ADVANCED GENETIC GENEALOGY:

TECHNIQUES AND CASE STUDIES

ADVANCED GENETIC GENEALOGY:

TECHNIQUES AND CASE STUDIES

Editor
Debbie Parker Wayne

2019

Advanced Genetic Genealogy: Techniques and Case Studies
Debbie Parker Wayne, Editor

Cover design by Debbie Parker Wayne using family photos, family documents, and altered images licensed through Pixabay; Pixabay images created by Tolucreations (background) and mohamed_hassan (DNA helix).

Published by Wayne Research, Cushing, Texas

ISBN 978-1-7336949-0-2
LCCN 2019902034

Cataloging data

Wayne, Debbie Parker, 1954–

Advanced Genetic Genealogy: Techniques and Case Studies by Debbie Parker Wayne, Editor

Includes index

1. Genetic Genealogy–Handbooks, Manuals, etc.
2. DNA–Analysis
3. Genealogy

Printed in the United States of America

Book design by Iron Gate Publishing

CONTENTS

DNA and the Genealogical Proof Standard

LIST OF FIGURES AND TABLES

Lessons Learned from Triangulating a Genome

Visual Phasing Methodology and Techniques

X-DNA Techniques and Limitations

Y-DNA Analysis for a Family Study

Unknown and Misattributed Parentage Research

The Challenge of Endogamy and Pedigree Collapse

Parker Study: Combining atDNA & Y-DNA

Would You Like Your Data Raw or Cooked?

Writing about, Documenting, and Publishing DNA Test Results

Preface

Debbie Parker Wayne, CG, CGL

Ask a question about using DNA for genealogical research in any online forum or on social media and dozens of immediate responses are received. Some of those responses will be from people with limited experience and expertise. There is little chance for a newcomer to DNA analysis to know which answers are right, which are wrong, which are partially right in a limited number of situations, or which responders are knowledgeable, and which are less so. Many responders have only used one type of DNA and may have looked only at the most basic results from that one type of test or have researched only one family that is different from the questioner's family in many ways. Too many researchers accept a response that fits with preconceived notions of what "seems" right.

What if that answer that "seems" right is wrong or incomplete? How many will use Google to try to learn more? Readers of this book are likely to be the ones who know there is much more to learn.

Genealogical Proof and Standards

All genealogical research should follow standards to achieve the best results. *Genealogy Standards*[1] can be applied no matter which region or which timeframe is under study. Because the focus of this book is case studies illustrating advanced DNA analysis techniques, some documentary evidence is omitted, especially for recent generations, to make room for more DNA analysis.

Researchers *should* include all necessary documentary evidence, as standards and best practices indicate, when writing proof arguments and essays that include DNA evidence. Three chapters focus specifically on standards: "Drowning in DNA? The Genealogical Proof Standard Tosses a Lifeline" by Karen Stanbary, "Correlating Documentary and DNA Evidence

1. Board for Certification of Genealogists (BCG), *Genealogy Standards*, 2nd ed. (Nashville, Tenn.: Ancestry.com, 2019).

to Identify an Unknown Ancestor" by Patricia Lee Hobbs, and "Writing about, Documenting, and Publishing DNA Test Results" by Thomas W. Jones.

This book was written while the Board for Certification of Genealogists (BCG) was considering and modifying guidelines for using DNA for genealogical research to be added to *Genealogy Standards*.[2] The new standards have been approved but not published as we go to press. Some chapter authors were involved in the standard drafting and adoption process and have access to the revisions approved by BCG. Therefore, some chapters have fully incorporated those standards, some partially, and some, written before the proposed standards were publicly available, may not mention the standards. Some chapters include references to the standards that the author believes apply to the described analysis. That does not mean that BCG agrees with the author's judgement.

All of the authors may not fully agree with all of the standards themselves, although most do. Five authors were members of the BCG committee working to define the DNA-related standards. All authors provide detailed instructions on how to apply techniques to specific situations. These techniques and methods should help researchers understand how to incorporate DNA analysis into a case study that can meet the criteria of the Genealogical Proof Standard.[3] Just as with documentary research, each individual researcher must make a judgment call on how much DNA evidence, if any, is required for a given problem. No one can say that every situation needs X number of test takers or Y number of lines tested or Z matching DNA segments or markers. A common phrase in the genealogical community is "it depends." Research requirements depend on the specifics of any given case. This statement is true for DNA and other types of evidence.

Several case studies illustrate the importance of combining DNA and documentary evidence, within the space limitations imposed by this publication. DNA evidence alone solves no problems. Even a parent-child link, which results in a fairly unique chromosome comparison, requires the addition of test-taker ages to determine which is the parent and which the child.

Many will ask what BCG requires for portfolios and whether DNA evidence is required for every case. This book is not a publication of nor is it sponsored by BCG, even though many of the authors hold credentials licensed from BCG. Contact BCG with any questions specific to using DNA in application and renewal portfolios or interpretation of standards.

The standards received rigorous review to achieve precise and accurate wording. However, emerging uses of DNA and the potential for individuals to interpret phrases differently may result in refinement of best practices in the future.

---

2. BCG, "Proposed DNA Standards: For Public Comment," *SpringBoard*, 23 May 2018 (https://bcgcertification.org/proposed-dna-standards-for-public-comment/).

3. BCG, *Genealogy Standards*, 2nd ed., 1–3.

Background Level of Audience

This book primarily includes intermediate to advanced level analysis techniques, although there is recapitulation of some basic information. There are many notes referencing material to help readers understand statements included here and learn more about the topic under discussion.

Readers who have trouble understanding the advanced methods and techniques described here may need to study other sources to build their basic understanding. This book is considered a follow-on to the DNA basics covered in *Genetic Genealogy in Practice,*[4] *The Family Tree Guide to DNA Testing and Genetic Genealogy,*[5] and similar books. The basic knowledge can also be obtained through sources in the Recommended Reading list. For general genealogical research knowledge *Mastering Genealogical Proof*[6] and other basic books are available.

Most genealogical researchers now recognize the need to effectively use DNA for family history. The need to understand math and scientific principles can be daunting for some. However, anyone who can mine documentary sources for relevant information, logically analyze that information, and correlate all of the evidence to reach a credible conclusion can likely learn to use DNA information to help solve genealogical problems. Some complex concepts may require dedicated study to achieve mastery.

Chapter Order and Contents

Chapter arrangement follows a logical sequence for those who read from front cover to back. However, each chapter stands alone. The reader can study chapters in order of preference. There are three major sections. The first covers analysis: advanced methods, tools, and techniques, with cases applying those techniques to genealogical problems. The second section links steps in DNA analysis to *Genealogy Standards* using a hypothetical case concentrating on the DNA instead of documents, includes a case study of a real family illustrating correlation of documentary and DNA evidence to solve a problem, and ends with a chapter on considerations when writing about DNA test results. The third section covers thought-provoking issues that every researcher should consider to be prepared for the effect of DNA testing on family dynamics, now and in the future. Figures and tables within some chapters use informal styles helpful during the research, analysis, and write-as-you-go phase. The chapter on writing helps to convert those items to a formal style for publication.

Several chapters use hypothetical cases or fictionalized names for real cases. This protects the privacy of those who do not wish to be identified publicly. Importantly, it also allows the

4. Blaine T. Bettinger and Debbie Parker Wayne, *Genetic Genealogy in Practice* (Falls Church, Va.: National Genealogical Society, 2016).
5. Blaine T. Bettinger, *The Family Tree Guide to DNA Testing and Genetic Genealogy* (Blue Ash, Ohio: Family Tree Books, 2016).
6. Thomas W. Jones, *Mastering Genealogical Proof* (Falls Church, Va.: National Genealogical Society, 2013).

authors to incorporate elements from multiple cases to illustrate techniques that genealogists may apply in their own research. For examples correlating documentary and DNA evidence in real-life case studies see the "Y-DNA Analysis for a Family Study," "Parker Study: Combining atDNA & Y-DNA," "Correlating Documentary and DNA Evidence to Identify an Unknown Ancestor" chapters. Some examples in other chapters are based on actual studies but use hypothetical names. Some chapters include links to online sources so the reader can access enlarged images and additional family information.

Terminology and Styles

In most cases, this book uses the term *genealogists* with only a few statements using the phrase *genetic genealogists*. Some authors and readers have strong preferences for which term they apply to themselves and colleagues. No matter which term is preferred, the techniques discussed here will be useful to any researcher using DNA test results to study family history.

The discussion within the community about which term should be used in which circumstances will probably go on for several more years. "Advanced genetic genealogy" is used in the title of this book, not because using DNA test results for genealogical research should be a practice of only a few "genetic genealogy" experts, but to make it clear that the chapters cover advanced techniques and methods that require the researcher to have the underlying basic understanding for using DNA accurately for genealogical research. Besides, "using DNA test results for genealogical research" is a more awkward phrase than "genetic genealogy." "Census genealogy" and "deed genealogy" are not common phrases in genealogical circles. However, "German genealogy" and "African American genealogy" are frequently used phrases for techniques that are useful for all genealogists when researching those groups. In the same way, genetic-genealogy techniques should be followed by all genealogists using DNA test results.

Most genealogical writing adheres to *Chicago Manual of Style* recommendations. However, writing about genetics includes more technical and scientific information than that typically seen in genealogical writing before DNA test results came into common use. Where it helps make the meaning more clear, in the view of the editors of this book, a technical style is followed instead of a humanities style in the handling of numbers and scientific data.

Accessing Sources and Tools

Where applicable, authors cite scientific and technical papers as sources to support their statements. Readers can greatly enhance their genetics knowledge by studying these sources. Some of the web pages cited may no longer be available at the same place as when the author accessed the pages. This is a problem faced by all authors in our fast-changing internet world. Two techniques can help the reader find the sources in a new location. First, try searching for the title of the article, web page, or website in a search engine, such as *Google*.

If that is not successful, use the *Wayback Machine*[7] to search for an archived copy of the web page. In rare cases, some scientific and technical papers may be behind a paywall. A university library or large public library may have these journals available.

Writers about DNA are often frustrated—in both good and bad ways. It is good that so many new tools are making DNA analysis easier. It is bad that new and important tools frequently appear just as a book goes to press. Many examples in this book are based on *GEDmatch* utilities. At the time of this writing, *GEDmatch* Classic is in the process of switching to *GEDmatch* Genesis; the names and locations of some tools may change and some functionality in Genesis is not finalized. Also, FamilyTreeDNA just announced new Y-DNA and mtDNA trees, tools, and tests. Such changes do not render the discussions in this book obsolete, but any screen shots used may have a different appearance than readers will see when accessing the tools. It is important to focus on the DNA analysis concepts, techniques, and methodologies used and not on the specifics of any given tool when options and appearances change.

Conclusion

Many areas of genetic genealogy need more study. These areas include multi-generational family studies for recombination models in autosomal and X-DNA, typical amounts of shared DNA and number of shared segments for specific relationships, comparisons between test results at different companies, and comparing whole genome sequences to the microarray genotypes prevalently used now.

While this book does not cover every DNA tool available or every analysis technique, a large number of these are covered. Researchers of all experience levels should learn something. Researchers who are at an intermediate level will understand more of the information at first read. Less experienced readers may need more than one reading or need to study basic information and come back to this book.

Everyone who contributed to this project hopes readers—citizen scientists—will use these tools, techniques, and methods to study their own families or regional groups, make fascinating discoveries, answer genealogical questions, and share the findings so all genealogists can learn more of our shared history. We will then all be better able to respond to those questions on social media and forums with answers that not only seem right but that are right.

7. Wayback Machine, *Internet Archive* (https://archive.org/).

Acknowledgments

Many people, beyond just the authors, contributed to this book. My thanks to everyone who helped and my sincere apologies if I forgot to include anyone here. Blaine T. Bettinger and I planned this book as another collaborative effort. When Blaine decided to focus on his speaking commitments, I decided to continue with the book. The author selection was a collaborative effort. Some issues with availability changed the main focus of the book, but the final product should help all genealogists using DNA evidence to answer questions.

Many people contributed time to provide feedback on the book and individual chapters. Patricia Lee Hobbs reviewed every chapter for scientific and biological accuracy as well as general readability. Pat Gordon and Susan Ball provided editorial expertise. Laurel Baty, Claire Mire Bettag, Blaine T. Bettinger, Janis Walker Gilmore, Thomas W. Jones, Kimberly Powell, Ann Raymont, Patricia Roberts, Rick Sayre, and Karen Stanbary commented on content and structure. Harold Henderson and Amy K. Arner gave editorial advice early in the process. Allen Peterson answered questions related to English research. All of the authors provided suggestions, in addition to writing their own chapters, that led to a better end product. The blame for any remaining errors is my own.

My husband also deserves my thanks. He kept me fed and was understanding when I delayed household responsibilities to work on the manuscript. I am grateful to all of my family members and clients who have allowed me to learn about DNA while analyzing their genetic data and to use that research in presentations, articles, and books.

The authors of this book were selected for expertise in the chapter subject matter. That expertise was sometimes gained through good experiences, some less good, and some downright bad. All have been using DNA for genealogical analysis for many years. Some are autosomal DNA experts, some Y-DNA experts, some experts in specific tools, concepts, or ancestral situations. A few are experts in all or almost all areas that a genealogist needs to understand.

All of these experts may not agree with each other 100% on the usefulness of certain methodologies, tools, or even terminology. Each sees differing situations in their own research. Genetic genealogy is still new enough that none of us has seen every situation that may occur. However, all authors offer in-depth understanding in using DNA to answer genealogical questions.

Some authors are more comfortable writing in first-person when using families as case studies and examples. Some authors prefer the more formal third-person. We have tried to leave as much as possible in the voice and words of the author while editing for clarity and consistency. Every editor has his or her own preferences. When editors and readers disagreed, my own preferences prevailed, with clarity for readers as a motivating factor.

Author Biographies

Jim Bartlett, PE

Chapter 1: Lessons Learned from Triangulating a Genome

Jim Bartlett has been a genealogist since 1974, and he still visits courthouses and scrolls microfilms, in addition to doing online research. He has been using DNA for genealogy since 2002 when he started the BARTLETT Y-DNA project at FamilyTreeDNA. Since 2010 he has tested at all of the major DNA companies, and has become an avid fan of autosomal DNA as a genealogy tool. Jim is a retired Navy Captain and a Registered, Professional Engineer who loves puzzles—he considers a chromosome map the ultimate puzzle, with very practical implications. He regularly gives beginner, intermediate, and advanced talks on DNA to genealogy groups and at International Conferences. He writes the plain English *Segment-ology* blog for genealogists at www.segmentology.org.

Blaine T. Bettinger, JD, PhD

Chapter 2: Visual Phasing Methodology and Techniques

Blaine T. Bettinger is a professional genealogist specializing in DNA evidence. In 2007 he started *The Genetic Genealogist* (http://www.thegeneticgenealogist.com), one of the earliest blogs on the topic. Blaine is the author of *The Family Tree Guide to DNA Testing and Genetic Genealogy*, and coauthor with Debbie Parker Wayne of the award-winning *Genetic Genealogy in Practice*, the world's first genetic genealogy workbook. He also coauthored "Genetics for Genealogy" with Judy G. Russell in 2018's *Professional Genealogy: Preparation, Practice & Standards* (ProGen PPS) edited by Elizabeth Shown Mills. Blaine is or has been an instructor for genetic genealogy courses at the Institute of Genealogy and Historical Research, Salt Lake Institute of Genealogy, Genealogical Research Institute of Pittsburgh, and Virtual Institute of Genealogical Research. Blaine is a graduate of ProGen Study Group 21, a trustee for the New York Genealogical and Biographical Society's Board of Trustees, and a board member for the Association of Professional Genealogists.

Patricia Lee Hobbs, CG

Chapter 10: Correlating Documentary and DNA Evidence to Identify an Unknown Ancestor

Patricia Lee Hobbs, Certified Genealogist®, with a BA in biology, works as a professional genealogist and occasionally at the Springfield-Greene County Library in Springfield, Missouri. Patti instructs on the use and analysis of DNA testing and how it complements traditional research in original records. She has presented numerous workshops on DNA locally and regionally as well as at the week-long "Practical Genetic Genealogy" course at the Genealogical Research Institute of Pittsburgh. Patti serves as a Board for Certification of Genealogists trustee, a BCG Education Fund trustee, the chair of the Ozarks Genealogical Society Education Committee, and the course coordinator for the Institute of Genealogy and Historical Research's "Genetics for Genealogists: Beginning DNA."

Melissa A. Johnson, CG

Chapter 5: Unknown and Misattributed Parentage Research

Melissa A. Johnson, Certified Genealogist®, specializes in genealogical research; family history writing, editing and publishing; and using DNA to solve difficult research problems and identify unknown parentage. She has expertise in researching families from New Jersey, New York, Pennsylvania, and the British Isles. Melissa is an instructor in Boston University's Genealogical Research Certificate Program, and is on the faculty of several genealogical institutes, including the Genealogical Research Institute of Pittsburgh, Institute of Genealogy and Historical Research, and Salt Lake Institute of Genealogy. She serves on the Board of Trustees of the Genealogical Society of New Jersey and the International Society for British Genealogy and Family History. Melissa owns and operates Johnson Genealogy Services and the New Jersey Family History Institute. Her work has been published in several genealogical journals, including the *National Genealogical Society Quarterly*, the *New York Genealogical and Biographical Record* and the *Genealogical Magazine of New Jersey*, as well as several other publications.

Kathryn J. Johnston, MD

Chapter 3: X-DNA Techniques and Limitations

Kathryn J. Johnston is a retired dermatologist who has been doing genealogical research for more than twenty-five years and genetic genealogy for ten years. She has been researching the X chromosome since 2008. She is one of the administrators for the Southern California Genealogical Society DNA Interest Group and has been involved in developing techniques for chromosome mapping for genealogists. When she is not home in Palos Verdes, California working on her hobby, she is volunteering at the local county hospital or traveling with husband, Matt. They are the proud parents of four children and five grandchildren.

Thomas W. Jones, PhD, CG, CGL, FASG, FUGA, FNGS

Chapter 11: Writing about, Documenting, and Publishing DNA Test Results

Thomas W. Jones has pursued his family's history since 1963. For the first twenty-five years he was clueless about what he was trying to accomplish and how to do it. When he started climbing the genealogy learning curve he repeatedly experienced the challenges, joys, and rewards of tracing ancestors reliably and of fully understanding their lives. Tom eventually became an award-winning genealogical researcher, writer, editor, and educator. He also is a professor emeritus from Gallaudet University, where he designed and managed graduate programs, conducted research, and taught and mentored graduate students for twenty-seven years. Tom is a former trustee and past president of the Board for Certification of Genealogists and co-edited the *National Genealogical Society Quarterly* from 2003 through 2018. He wrote the textbooks *Mastering Genealogical Documentation* and *Mastering Genealogical Proof* and three chapters in *Professional Genealogy: Preparation, Practice & Standards*. A popular speaker at national and local genealogical society conferences, he teaches genealogy at the Genealogical Research Institute of Pittsburgh and the Institute of Genealogy and Historical Research. He previously taught genealogy at Boston University, the Salt Lake Institute of Genealogy, and elsewhere.

Debbie Kennett, MCG

Chapter 14: The Promise and Limitations of Genetic Genealogy

Debbie Kennett is a U.K.-based genealogist, writer, and lecturer. She has been researching her family history for nearly twenty years and has a particular interest in her rare maiden name Cruwys, which can be traced back to the 1200s in Devon. Debbie has been a long-time member of the Guild of One-Name studies and runs the Guild's social media accounts. She was recognized as a Master Craftsman of the Guild (MCG) in 2016. Debbie is an Honorary Research Associate in the Department of Genetics, Evolution and Environment at University College London. She is the author of two books for the History Press: *DNA and Social Networking* and *The Surnames Handbook*. She has written articles for many family history magazines including *Who Do You Think You Are? Magazine* and *Family Tree Magazine* (U.K.). Her highly regarded blog *Cruwys News*, although originally set up to focus on her Cruwys one-name study, has evolved into the leading genetic-genealogy blog in the U.K. She is a regular speaker at genealogy events in the U.K. and Ireland and has been an invited speaker at conferences in the U.S., Canada, and Portugal. Debbie is the co-founder with Tom Hutchison of the *ISOGG Wiki*, and provides much of the content. She is the administrator of the Cruwys DNA project, Devon DNA Project, Haplogroup U4 Project, and a co-administrator of the Haplogroup U106 Project.

Michael D. Lacopo, DVM

Chapter 13: Uncovering Family Secrets: The Human Side of DNA Testing

Michael Lacopo is a life-long resident of northern Indiana. He began his genealogical research in 1980. He has contributed to numerous periodicals and has helped numerous people in their quests to locate their relatives—living and dead. He appeared in *USA Today* in 2000 discussing genealogy and the proposed destruction of the federal census tabulated in that year. His national lecturing began in Sacramento, California, at the National Genealogical Society's conference in 2004, and has continued with local, state, national, and international conference speaking engagements to this present day. Michael was an early adopter of the use of DNA in practical genealogical research and lectures; he teaches extensively on a number of DNA and genetic genealogy topics. He believes that genealogists should tell the tales of our ancestors and encourages learning the social history that interweaves our ancestors into the fabric of the past.

James M. Owston, EdD

Chapter 4: Y-DNA Analysis for a Family Study

James M. Owston has been interested in genealogy since being assigned a family tree project in an eighth grade English class in 1968. His interest in genetics for genealogy began in late 2007, and to date, he has tested over 70 members of his family. Owston serves as an academic administrator and a tenured professor at Alderson Broaddus University. He holds the following academic degrees: Doctor of Education, Specialist in Education, two Master of Arts, Bachelor of Theology, Bachelor of Arts, and two Bachelor of Science degrees. He was the 2008 recipient of the Leo and Margaret Goodman-Malamuth Outstanding Higher Education Administration Research Award from the American Association of University Administrators and received international recognition as the 2009 winner of the Alice L. Beeman Outstanding Research Award for Communication and Marketing for Educational Advancement from the Council for Advancement and Support of Education. He is a member of numerous organizations including being a charter member of the John Beckley chapter of the SAR. He additionally serves as a peer reviewer for the Higher Learning Commission and the Journal of Marketing in Higher Education and is a board member for the American Association for University Administrators.

Kimberly T. Powell

Chapter 6: The Challenge of Endogamy and Pedigree Collapse

Kimberly T. Powell is a professional genealogist, author, educator, and volunteer. She is a past president of the Association of Professional Genealogist, serving seven years as a Director and on APG's Professional Development Committee. She is on the faculty of the Genealogical Research Institute of Pittsburgh, the Institute of Genealogy and Historical Research (IGHR), and the Salt Lake Institute of Genealogy, and a facilitator for Boston University's Genealogical Research Certificate Program. Kimberly was the genealogy expert for About.com from 2000 to 2016 and is the author of several books, including *The Everything Guide to Online Genealogy*, 3rd ed. (Adams Media, 2014). She has published articles in BBC's *Who Do You Think You Are? Magazine*, *Family Tree Magazine*, and the *Association of Professional Genealogists Quarterly*. She is the proud recipient of the Utah Genealogical Association's Silver Tray award for genealogical publishing (2012) and the Grahame T. Smallwood Jr. Award of Merit from the Association of Professional Genealogists (2017). You can find her online at Learn Genealogy (http://www.learngenealogy.com/).

Judy G. Russell, JD, CG, CGL

Chapter 12: Ethical Underpinnings of Genetic Genealogy

The Legal Genealogist Judy G. Russell is a genealogist with a law degree. She writes, teaches, and lectures on a wide variety of genealogical topics, ranging from using court records in family history to understanding DNA testing. A Colorado native with roots deep in the American south on her mother's side and entirely in Germany on her father's side, she holds a bachelor's degree in journalism with a political science minor from George Washington University in Washington, D.C., and a law degree from Rutgers School of Law-Newark. She has worked as a newspaper reporter, trade association writer, legal investigator, defense attorney, federal prosecutor, law editor and, for more than 20 years before her retirement in 2014, was an adjunct member of the faculty at Rutgers Law School. Judy has written for the *National Genealogical Society Quarterly* (from which she received the 2017 Award of Excellence), the *National Genealogical Society Magazine*, the *FGS Forum*, BCG's *OnBoard*, and *Family Tree Magazine*, among other publications. She is on the faculty of the Institute of Genealogy and Historical Research, the Salt Lake Institute of Genealogy, the Genealogical Research Institute of Pittsburgh, the Midwest African American Genealogy Institute, and the Genealogical Institute on Federal Records. She also serves as a member of the Board of Trustees of the Board for Certification of Genealogists®, from which she holds credentials as a Certified Genealogist® and Certified Genealogical Lecturer℠. Her award-winning blog appears at *The Legal Genealogist* website (http://www.legalgenealogist.com).

Karen Stanbary, CG

Chapter 9: Drowning in DNA? The Genealogical Proof Standard Tosses a Lifeline

Karen Stanbary holds a Master's Degree in Clinical Social Work from the University of Chicago. She specializes in complex problem-solving using genetic and documentary sources and the Genealogical Proof Standard. Karen's regional specialties include Midwestern United States, Chicago, and Mexico. She is bilingual (Spanish). She is a coordinator and faculty member at the genealogical institutes— Genealogical Research Institute of Pittsburgh, Institute of Genealogy and Historical Research, and Salt Lake Institute of Genealogy. The National Genealogical Society honored her with the *NGSQ* Award for Excellence for her complex-evidence case study incorporating traditional documentary research and autosomal DNA analysis in the June 2016 issue of the *National Genealogical Society Quarterly*. She holds the credential Certified Genealogist® from the Board for Certification of Genealogists® where she serves as a Trustee and chair of the DNA Standing Committee.

Ann Turner, MD

Chapter 8: Would You Like Your Data Raw or Cooked?

Ann Turner was an early adopter of DNA testing for genealogy. She began in 2000 with a mitochondrial DNA test from Oxford Ancestors. This inspired her to create the GENEALOGY-DNA mailing list on RootsWeb, where people could discuss how to analyze all manner of tests as they appeared on the market. Witnessing these developments enabled her to learn gradually, one step at a time. She and Megan Smolenyak met online and decided it was time for a book-length presentation, which culminated in 2004 with *Trace Your Roots with DNA*. Ann was among the first to take an autosomal test in 2007 at a price point of $1000. As the cost dropped, she tested more cousins who shared her inherited hearing impairment and identified a segment common to them all. She was then among the first to obtain whole exome sequencing when 23andMe offered it in 2011, again at a cost of $1000. A novel mutation in a gene not even known to be important for hearing was discovered in this segment.

Debbie Parker Wayne, CG, CGL

Editor, Advanced Genetic Genealogy Techniques and Case Studies

Chapter 7: Parker Study: Combinging atDNA & Y-DNA

Debbie Parker Wayne, Certified Genealogist® and Certified Genealogical Lecturer℠, is experienced using DNA analysis and traditional techniques for genealogical research. Debbie is the coauthor with Blaine T. Bettinger of the award-winning DNA workbook, *Genetic Genealogy in Practice*, published by the National Genealogical Society (NGS). She is the author of the online, self-paced course *Continuing Genealogical Studies: Autosomal DNA*, offered by NGS. She is the DNA Project Chair for the Texas State Genealogical Society (TxSGS) and the Early Texans DNA Project. Debbie serves as a trustee of the Board for Certification of Genealogists®. Her publications include a column on using DNA analysis for genealogical research in *NGS Magazine* and in TxSGS's *Stirpes* journal. Debbie was the course coordinator for the first beginner and intermediate DNA courses offered at major U.S. genealogy institutes: Genealogical Research Institute of Pittsburgh, Salt Lake Institute of Genealogy, Institute of Genealogy and Historical Research, and Forensic Genealogy Institute. See http://debbiewayne.com/ for more information and for archived versions of many of her articles.

Lessons Learned from Triangulating a Genome

Jim Bartlett, PE

Introduction

In this chapter, in order to make broad statements, the focus will be autosomal DNA—the twenty-two numbered chromosomes in a genome—except as noted otherwise. All reference to segments shall mean IBD (Identical By Descent) segments inherited from a shared ancestor with a Match—except as noted otherwise. In this chapter, the capitalized words "Match" and "Ancestor" refer to a real person and "Triangulation" refers to a specific process.

My overall objective in genetic genealogy is chromosome mapping—determining which parts of my DNA came from which Ancestors—to assist in determining the Ancestor shared with new matches. The "parts" are specific DNA segments. Since the Ancestor who first formed each segment may be beyond my known genealogy, I identify ancestral lines from me back toward those Ancestors. Each segment of DNA came from a specific Ancestor, down a specific path of descendant Ancestors (the ancestral line), to a paternal or maternal chromosome. This is the top-down viewpoint. The bottom-up viewpoint is each chromosome (composed of many segments) came entirely from one of two parents; they got their chromosomes from their parents, who got them from their parents, and so on. In other words, each segment of DNA came from one parent, one grandparent, one great-grandparent, and so on up the ancestral line.

The most basic solution to chromosome mapping is twenty-two large segments (full chromosomes) from each parent. The next step is to determine which segments of each chromosome came from which grandparent. Then, at the great-grandparent level, some, but not all, of these grandparent segments are further subdivided into smaller segments. This happens again at each generation going back—some of the segments are subdivided, some are not (some segments are passed intact without subdivision).

We cannot "see" any segments of our own DNA—the raw DNA file consists of values at approximately 700,000 locations (called SNPs). When we compare our DNA to another's DNA, we then "see" shared segments in a chromosome browser or spreadsheet—segments of DNA that are the same for both people being compared. We do not really know if a shared segment represents all or part of a segment from one of our ancestors. In fact, we do not really know if a shared segment represents all or part of a larger segment from a Common Ancestor between us and a Match.

Triangulated Groups are helpful in determining which segments of DNA came from which of a person's ancestors. A Triangulated Group is composed of three or more segments shared between Matches, where each shared segment overlaps with the others.[1] The start and stop points of a Triangulated Group occurs at natural breakpoints in the DNA (crossover points[2]). Each shared segment identifies part of the DNA that came from a specific Ancestor. The Triangulated Group represents an ancestral segment. The Matches who have the shared segments with you, will be cousins somewhere on your ancestral line to (or through) the Triangulated Group Ancestor.

Background Terminology

The following definitions and concepts apply to Triangulation and chromosome mapping.

SEGMENTS of DNA on your chromosomes come from your Ancestors. A SHARED SEGMENT with a Match represents an overlap of your DNA and your Match's DNA. A computer algorithm compares your DNA SNP values with those of a Match to determine which segments are shared. The algorithm reports those segments that are deemed to "match"—they have the same DNA sequence. Some companies incorporate other criteria in matching algorithms to increase the odds that the segment will be Identical By Descent (IBD), meaning it really came from a Common Ancestor shared by you and your Match.

TRIANGULATION actually has two meanings in genetic genealogy. In one meaning, it is a process of determining that three or more people share the same segment (or significantly overlapping segments) with each other. Technically, the Matches' shared segments form a TRIANGULATED GROUP (TG). This is called SEGMENT TRIANGULATION. You and two or more Matches share the same segment and your Matches share this same segment with each other. In other words, you have three comparisons on the same segments which match: you to Match A; you to Match B; and Match A to Match B. A Triangulated Group can exist whether or not the Ancestor or ancestral line is identified.

All URLs were accessed 27 January 2019 unless otherwise noted.

1. Jim Bartlett, "Anatomy of a TG," *segment-ology*, 5 February 2016 (https://segmentology.org/2016/02/05/anatomy-of-a-tg/).

2. "Recombination," *International Society of Genetic Genealogists (ISOGG) Wiki* (https://isogg.org/wiki/Recombination).

The other meaning of Triangulation is called GENEALOGY OR TREE TRIANGULATION. In this sense, at least three Matches identify the same Most Recent Common Ancestor (MRCA)—sometimes just abbreviated to Common Ancestor (CA). The MRCA is usually a couple—two parents—and you descend from one of their children and your Match descends from a different child of that couple. In a few cases, where one parent is married more than once, the MRCA may be a single Ancestor with you and your Match descending from different spouses or mates. Genealogy Triangulation is almost all about genealogy—collaborating with your Matches to determine MRCAs, and forming a consensus among several Matches. Sometimes several Matches in a Triangulated Group agree on the same MRCA, and sometimes we "Walk the Ancestor Back" as will be illustrated later.

Sometimes I refer to *segments* in a TG, sometimes I refer to the human *Matches* in a TG. That is because forming a TG is a mechanical process with segment data; the human Matches, their names, and their ancestry have nothing to do with this phase of Triangulation. Determining which "side" (paternal or maternal) a TG is on or finding MRCAs requires collaboration between you and your Match. The collaboration might be an email, a message, a phone call, or a review of a Match's Ancestors (in a tree, pedigree chart, fan chart, Ahnentafel, descendants list, family group sheet, list of surnames, or even a long rambling story about their family).[3] Many more Common Ancestors will be found by sharing information with Matches.

A Match may share more than one DNA segment with you, and you may have different MRCAs for each of these segments. A Match may share only one segment, but have multiple MRCAs. This is rather common in many populations, such as in Colonial America. In this case, you would be genetic cousins on only one MRCA (the MRCA linked to the segment), but you would still be genealogy cousins on the other MRCA(s).

In genetic genealogy and chromosome mapping, we deal with a lot of DNA data. Triangulation lets us group Match-segments that "go together." A TG represents a real segment of your DNA that comes from a specific Ancestor down one ancestral line to you. In general, all of the Matches in a TG who share that full segment will share the same Common Ancestor with you and usually are more closely related than those who share only a portion of that segment.[4]

> *Takeaway*: Every IBD segment belongs to a specific TG. This is a major outcome and benefit of Triangulation.

CHROMOSOME MAPPING combines Segment Triangulation, which defines each segment in the map, with Genealogy Triangulation which links an Ancestor to each segment. In other words, specific ancestral lines are mapped (linked) to specific DNA segments. TGs tend to

3. Readers unfamiliar with charts or Ahnentafel numbers can find examples using a Google search. See also "Ahnentafel," *Wikipedia* (https://en.wikipedia.org/wiki/Ahnentafel).

4. Matches with smaller segments than the full TG may share a more distant MRCA with you, that is, an ancestor of the CA who created the TG segment.

be adjacent to each other and cover each chromosome from one end to the other. Mapping ancestral lines to specific DNA segments is easier to visualize when the "side," paternal or maternal, is known. This is comparable to working a jigsaw puzzle upside down and without the picture—just by putting the pieces in place where they fit. Identified MRCAs are the pictures for each jigsaw puzzle piece (segment or TG).

Using Triangulation

There are several benefits of Triangulation.

1. *Grouping your Match-segments.* Although no genealogy is required to sort segments into TGs with Triangulation, some genealogy is required to determine whether a TG is on the maternal or paternal side. These groups definitely help organize your data and there can be synergy in each group collaborating to find the CA.
2. *Research Confirmation.* A TG with Genealogy Triangulation reinforces that the genealogy research is correct and builds confidence in conclusions.
3. *IBD Segments.* Some shared segments under about 15 cM will be false. Triangulation will identify virtually all of the false segments, because they will not Triangulate with established TGs—they will not match an overlapping TG on either side. You can be confident that virtually all of the segments in a TG are IBD. This statement has been contested because it has not been proved or published. However, after five years of Triangulating, I have not found any evidence to the contrary, and I have culled out thousands of false segments that do not Triangulate.
4. *Brick walls.* All of your TGs will pretty much cover your DNA. Some of the TGs will represent segments from brick wall Ancestors. They are identified through cousins who are descendants of the brick wall Ancestor. Two of the Matches in one of these TGs may have a Common Ancestor at, or ancestral to, the brick wall Ancestor. This is a very important clue to resolving your brick wall.
5. *Finding NPEs or Misattributed Parentage.* If some TGs have plenty of Matches with long, shared segments, but no or few MRCAs are found with these Matches, then perhaps some of the ancestral links are incorrect. This may be through misattributed parentage, faulty genealogy, or incomplete trees.
6. *Chromosome Mapping.* A major outcome of Triangulation is a map assigning DNA segments to the Ancestor who passed the segment to you. Then DNA Matches can be compared to the map to indicate an ancestral line, confirm genealogy, or provide clues for further research.

Insight: When you form TGs with almost all of your segments, you will have TGs that "cover" most of your chromosomes. This provides a chromosome map of the segments from your ancestors, which can then be linked to the Ancestor who passed each segment down; as well as providing genealogy insights to your Matches.

My experience

Numbers of Matches. Since 2010, my total number of Matches doubled about every fourteen months up to 2015 and is increasing at a faster rate now. This increase means we have a lot of data to work with, and new data is pouring in every day. We need to plan for this increase and determine a way to capture, organize, manage, analyze, and make effective use of this data. Many reported segments smaller than 7 cM to 15 cM in size will either be false or pseudo-segments. These should be given low priority in the absence of evidence to the contrary.

> *Lessons Learned*: To budget your time, select a minimum segment cM threshold that works for you and leaves time to communicate with Matches and determine MRCAs. Sort your data on the cM size column and remove all rows with segments under, say, 15 cM. The remaining Matches would tend to be closer cousins. You would have fewer, but larger, TGs spread over your DNA. Find a balance between collecting and Triangulating segments and working with Matches to find the MRCAs. You can always add the next lower range back into your data later. The objective is to form and fill out TGs *and* find MRCAs with the Matches. As you adjust the cM size of segments, look for your sweet spot where you can keep up. In my case, I use a 7 cM threshold, but am now getting behind on the genealogy phase of determining MRCAs. My solution is to remove the segments under 12 cM and focus on my remaining Matches.

Number of TGs. I have 377 TGs. These are broken down into 274 TGs with an MRCA at the grandparent level or more distant; 88 TGs with at least two IBD segments in each one but no credible MRCA yet (although the TGs *are* identified as to parental side); and 15 gaps with a "filler" TG but no segments over 7 cM from any Match. (There are also five areas of our DNA for which the companies do not report any segments. These are at the beginning of chromosomes 13, 14, 15, 21, and 22 on each side.) It is hard to determine the cMs for each TG, but I can easily subtract the End Point from the Start Point and get the TG "size" in Megabase pairs (Mbp). Mbp very roughly equates to cMs. The 362 TGs, not counting the 15 gaps, with Match-segments are broken down as shown in figure 1.1.

The bulk of my TGs are in the 10–20 Mbp (or cM) range and a few range over 40 Mbp. Some of these larger TGs will likely be subdivided later as more data is available. The segments in the few 5–6 Mbp TGs are over 7 cM in size.

> *Lessons Learned*: When using a 7 cM threshold, you can expect a wide range of TG sizes. DNA inheritance is random; go with the flow and follow the natural break points in the data. With enough data, you should expect TGs that are adjacent to each other from the beginning of each chromosome to the end. Expect Matches with shared segments about the same size as a TG to be closer cousins and Matches with smaller shared segments

within a TG to be cousins that are more distant—to have an MRCA beyond the Common Ancestor for the whole TG. Expect large TGs to be subdivided into smaller ones.

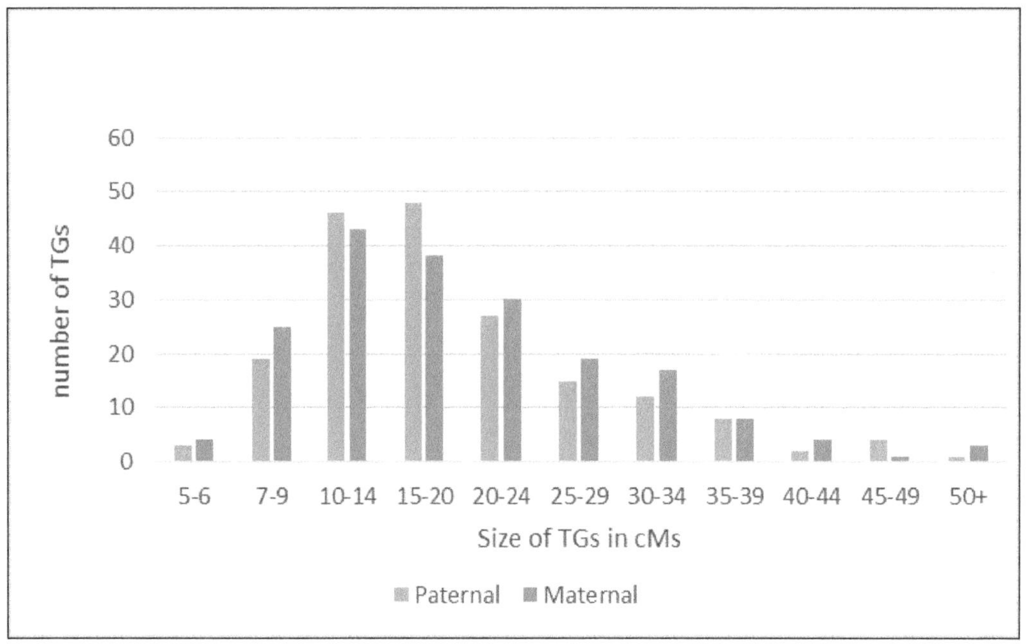

Figure 1.1. Number of TGs in each Mbp size range

Spreadsheets

Although other tools are available, I use spreadsheets to track my Matches and segments.

Spreadsheets. It is your decision whether to use a spreadsheet or a third-party tool.[5] You need some method to manage data from different companies unless you plan to use only one company and its data as presented. This is a very personal decision. I recommend a spreadsheet to have full control, to add, delete, or hide any data column(s) and to manipulate, analyze, sort, and search the data in any way desired. At the companies, lists of Matches and segment data can be downloaded in spreadsheet or Comma-Separated-Value (CSV) formats. Creating and updating a spreadsheet is labor intensive and reduces the time for actual genealogy.

> *Insight*: Third-party tools may be easier (and more accurate), but they, too, require a learning curve. Also, they are often non-functional and not always reliable, often because companies change the way data is presented. This chapter focuses on using spreadsheets.

5. "Autosomal DNA Tools," *ISOGG Wiki* (https://isogg.org/wiki/Autosomal_DNA_tools), see AncestryDNA Helper, *DNAGedcom, GEDmatch,* Genome Mate Pro, and other tools.

Data to Download. Each company provides different data, and some data only you can provide such as remarks, MRCA, pedigree charts, and so on. Each piece of data you want to add to a spreadsheet must be input by typing or downloading. Choose wisely.

> *Lessons Learned*: Use different columns for each company's version of some data. For example, I have separate columns for full name, surname, AncestryDNA name, *GEDmatch* name, and administrator or Point of Contact (POC) name, as they may all be different. Also, companies have different criteria for determining total amounts of shared DNA and estimated relationships. However, individual segment data is more standard and equivalent. The same is true with email and haplogroups.

Manually Added Data. See table 1.1 for data to be manually added to a spreadsheet. These key data points, often not in a download, make life easier. Add a separate column for each.

> *Lessons Learned:* Add any other columns you want; just be prepared to enter the data for each added column if it is not part of the downloaded file.

Matches Tested at Multiple Companies. Create one row for each segment for each Match from each of the testing companies. This allows forming TGs within each company and across multiple companies into one global TG. It is easier to keep track of the data if the Match uses a different name at different companies.

> *Lessons Learned:* It is important to be able to "find" a Match at each company from the name in your spreadsheet.

Table 1.1. Data added manually to spreadsheet		
Column Heading	**Data**	**Remarks**
Co	text	Company name. FF for Family Finder; AD for AncestryDNA; 23 for 23andMe; MH for MyHeritage; LD for LivingDNA; Gm for GEDmatch; DL for DNA.Land – this is a narrow column. Initially you will not have any AD rows with segment data, but gradually you will find Gm Matches you can pair up with AD kits (then change the Gm column to AD). Company name helps in several ways: remembering where the Match tested; sorting by company and date to find your most recent Match, knowing which new ones to add; including statistical info; etc.
Donor	text	Name of Donor (Test taker). Use initials—I am jvb; my brother is dmb. An autosomal DNA spreadsheet should be for one person and that person's matches! This is very important: do not combine spreadsheets with multiple kits—it is just too confusing. From time to time you might want to combine two spreadsheets to compare data, but then separate them again using the Donor column.
POC	text	Point of Contact. Often this is not in a download
Side	text	Side. When you determine that a segment is Paternal or Maternal, enter a P or M in this column. Or use Ahnentafel numbers. We often say a Match or relationship is on our Maternal "side" meaning on our maternal chromosome or is from a maternal Ancestor.
TG ID	text	ID for each TG. As you identify Segment TGs it is very helpful to give each a name or ID. Example: the 06C36 TG is on Chr 06; starts roughly at "C" (which is between 20 and 30Mbp—every 10Mbp gets a letter); and is on my Ahnentafel number 3 (or M = Maternal) side; and, in this case, it is on my mother's father's (Ahnentafel 3-6) line. Presently, TG ID is only carried out two generations, but it could be extended.
Date	date	Date. If your downloaded data does not include a date, add one—your best guess as to when the Match occurred. This will help you—trust me.
Tree	hyper-link	URL. Whenever you find an online tree or pedigree, copy the URL and paste it into this column of your spreadsheet. It will then be handy when you are reviewing Matches in a TG.
Kin	text	A close relative in this TG. Add initials of close kin who also match this Match on the same segment; this indicates which side the segment is on. Some of the time, a Match with two shared segments will have one segment on each side, so be sure to check each shared segment separately.
Email	text	Email. Sometimes provided by a Match; needs to be carefully typed in or pasted.
GEDmatch	text	GEDmatch ID. Often this will be provided by a Match; needs to be carefully typed in or pasted.
MRCA	text	MRCA Couple. I usually use the two surnames of a couple, such as BARTLETT/NEWLON
Cousin (Czn)	text	Cousinship. The exact relationship between you and a Match (determined by you, NOT the company estimate). Use abbreviations such as 3C1R = 3rd Cousin once removed; 3Cx2 = double 3C; 1C/2 = half 1C; etc.
Q	text	Quality Factor of MRCA. Use a Scale of 1 to 9; use 1 if you are absolutely sure of the *genealogy* link to an MRCA, 2 = pretty sure, down to 8 = pretty iffy, 9 = guess. You do not really know, initially, if an MRCA is the correct *genetic* link or not—you have to wait until you are sure of Genealogy Triangulation before you can become confident of the genetic link.
Notes	text, etc.	Notes or Remarks. Enter anything and everything you want.

Lessons Learned from Specific Case Studies

This section includes specific examples that outline concepts and offer Lessons Learned, Techniques, and Tips.

Case 1: genealogy triangulation

Triangulation of segments is the first step to form TGs. Each TG represents a specific segment from an Ancestor. The second step is linking the TGs to *the* specific Ancestor. We cannot find an MRCA with one Match and be positive *that* MRCA is the one who passed down the DNA segment to us.[6] Particularly with smaller segments, say under 20 cM or 30 cM, there are usually just too many MRCA possibilities to be confident that any particular one is the correct MRCA. One of the ways to build confidence in the correct MRCA is through Genealogy Triangulation—finding at least three Matches in a TG who all agree on the same MRCA.

A person and that person's parent or child can only be considered as one of the three Matches in Genealogy Triangulation. The DNA of a parent and child matches over the entire genome, so this does not form a *test* for matching, because it is always a match. The three Matches should be no closer than first cousins to each other. This is the absolute minimum criteria for Genealogy Triangulation—more cousin Matches increases the confidence that the MRCA has been correctly identified.

										1C	2C	3C	4C	5C	6C	7C	8C	9C	10C		
Co	Don	MATCHNAI	Chr	Start	End	cMs	SNPs	TG ID	**G2**	**G3**	**G4**	**G5**	**G6**	**G7**	**G8**	**G9**	**G10**	**G11**	**G12**	**Czn**	**MRCA Couple**
	jvb	--------------	1	0.0	249.3	**CHROMOSON**		1	2												
	jvb	**SEGMENT -**	1	0.0	6.1	*17.0*	1,686	01A25	2	5	11	22	44	88	**176**					6C	**BUTCHER/BUSH**
AD	jvb	Match A1	1	0.1	7.0	17.3	1,788	01A25	2	5	11	22	45	91	182	**364**				6C1R	**BUSH/Susannah**
AD	jvb	Match A2	1	0.1	7.0	17.3	1,788	01A25	2	5	11	22	44	88	**176**					5C1R	**BUTCHER/BUSH**
AD	jvb	Match B	1	0.1	6.2	15.7	1,615	01A25	2	5	11	22	44	88	**176**					6C1R	**BUTCHER/BUSH**
Gm	jvb	Match C	1	0.1	3.6	9.3	633	01A25	2	5											
Gm	jvb	Match D	1	0.7	4.1	10.8	561	01A25	2	5											
Gm	jvb	Match E	1	1.0	3.4	8.7	534	01A25	2	5											
FF	jvb	Match F	1	1.2	6.1	12.2	1,686	01A25	2	5	11	22	44	88	**176**					6C	**BUTCHER/BUSH**

Figure 1.2. A simple example with three matches in a genealogy triangulation

Figure 1.2 shows six Matches that triangulate; they match me and each other in TG-01A25 which is my identification code for this segment. These widely separated Matches are greater than third cousins apart. Match A has two MRCAs with me, both listed. Three of these Matches (A, B, and F) have the same MRCA with me: my fifth-great-grandparents Paulser Butcher (1753–1829) and Elizabeth Bush (1758–1839) who lived in Lewis County, Virginia. The abbreviation BUTCHER/BUSH is used in the MRCA Couple column.

6. The exception is a very long segment, say 100cM, which is part of a total of 900cM from a first cousin. Because of the amount of shared DNA, we can be fairly confident that the MRCA is a grandparent. I have found instances where a second cousin, who shared six DNA segments with me, was also a sixth cousin, and the smallest of the segments was from this more distant MRCA.

I have columns for each generation out to the MRCA couple with Ahnentafel numbers in these cells. Column G2 is my parent, in this case my father, and "2" is the Ahnentafel number for him. Paulser Butcher is Ahnentafel 176; this number is bold to indicate it represents the couple. At this point, we do not know which one of these two fifth-great-grandparents is the one who passed this segment down to me or the overlapping segments down to each of the Matches. Each Ahnentafel number represents a specific person in my tree. This system offers some advantages in analyzing TGs as we will see in other examples.

You can see the "removes" in the Czn column. This shows that Match A is a generation closer to the MRCA than I am and Match B is a generation further. Match F and I are on the same generational level, both of us being fifth-great-grandchildren of BUTCHER/BUSH. Note the "CHROMOSOME" bar (row 3) that provides a visual break between chromosomes in my spreadsheet; the "SEGMENT" bar (row 4) that summarizes the TG-01A25 segment spanning from the beginning of chromosome 1 to about 6.1 Mbp, at which the next TG-01Aa25 starts. There is often fuzziness at the end of each TG, but it is nothing to worry about. The focus is on the 0.0-6.1 Mbp segment which is about 17 cM long. I would prefer two more Matches to agree on the MRCA in this TG. Six Matches for the sixth-great-grandparent level would provide a higher level of confidence for this MRCA identification.

For convenience, I color some paternal data shades of blue and maternal data shades of red. I use green when I develop a hypothesis for the MRCA of a TG.

> *Lessons Learned*: In Triangulation and chromosome mapping we are dealing with a lot of data. Develop methods that suit your style of thinking and analyzing. The goals are to group data and highlight the evidence gleaned from it. A single MRCA in a TG is just one clue; we should have multiple pieces of evidence before we can reach conclusions.

Case 2: walking the ancestor back

The word "prove" should not be used when we link ancestors to TGs; the DNA is randomly recombined with many possibilities. We can increase confidence that we are on the right track using several methodologies. Genealogy Triangulation was illustrated above. Another method is "Walking the Ancestor Back." This means having a TG which includes several different levels of cousins to different MRCAs, all on the same ancestral line. For example a second cousin (2C), a fourth cousin once removed (4C1R), a sixth cousin (6C), and a seventh cousin (7C), all on the same line of descent from a sixth-great-grandparent couple form a TG. Together these offer corroborating evidence that the sixth-great-grandparents are a CA for this TG. All the Matches descend from this CA.

										1C	2C	3C	4C	5C	6C	7C	8C	9C	10C		
Co	Don	MATCHNAM	Chr	Start	End	cMs	SNPs	TG ID	G2	G3	G4	G5	G6	G7	G8	G9	G10	G11	G12	Czn	MRCA Couple
	jvb	**SEGMENT -**	11	115.8	131.3	*24.0*	4,178	11L25	**2**	**5**	**11**	**23**	**47**	**95**	190	380				7C	**HENRY/ROBINSON**
	jvb	**SEGMENT -**	11	115.8	131.3	*24.0*	4,178	11L25	**2**	**5**	10	21	42	84	168	337	**674**			7C1R	**HARRIS, Hugh**
	jvb	**SEGMENT -**	11	115.8	131.3	*24.0*	4,178	11L25	**2**	**5**	**11**	**23**	**47**	**94**						5C	**MEANS/HENRY**
AD	jvb	Match A1	11	115.8	129.2	23.8	3,942	11L25	2	5	11	23	**46**							3C	GLAZE/MEANS
AD	jvb	Match A2	11	115.8	129.2	23.8	3,942	11L25	2	5	10	21	42	84	168	337	**674**			7C1R	HARRIS, Hugh
Gm	jvb	Match B	11	115.9	122.2	10.4	1,903	11L25	2	5	11	23	**47**	**95**	190	**380**				7C	HENRY/ROBINSON
FF	jvb	Match C	11	116.0	128.8	23.8	4,178	11L25	2	5	11	**22**								2C-1	BUTCHER/GLAZE
FF	jvb	Match D	11	116.0	128.8	23.8	4,178	11L25	2	5	11	23	**46**							3C-1	GLAZE/MEANS
FF	jvb	Match E	11	116.0	128.8	23.8	4,178	11L25	2	5											
23	jvb	Match F	11	116.6	125.9	15.2	2,297	11L25	2	5											
AD	jvb	Match G	11	116.6	124.1	13.0	2,286	11L25	2	5	11	23	**47**	**94**						5C1R	MEANS/HENRY
FF	jvb	Match H	11	117.1	129.2	22.6	3,878	11L25	2	5											

Figure 1.3. A few different MRCAs in genealogy triangulation

Figure 1.3 illustrates TG-11L25, which is a little more elaborate than the Genealogy Triangulation example. Match C is a close cousin with an MRCA on our BUTCHER/GLAZE ancestors. He shares several segments with me. The MRCA is bolded Ahnentafel 22, meaning Ahnentafel 22 or 23. Match D is another generation out; our MRCA is bolded 46, the ancestral couple is GLAZE/MEANS. Ahnentafel numbers indicate that 46 is the father of 23, so this is on the same line as Match C. In other words either GLAZE (46) or MEANS (47) could have passed an overlapping segment in TG-11L25 to Match D and to Match C and also to me and Match A.

Now look at Match G. Our MRCA is bolded 94, MEANS/HENRY, whose daughter was 47, the wife of 46. This indicates either MEANS (94) or HENRY (95) could have passed overlapping DNA segments down to me, Match D, Match A, and Match C. Finally, we have Match B, with a smaller segment (10.4 cM) who is a seventh cousin on our MRCA 380/381, HENRY/ROBINSON, two generations further out. From the Ahnentafel numbers we can immediately see their granddaughter, 95, was the wife of the Match G MRCA. Either 380 or 381 could have passed overlapping DNA segments down to me, Match G, Match D, Match A, and Match C.

The Match B MRCA of HENRY/ROBINSON is speculative; we have no corroboration yet. Green highlighted segments have some confirmation. Note that Match A2 is also a seventh cousin once removed on an unrelated HARRIS line. I have noted that clue, but in light of all the other evidence, I am pretty confident Match A and I are genetically linked through our GLAZE to MEANS to HENRY to HENRY-or-ROBINSON line.

TG-11L25 has me and five Matches all in agreement on the same ancestral line. We have "Walked the Ancestor Back" to at least MEANS/HENRY and probably to HENRY (380) or ROBINSON (381). That is my hypothesis for TG-11L25 in my chromosome map—subject to change as new data flows in.

Lessons Learned: The lesson is to communicate with your Matches. It takes luck to find several cousins in a TG with robust enough trees to find MRCAs that all align. It does not always work that way. It is hard work communicating with many Matches.

Case 3: the Higginbotham story

With over 50,000 Matches at AncestryDNA, over 100 had Higginbotham ancestry. I copied then pasted the list to a Word document and then analyzed each Match. The copied list preserves the hyperlink to each kit at AncestryDNA; clicking on the link brings up that Match's page. A quick review of each page found over fifty of the Matches were cousins on my maternal Higginbotham line.

My Higginbotham ancestors are

- Fourth-great-grandparent—Elizabeth Higginbotham, b. 1778 Amherst County, Virginia; m. 1792 Amherst County, Virginia, to James Shields, b. 1772 Nelson County, Virginia
- Fifth-great-grandparent—Aaron Higginbotham Jr., b. 1752 Amherst County, Virginia; m. 1775 Amherst County, Virginia, to Nancy Croxton, b. 1756 Amherst County, Virginia
- Sixth-great-grandparent—Aaron Higginbotham Sr., b. 1720 Amherst County, Virginia; m. 1740 to Clara Graves, b. 1723 Accomack County, Virginia
- Possibly Seventh-great-grandparent—John Higginbotham, b. 1695 Barbados; m. to Frances Riley, b. 1696 Ireland (This generation is not proven and possibly is not correct, but most online trees list John and Frances this way. All of the children for this couple appear to be correct in most online trees. Sometimes speculative ancestors, not yet proven, are placed in research trees while working on a hypothesis.)

Because AncestryDNA does not provide shared segment information, it is necessary to use other tools, such as those on *GEDmatch,* for segment analysis. I crafted a short message outlining my quest to find the DNA segment(s) for this Higginbotham line, explained how important *GEDmatch* is, and requested that the Matches upload to *GEDmatch.* I provided a link to my blog post with detailed steps to upload from AncestryDNA[7] and promised a report of my DNA analysis and findings. I was helpful and upbeat; I promised to do the work if they just uploaded. I sent about ten of these messages each night (to avoid the Ancestry spam detectors). At least thirty-three Matches uploaded and provided their *GEDmatch* kit number. Four kits proved to be on my paternal side so clearly my maternal Higginbotham ancestors were not the MRCA. Thirteen of these Matches from ten different families were all included in TG-04P36. On TG-04E36 there were seven Matches from six families. The rest of these Higginbotham Matches were spread over twelve other TGs, with no more than two per TG.

Is this outcome reasonable? We have 256 seventh-great-grandparent *couples* and an estimated 377 TGs across all of our DNA. We might expect, on average, each seventh-great-grandparent couple to "have" one or two TGs. So having two Higginbotham TGs is reasonable. Since

7. Jim Bartlett, "Uploading to GEDmatch," *segment-ology*, 19 January 2017 (https://segmentology.org/2017/01/19/uploading-to-gedmatch/).

most of my Matches share my Colonial Virginia ancestry, they often share several MRCAs with me. I expect that the other Higginbotham Matches will have other MRCAs in those TGs, and they often do. However, those Matches will remain *genealogy* Higginbotham cousins based on our paper trails but probably not *genetic* Higginbotham cousins.

										1C	2C	3C	4C	5C	6C	7C	8C	9C		
Co	Don	MATCHNAM	Chr	Start	End	cMs	SNPs	TG ID	**G2**	**G3**	**G4**	**G5**	**G6**	**G7**	**G8**	**G9**	**G10**	**G11**	**Czn**	**MRCA Couple**
	jvb	**SEGMENT -**	4	154.2	179.6	*28.0*		04P36	3	6	13	26	53	107	214	428	**856**		8C	HIGGINBOTHAM/RILEY
23	jvb	Match A	4	154.2	175.1	22.6	3,733	04P36	3	6										
AD	jvb	Match B	4	154.2	177.8	25.8	4,637	04P36	3	6	13	26	53	107	214	**428**			7Cc	HIGGINBOTHAM/GRAVES
AD	jvb	Match C1	4	154.2	165.9	13.1	2,262	04P36	3	6	13	26	53	106	**212**				6C	SHIELDS/FINLEY
AD	jvb	Match C2	4	154.2	165.9	13.1	2,262	04P36	3	6	13	26	53	107	**214**				6C1R	HIGGINBOTHAM/CROXTON
AD	jvb	Match D	4	154.2	162.4	8.5	1,615	04P36	3	6	13	26	53	**106**					4C1R	SHIELDS/HIGGINBOTHAM
AD	jvb	Match E1	4	154.2	164.0	10.1	1,907	04P36	3	6	13	26	53	107	214	**428**			6C1R	HIGGINBOTHAM/GRAVES
AD	jvb	Match E1	4	154.2	164.0	10.1	1,907	04P36	3	6	23	26	52	104	208	**416**			7C1R	PLUNKETT/FRANK
AD	jvb	Match F	4	154.2	179.7	28.3	3,525	04P36	3	6	13	26	53	107	214	**428**			7Cc	HIGGINBOTHAM/GRAVES
Gm	jvb	Match G	4	154.2	167.2	14.4	1,786	04P36	3	6										
FF	jvb	Match H	4	154.3	169.8	17.2	3,400	04P36	3	6										
23	jvb	Match J	4	154.5	177.4	26.0	4,161	04P36	3	6	13	26	53	**106**					5C	SHIELDS/HIGGINBOTHAM
AD	jvb	Match K	4	154.5	164.6	10.4	1,971	04P36	3	6	13	26	53	107	214	428	**856**		8C	HIGGINBOTHAM/RILEY
Gm	jvb	Match L	4	156.7	166.7	10.3	1,287	04P36	3	6										
FF	jvb	Match M	4	156.8	179.8	27.0	5,000	04P36	3	6	13	26	53	107	214	428	**856**		8C	HIGGINBOTHAM/RILEY
AD	jvb	Match N1	4	158.1	171.1	13.5	2,472	04P36	3	6	13	26	52	105	211	422	**844**		7C1Rx	RUCKER/FIELDING
AD	jvb	Match N2	4	158.1	171.1	13.5	2,472	04P36	3	6	13	26	53	107	214	428	**856**		8C1Rx	HIGGINBOTHAM/RILEY
AD	jvb	Match P	4	161.1	179.9	21.2	2,661	04P36	3	6	13	26	53	107	214	428	**856**		8C	HIGGINBOTHAM/RILEY
23	jvb	Match Q	4	161.8	175.6	15.5	2,531	04P36	3	6										
AD	jvb	Match R1	4	161.9	169.1	8.0	1,482	04P36	3	6	13	27	54	**108**					5C1R	PLUNKETT/HILL
AD	jvb	Match R2	4	161.9	169.1	8.0	1,482	04P36	3	6	13	26	53	**106**					5C1R	SHIELDS/HIGGINBOTHAM
23	jvb	Match S	4	165.9	179.4	16.3	2,463	04P36	3	6										
Gm	jvb	Match T	4	167.0	179.8	14.2	2,590	04P36	3	6										
AD	jvb	Match U	4	167.2	179.7	14.0	2,516	04P36	3	6	13	26	53	107	214	428	**856**		8C	HIGGINBOTHAM/RILEY
23	jvb	Match V	4	169.5	179.6	13.3	1,870	04P36	3	6										
AD	jvb	Match W	4	169.7	179.8	11.6	2,114	04P36	3	6	13	26	53	107	214	428	**856**		8C	HIGGINBOTHAM/RILEY

Figure 1.4. The Higginbotham story in segments

Figure 1.4 shows the HIGGINBOTHAM shared segments in TG-04P36. There are eleven different Matches (a few representing two or more family members) from AncestryDNA, plus others from 23andMe and FamilyTreeDNA (FTDNA). Three Matches are at the fifth cousin level, one at the sixth cousin level, three at the seventh cousin level, and six at the eighth cousin level. Matches C, E, N, and R were also cousins on a different line, as is common in Colonial Virginia, but the overwhelming evidence points to a Higginbotham TG. The TG and segments colored dark green are all on the same line.

Several of these MRCAs were found by extending my Matches' trees further out. In this case I knew exactly what I was looking for: Higginbothams in Amherst County, Virginia, in the 1700s. Several of these Higginbotham Matches showed up as AncestryDNA Hints. Because they are all beyond the fourth cousin range, they were generally not included in Shared Matches.

Lessons Learned: It is relatively easy to start with the surname of a fifth- to seventh-great-grandparent and see if several of them fall into a few TGs. Success partly depends on the ancestor having a large family, and thus many living descendants and DNA Matches.

Takeaway: Triangulation, both Segment and Genealogy, works well beyond fourth cousin level. However, higher levels of confirmation are required for the more distant MRCAs. A *rule-of-thumb* is that the number of Match/families with MRCAs in a TG needs to be at least the same number as the number of "Greats" in the MRCA. That is, to be confident of a seventh-great-grandparent MRCA, you should have seven or more different Match/families with MRCAs on that line in the TG.

For reference, here is a partial Outline of the Higginbotham line showing where the TG-04P36 shared DNA segments were from.

John Higginbotham c1690-1762; m Frances Riley
- **Joseph** Higginbotham
  - **Jacob** Morrison Higginbotham
    - **James C** Higginbotham
      - **4 DNA kits=> [04P36]**
    - **Benjamin G** Higginbotham
      - **1 DNA kit => [04P36]**
- **Moses** Higginbotham
  - **Joseph S** Higginbotham
    - **1 DNA kit => [04P36]**
  - **Robert** Higginbotham
    - **1 DNA kit => [04P36]**
- **Aaron** Higginbotham Sr.
  - **Samuel** Higginbotham
    - **2 DNA kits => [04P36]**
  - **Aaron** Higginbotham Jr .
    - **1 DNA kit =>** Me/Jim Bartlett **[04P36]**
- **Benjamin Aimsley** Higginbotham Sr.
  - **Caleb** Higginbotham
    - **1 DNA kit => [04P36]**
  - **Benjamin Graves** Higginbotham
    - **3 DNA kits => [04P36]**

This Higginbotham family was very prolific, with many living descendants, many of whom have taken an atDNA test and many who match me. However, is every one of my Matches who descends from this line genetically related to me through Higginbotham DNA? Almost certainly not.

This far back into Colonial Virginia, each of us has many possibilities for MRCAs and often share multiple MRCAs with our Matches. At AncestryDNA, I have over 50,000 Matches and only about 1,000 are fourth to sixth cousins or closer. By count, about one-third of my 704 Shared Ancestor Hints at AncestryDNA are at the eighth cousin level. By extrapolation, over 10,000 (20%) of my AncestryDNA Matches would be about eighth cousins. Spread over 400 TGs, this would be twenty-five eighth cousins in each group. Even if half of all AncestryDNA Matches are false, there are still many cousins in each TG. So having thirteen Matches with the same MRCA in a TG should be expected and is reasonable.

You might notice an almost impossible situation in the outline above. The same DNA segment, TG-04P36, is passed down to distant descendants from four siblings (Joseph, Moses, Aaron, and Benjamin, all sons of John Higginbotham). The odds of this are almost zero. The odds of the same segment from two siblings passing down to two eighth cousins is fairly small, but it happens because we have so many eighth cousins. Two siblings share about 50% of their DNA with each other. Three siblings share only about 25% of their DNA and the odds of the same segment from three siblings passing down to three eighth cousins is small, but possible. We cannot reconcile an almost zero probability of four siblings sharing the TG-04P36 segment with various eighth cousins as shown in the outline above and in Figure 1.4. Probably the two kits from Moses are not genetically linked to this segment. Another possibility is that some of the genealogies are incorrect; the Matches descend from this Higginbotham line, but sometimes, particularly with the same given names used multiple times, the wrong Higginbotham might be selected as the ancestor. A third possibility is that some of the same Matches who Triangulate on segment TG-04P36 descend differently—perhaps from an ancestor of one of the MRCA couple and the path for this segment is that line.

The result is a strong case for TG-04P36 being linked to the Higginbotham line; however, that does not mean every Match is correct as noted. Through Segment Triangulation, we can be confident that the TG-04P36 is strong and comes from some ancestor. Based on Genealogy Triangulation and Walking the Ancestor Back, the strongest case for the MRCA is the Higginbotham line.

Matches in this TG will be contacted for more information. Finding a third or fourth cousin in this TG with Higginbotham ancestry would corroborate this hypothesis.

Case 4: the Colonial Virginia headache

One part of my ancestry is particularly tangled while the rest of my ancestry is more straightforward. On the straightforward side, my father's parents both had Colonial Virginia ancestry, but they came from very different regions. Dad's paternal ancestry was from the Northern Neck of Virginia in the 1600s and 1700s; they married in Northern Virginia to lines from Pennsylvania and Maryland. Dad's maternal ancestry was from Southwestern Virginia. My father's parents, with different geographic backgrounds, met and married in West Virginia. My mother's maternal ancestors were immigrants in the 1850s from Scotland and Germany. My mother's paternal grandfather came from Pennsylvania to Central Virginia with small families resulting in fewer Matches than other lines. On the tangled side, my mother's paternal grandmother came from Central Colonial Virginia. Her ancestors were farmers with large families in the same area for several generations. The Ahnentafel numbers for this line are 3-6-13 and I see that pattern often in my spreadsheets. From this line, I get Matches with multiple MRCAs. I get TGs with Matches with conflicting MRCAs even within 3-6-13. One such group is TG-08L36.

										1C	2C	3C	4C	5C	6C	7C	8C		
Co	Don	MATCHNAM	Chr	Start	End	cMs	SNPs	TG ID	**G2**	**G3**	**G4**	**G5**	**G6**	**G7**	**G8**	**G9**	**G10**	**Czn**	**MRCA Couple**
	jvb	**SEGMENT -**	8	105.9	128.7	*22.0*		08L36	3	6	13	27	55	110	221	443	886	10C1R	**CHILES, Walter**
	jvb	**SEGMENT -**	8	105.9	128.7	*22.0*		08L36	3	6	13	27	54	108	217	435	870	10C	[TINSLEY]
	jvb	**SEGMENT -**	8	105.9	128.7	*22.0*		08L36	3	6	13	27	54	109	219	439	**878**	8C	**WALLER/CARR**
	jvb	**SEGMENT -**	8	105.9	128.7	*22.0*		08L36	3	6	13	27	54	108	217	434	868	10C	**RUCKER**
	jvb	**SEGMENT -**	8	105.9	128.7	*22.0*		08L36	3	6	13	26	52	104	209	418	836	10C	**HAWKINS, Sir**
	jvb	**SEGMENT -**	8	105.9	128.7	*22.0*		08L36	3	6	13	26	52	104	209	**418**		7C1R	**HAWKINS/WILLIS**
	jvb	**SEGMENT -**	8	105.9	128.7	*22.0*		08L36	3	6	13	26	53	106	213	427	854	9C	**CALDWELL/PHILLIPS**
	jvb	**SEGMENT -**	8	105.9	128.7	*22.0*		08L36	3	6	13	27	54	109	218	436	873	9C	**MARRIOTT/WARREN**
	jvb	**SEGMENT -**	8	105.9	128.7	*22.0*		08L36	3	6	13	27	54	109	219	438	876	10C	**OVERTON/GARDINER**
	jvb	**SEGMENT -**	8	105.9	128.7	*22.0*		08L36	3	6	13	27	54	109	219	438	877	10C	**JENNINGS**
	jvb	**SEGMENT -**	8	105.9	128.7	*22.0*		08L36	3	6	13	27	55	110	221	443	886	10C	**PAGE/LUKIN**
	jvb	**SEGMENT -**	8	105.9	128.5	*23.0*		08L36	3	6	13	26	52	105	211	423	**846**	8C	**RUCKER**
23	jvb	Match A	8	106.8	124.5	13.4	3,267	08L36	3	6	13	26	52	104	209	418	836	10C	**HAWKINS, Sir**
23	jvb	Match B	8	106.8	127.5	19.4	4,016	08L36	3	6	13	27	54	109	219	439	878	9C2R	**WALLER/CARR**
FF	jvb	Match C1	8	107.3	121.7	11.2	2,800	08L36	3	6	13	27	54	109	219	438	877	10C	**JENNINGS**
FF	jvb	Match C2	8	107.3	121.7	11.2	2,800	08L36	3	6	13	27	54	109	218	436	873	9C	**MARRIOTT/WARREN**
Gm	jvb	Match D	8	107.8	125.0	14.7	3,281	08L36	3	6	13	27	54	108	217	435	870	10C	[maybe TINSLEY]
Gm	jvb	Match E	8	108.3	126.9	17.7	3,639	08L36	3	6	13	27	55	110	221	443	886	10C1R	**CHILES, Walter**
FF	jvb	Match F	8	111.5	124.3	14.0	2,693	08L36	3	6	13	27	54	109	219	438	876	10C	**OVERTON/GARDINER**
FF	jvb	Match G	8	115.7	124.1	10.3	2,000	08L36	3	6	13	26	53	106	213	427	854	9C	**CALDWELL/PHILLIPS**
FF	jvb	Match H	8	115.7	124.1	10.3	2,000	08L36	3	6	13	26	53	106	213	427	854	9C	**CALDWELL/PHILLIPS**
23	jvb	Match J	8	116.7	127.0	13.8	2,336	08L36	3	6	13								[4gp: Australia]
FF	jvb	Match K	8	117.4	124.1	9.1	1,700	08L36	3	6	13								[maybe HAWKINS]
FF	jvb	Match L1	8	118.6	127.2	14.4	2,292	08L36	3	6	13	27	54	108	217	434	868	10C	**RUCKER**
23	jvb	Match L2	8	118.7	126.9	11.6	1,920	08L36	3	6	13	27	55	110	221	443	886	10C	**PAGE/LUKIN**
23	jvb	Match L3	8	118.7	126.9	11.6	1,920	08L36	3	6	13	26	52	105	211	423	**846**	8C	**RUCKER**
Gm	jvb	Match M	8	119.4	127.8	13.1	1,987	08L36	3	6	13								[maybe HIGGINBOTHAM]
				[TG [08L36] has n> 100 other Triangulated Matches															

Figure 1.5. The Colonial Virginia headache

Figure 1.5 highlights a recurring issue with some Colonial Virginia ancestry. TG-08L36 represents an ancestral DNA segment of about 22 cM. It comes down one ancestral line to me. However, each TG can only come down one way—through one ancestral line—in this example, from 836, 846, 854, 868, 870, 873, 876, 878, *or* 886. Notwithstanding the eighth, ninth, tenth, and eleventh cousin relationships, the Ahnentafel numbers are not very different. They all come from a very narrow part of my ancestry; they are all ancestral to 3-6-13. Match C has two known MRCAs with me and Match L has three MRCAs. This tells me two things: (1) I am almost certain this is on my 3-6-13 line, and (2) I do not have a clue who the correct MRCA is. I would have great confidence in the chromosome map at the great-grandparent level, but as I try to map more distant generations it turns into a guessing game. It could be either Ahnentafel number 26 or 27 at the second-great-grandparent level. There might be one of the MRCAs above for whom I have done a lot more research than for others, but the DNA is independently random. TG-08L36 could have come from any one of these MRCAs.

So what do we do? We keep researching, communicating with the Matches in this TG, and reaching out to new Matches pouring in every week. In particular, there is a need to search for "intermediate" cousins in this TG on a much closer MRCA. As a general rule, give more weight to the MRCAs with closer cousins. Of course, in the end we have to weigh all of the evidence from DNA and documents to come up with a good MRCA hypothesis. See also additional research steps in "The Challenge of Endogamy and Pedigree Collapse" chapter.

> *Lessons Learned*: Look at Figure 1.5 again. In the first column, you can see kits from 23andMe, FTDNA and *GEDmatch* (with the full list of Matches in this TG, there are also some AncestryDNA kits). This shows the value of testing at all the major companies and uploading to *GEDmatch*. You never know when a close cousin might show up or which Matches will turn out to have MRCAs in their tree.

Chromosome Mapping

Once all of the TG boundaries are determined with Segment Triangulation, the outlines of a chromosome map fall into place. The size and location of all the puzzle pieces become clear. Chromosome maps for the maternal and paternal sides are independent—there is no interrelationship between the two.

There are at least two fundamental ways to look at a chromosome map. In each case there is a separate map for the paternal chromosomes and the maternal chromosomes.

1. By *generation*. Looking at the grandparent generation shows the *paternal* map based on two grandparents. The grandparent segments alternate, unless a whole chromosome is passed down intact.[8] We could also look at the paternal great-grandparent generation for segments linked to each of the four paternal great-grandparents. Or we could look seven generations back and see how the sixty-four paternal fifth-great-grandparents link up with the segments on each chromosome. There is a high probability that only some of the fifth-great-grandparents would link on any given chromosome and some on none of the chromosomes at this distance.
2. By *TG segments*. TGs are formed between natural break points (crossovers) in the data. This creates a chromosome map of segments from various generations. Smaller TGs are often difficult to tease out of the data because some shared segments, from closer cousins, will overlap two or more ancestral segments. In practice, shared segments cover most of each chromosome and some crossovers are very clear. As MRCAs are linked to these TG segments, the picture becomes clearer. Chromosome mapping is an excellent tool to determine the correct MRCAs. With a chromosome map of TGs linked to MRCAs, we readily see the underlying map by generations.

8. Jim Bartlett, "Segments Bottom Up," *segment-ology*, 24 May 2015 (https://segmentology.org/2015/05/24/segments-bottom-up/) for more information and diagrams about ancestral segments and crossovers.

Chromosome mapping the grandparents on one side: quality control and parsimonious adjustments

My spreadsheet is based on "TG segments" with MRCAs from different generations. Ahnentafel numbers for the MRCA of each TG represent the different grandparents, great-grandparents, and so on. Of course, all my paternal TGs cover each of my paternal chromosomes. When I look at my 184 paternal TGs, I can consolidate them into sixty-two segments: twenty-five on my father's *paternal* side (Ahnentafel 2-4); twenty-seven on my father's *maternal* side (2-5); and ten that are uncertain with no high-confidence MRCAs identified for those segments. See a partial list of the paternal grandparent segments in figure 1.6.

		Chr	Start point	End poin	Mbp			MRCA	Mbp Summary Data		
TG3	Grandparent	1	0.0	35.9	35.8	**2**	**5**	MITCHELL		36	
TG3	Grandparent	1	35.9	102.1	66.2	**2**	**4**	BARTLETT	66		
TG3	Grandparent	1	102.1	158.3	56.2	**2**	**5**	MITCHELL		56	
TG3	Grandparent	1	158.3	238.1	79.8	**2**	**4**	BARTLETT	80		
TG3	Grandparent	1	238.1	249.3	11.3	**2**	**?**	Unknown			11
TG3	Grandparent	2	0.0	10.0	9.9	**2**	**?**	Unknown			10
TG3	Grandparent	2	10.0	120.4	110.4	**2**	**5**	MITCHELL		110	
TG3	Grandparent	2	120.4	220.8	100.4	**2**	**4**	BARTLETT	100		
TG3	Grandparent	2	220.8	243.2	22.5	**2**	**5**	MITCHELL		22	
TG3	Grandparent	3	0.0	13.8	13.7	**2**	**?**	Unknown			14
TG3	Grandparent	3	13.8	141.5	127.7	**2**	**4**	BARTLETT	128		
TG3	Grandparent	3	141.5	189.4	47.9	**2**	**5**	MITCHELL		48	
TG3	Grandparent	3	189.4	199.3	10.0	**2**	**4**	BARTLETT	10		
TG3	Grandparent	4	0.0	59.6	59.5	**2**	**4**	BARTLETT	60		
TG3	Grandparent	4	59.6	80.7	21.1	**2**	**?**	Unknown			21
TG3	Grandparent	4	80.7	143.6	62.9	**2**	**4**	BARTLETT	63		
TG3	Grandparent	4	143.6	191.1	47.6	**2**	**5**	MITCHELL		48	
		TRUNCATED									
TG3	Grandparent	18	0.0	48.1	48.0	**2**	**4**	BARTLETT	48		
TG3	Grandparent	18	48.1	78.1	30.1	**2**	**5**	MITCHELL		30	
TG3	Grandparent	19	0.0	63.8	63.8	**2**	**5**	MITCHELL		64	
TG3	Grandparent	20	0.0	56.1	56.0	**2**	**4**	BARTLETT	56		
TG3	Grandparent	20	56.1	63.0	7.0	**2**	**5**	MITCHELL		7	
TG3	Grandparent	21	9.8	34.1	24.3	**2**	**4**	BARTLETT	24		
TG3	Grandparent	21	34.1	48.1	14.1	**2**	**5**	MITCHELL		14	
TG3	Grandparent	22	14.5	51.3	36.9	**2**	**5**	MITCHELL		37	

	BART	MTCH	Unk	Total
Mbp	635	472	56	1163
# Segs	25	27	10	62
Chr				22
Crossovers				40
% Mbp	55	41	5	
% Segs	40	44	16	

Figure 1.6. Initial paternal grandparent distribution of 62 segments over 22 chromosomes

In the summary data at the lower right, subtract the number of chromosomes (22) from the number of segments (62) to get the number of crossovers (40). A total of 2,808 Mbp is contained in 22 paternal chromosomes (minus the untested portions on chromosomes 13, 14, 15, 21, and 22). This is the DNA that is tested and reported in shared segments and is the data available for analysis.

Is this outcome reasonable? The Mbp total is correct, so the individual segments are probably correct. The number of segments on each side is about the same, as expected. The percentage of the total between the two grandparents is not exactly half and is skewed a little more than expected. The total number of crossovers is high; forty is roughly 50% higher than the average twenty-seven crossovers expected for males in any one generation.[9]

The "Unknown" segments are analyzed and assigned in the most likely way. That is, if an unknown grandparent segment is at the beginning or end of a chromosome, it is assigned to the same grandparent as the adjacent TG (resulting in one larger segment from one grandparent, rather than two segments from different grandparents). If an unknown segment is between two segments from the same grandparent (both Bartlett or Mitchell in the example above), it is assigned to the same grandparent. This reduces the number of crossovers bringing the total closer to the average. Figure 1.7 is the resulting summary of paternal grandparent segments.

This process is a good quality control check for the conclusions. The Unknown segments have been eliminated. The total Mbp remains the same. The total number of crossovers from my father to me is reduced to twenty-seven, which is the average for a male. The ratio of the amount of DNA from each of my father's parents is tilted to the paternal side, but not as much as before. The ratio of the number of segments is tilted to the maternal side, as expected. Does this then mean that the parsimonious guesses are correct? Of course not, but they are insightful clues and probably more right than wrong.

This can be viewed in a chromosome map created from the spreadsheet data using Kitty Cooper's Chromosome Mapper.[10] Be sure to use her column headings, save the spreadsheet in CSV format, and navigate to that file from Cooper's Mapper page to get output similar to that seen in figure 1.8.

---

9. Chowdhury et al., "Genetic Analysis of Variation in Human Meiotic Recombination," *PLOS Genetics* 5 (September 2009): e1000648 (http://dx.doi.org/10.1371/journal.pgen.1000648), 2.

10. Kitty Cooper, "Chromosome Mapper - Make a graphic chromosome map from a CSV of your ancestral segments" (http://kittymunson.com/dna/ChromosomeMapper.php). See also the chromosome map tools on *DNA Painter* (http://www.dnapainter.com/).

		Chr	Start point	End poin	Mbp			MRCA	Mbp Summary Data	
TG3	Grandparent	1	0.0	35.9	35.8	2	5	MITCHELL		36
TG3	Grandparent	1	35.9	102.1	66.2	2	4	BARTLETT	66	
TG3	Grandparent	1	102.1	158.3	56.2	2	5	MITCHELL		56
TG3	Grandparent	1	158.3	249.3	91.1	2	4	BARTLETT	91	
TG3	Grandparent	2	0.0	120.4	120.3	2	5	MITCHELL		120
TG3	Grandparent	2	120.4	220.8	100.4	2	4	BARTLETT	100	
TG3	Grandparent	2	220.8	243.2	22.5	2	5	MITCHELL		22
TG3	Grandparent	3	0.0	141.5	141.4	2	4	BARTLETT	141	
TG3	Grandparent	3	141.5	189.4	47.9	2	5	MITCHELL		48
TG3	Grandparent	3	189.4	199.3	10.0	2	4	BARTLETT	10	
TG3	Grandparent	4	0.0	143.6	143.5	2	4	BARTLETT	144	
TG3	Grandparent	4	143.6	191.1	47.6	2	5	MITCHELL		48
		ROWS HIDDEN								
TG3	Grandparent	18	0.0	48.1	48.0	2	4	BARTLETT	48	
TG3	Grandparent	18	48.1	78.1	30.1	2	5	MITCHELL		30
TG3	Grandparent	19	0.0	63.8	63.8	2	5	MITCHELL		64
TG3	Grandparent	20	0.0	56.1	56.0	2	4	BARTLETT	56	
TG3	Grandparent	20	56.1	63.0	7.0	2	5	MITCHELL		7
TG3	Grandparent	21	9.8	34.1	24.3	2	4	BARTLETT	24	
TG3	Grandparent	21	34.1	48.1	14.1	2	5	MITCHELL		14
TG3	Grandparent	22	14.5	51.3	36.9	2	5	MITCHELL		37

	BART	MTCH	Unk	Total
Mbp	681	482	0	1163
# Segs	23	26	0	49
Chr				22
Crossovers				27
% Mbp	59	41	0	
% Segs	47	53	0	

Figure 1.7. Revised paternal grandparent distribution of 49 segments over 22 chromosomes

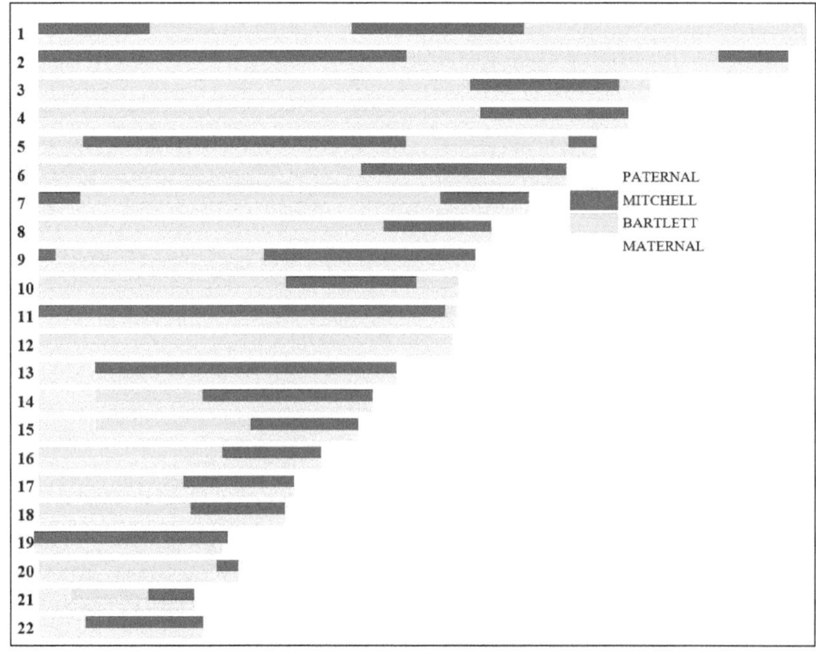

Figure 1.8. Chromosome map of paternal grandparents

On the chromosome map in figure 1.8 all, or nearly all, of chromosomes 11, 13, 19, and 22 were inherited from MITCHELL and all, or nearly all, of chromosomes 12 and 20 were inherited from BARTLETT. This is fairly unusual. Often one of the smaller chromosomes (19–22) will be inherited from one grandparent, but six is an unexpectedly high number. Probably a few of the parsimonious "adjustments" were incorrect, providing all the more reason to always stay flexible and look for more "intermediate" cousins to indicate the proper MRCAs in all TGs. As more Matches are analyzed, this Map will almost certainly change.

Looking at my 377 TGs, 113 paternal and 116 maternal TGs have MRCAs at the second-great-grandparent generation or more distant, forty have great-grandparent MRCAs, and thirty-seven more have grandparent MRCAs. This leaves seventy-one TGs with only the parent known. Figure 1.9 is a chromosome map showing the second-great-grandparent generation.

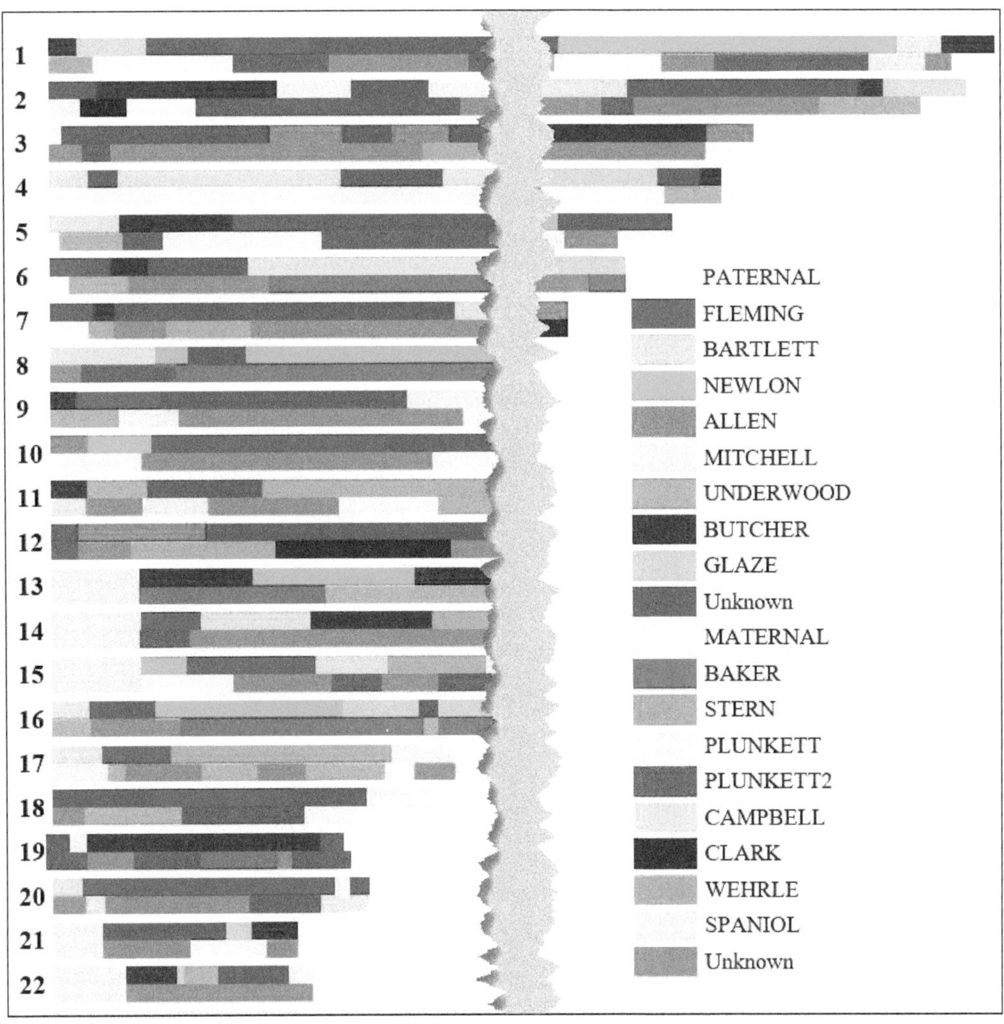

Figure 1.9. Chromosome map at the second-great-grandparent generation

This is what chromosome mapping is all about. Not all of my TGs are linked to this level, but some are linked to the great-grandparent level. This means TGs can only have one of two second-great-grandparents as an MRCA—the parents of the linked great-grandparent. For every generation where you are confident of the MRCA, there are only two possibilities for the next generation.

More Lessons Learned, Insights, Tips, and Techniques

Segment in*sights* from five years of experience

- All segments over 7 cM will form about 400 TGs in our DNA, with a little more than half on the maternal side. Each shared IBD segment will go into one of these TGs. This use of Triangulation arranges your segments into groups. New segments are easily matched to a paternal or a maternal TG. When not matched to a paternal or maternal TG, they are discarded as not being IBD.

- When a parent and a child both share an overlapping segment with a Match, it is virtually the same size and location most of the time, particularly in the 7 cM to 15 cM range. This indicates that such segments are not usually divided in the generation between the parent and child. If they are divided, the child often does not show up as a Match. This is normal.

- Some TGs with a lot of segments are due to known pileup areas.[11] Other TGs with many segments may be due to the MRCA and their descendants having large families or a very distant MRCA creating many descendants.

- Close cousins may have shared segments that span two or more TGs.

- Most Matches with a single IBD segment in the 5 cM to 20 cM range will have an MRCA well beyond ten generations back;[12] but many will be closer. The more Matches communicated with, the more MRCAs that will be determined.

- TGs are the effective equivalent of phased data—both have the same long segment of specific SNPs from an ancestor.

- Every shared segment has to be on the paternal chromosome or on the maternal chromosome or it is not IBD (meaning not from an ancestor). Every IBD segment should Triangulate on one side or the other.

---

11. "Excess IBD sharing," *ISOGG Wiki* (https://isogg.org/wiki/Identical_by_descent).
12. "Statistics of IBD genomic regions," *ISOGG Wiki* (https://isogg.org/wiki/Identical_by_descent).

Tips for TGs and spreadsheets

- Triangulated segments often have fuzzy boundaries. Focus on the fact that this segment came from a specific Ancestor rather than the precise boundaries. Look for a logical place to start a TG, and then arbitrarily end the previous TG summary bar at that same location. Some shared segments in the previous TG will "overrun" the next TG by 1–3 Mbp. Do not worry about it. For a variety of reasons, the data is not that precise.[13] Focus on the MRCA.

- For shared segments that span more than one TG, leave the whole shared segment in the first TG. The segment End position will be well past the TG End position. Then make a copy of that row, insert it in the following TG in the spreadsheet, and change the Start Point to the Start Point of that second (following) TG so it sorts with that TG. Often the MRCA is known with a long segment with a close Match, and this MRCA carries over to the second TG.

- Watch for potential misattributed parentage in your tree. Statistically, we likely all have some. As you determine grandparent TGs and great-grandparent TGs in your chromosome map, watch for some ancestors with no, or very few, assigned segments. This may be due to recent immigration or small families, but could be the result of an incorrect ancestor in your tree.

- Show Segment Start and End locations in Mbp with no decimal places (even though the examples in this chapter generally include one decimal place). These numbers are easier to use and just as accurate for our purposes.

- Triangulate the largest segments first, then lower the threshold and make another pass. With each pass, you will be adjusting the TG boundaries as the data indicates. Once you have some TGs established, adding new segments to them is much easier and areas for other TGs become apparent.

- Every TG will be a "work in process" in finding MRCAs with Matches and in refining the TG boundaries. Keep an open mind as new data becomes available.

- In Common With (ICW) lists provide *potential* TG candidate Matches. Triangulated segments will generally show up in each Match's ICW list, but many ICW groups will not triangulate (match on the same location of the same chromosome). Segment Triangulation depends on matching shared segments, which are not a criteria for an ICW list.

---

13. Jim Bartlett, "Fuzzy Data, Fuzzy Segments, No Worry," *segment-ology*, 30 May 2015 (https://segmentology.org/2015/05/30/fuzzy-data-fuzzy-segments-no-worry/).

- Maintain one spreadsheet for each test taker. TGs are based on the perspective of one test taker, the base person. Each test taker, including close relatives, will have uniquely different TGs. Spreadsheets can be combined for different kinds of analyses, but separate them again when done.

Quality control techniques

- Periodically, compare your data to a fresh download from each company. The companies sometimes "update" the segment data and the Matches sometimes change their names. This also catches mistakes made during data entry.

- Compare the Match MRCA relationships with the expected cM ranges.[14] For example, if you share 60 cM with a Match, the relationship is almost certainly closer than an eighth cousin. Conflicts and inconsistencies must be investigated.

- Develop a chromosome map at each generation and determine if the number of crossovers is reasonable. Confirm that the total amount of DNA for each Ancestor is reasonable. On average, there are about twenty-seven crossovers for males and forty-one for females in each generation, trending toward an average of thirty-four crossovers in later generations. There should be a roughly balanced distribution of total DNA among the ancestors in each generation (recognizing that some ancestral lines will gradually drift to zero in distant generations).

More lessons learned

- Test at as many companies as you can afford. Start with the Big Three: 23andMe, AncestryDNA, and FamilyTreeDNA. Then add MyHeritage, then Living DNA (now partnered with FindMyPast). Upload one kit to *GEDmatch* and DNA.Land. You never know where you will find a close cousin, a cousin with a lot of information to share, a cousin that will help define a TG, or two cousins with a Common Ancestor just beyond your brick wall. Each of these companies has something unique to offer.

- Test as many close relatives as you can afford. Start with the older generations who are closer to the MRCAs and will share more segments with the MRCAs than you do.

- Bite off small tasks that fit your time budget; get them completed so you have a working spreadsheet at the end of each task. Examples of such tasks include (1) add Match and segment data from one company and Triangulate them and (2) form all the TGs you can using only 20 cM and higher segments.

14. See the Shared cM Project in "Autosomal DNA Statistics," *ISOGG Wiki* (https://isogg.org/wiki/Autosomal_DNA_statistics).

Conclusion

Triangulation is a process and a powerful tool. Triangulated Groups can be a very valuable way to organize your work with a focus on a specific genealogical line in each TG. TGs form a chromosome map—combining and integrating genealogy and DNA. TGs and a chromosome map are invaluable sources of evidence in your research.

Autosomal DNA analysis has been quite a journey. A journey of learning about DNA and how atDNA segments are inherited. A journey that has connected me to many more cousins in the past seven years than I found in the previous thirty-six years. This has been, and continues to be, a journey on the frontier of genetic genealogy. All of us genealogists are pulling ourselves up by our bootstraps, learning from each other's experiences, and pushing the envelope every day. We have learned how to tell which side a Match is on, to do visual phasing to determine segments from our grandparents, and to form TGs that link to more distant ancestors. We are developing our own vocabulary (fuzzy data, sticky segments,[15] TGs, sides, pileups), and learning to describe genetics in plain language to millions of researchers with no genetics background. What a hoot for this old genealogist!

What of the future? Mapping our chromosomes means we know which segments of DNA came from which ancestral lines. From a genealogy standpoint, we use shared segments and family trees to identify Ancestors, break down brick walls, highlight potential misattributed parentage, resolve research disputes, find birth parents, and more. We can now know which genes came from which Ancestors. We may someday be using our genealogy hobby to improve our medical knowledge.

When all of this genetic genealogy stuff turns to mush, we can just accept our Match lists as our cousins and contact them to work out our shared ancestry. We do not need to know how to spell D-N-A. Just have fun with our genealogy research. And remember, this is *your* genealogy. Use your own judgment to come to conclusions with which you are confident, as long as you base your theories on accurate analysis techniques and accepted standards.

15. The term "sticky segments" is disliked by some genealogists because novices may misunderstand the meaning of the term. There is no inherent segment quality that results in a segment remaining intact for more generations than is typical. However, random recombination, location in an area known to experience less recombination, and small segment size may result in a segment that remains intact through multiple generations.

Visual Phasing
Methodology and Techniques

Blaine T. Bettinger, PhD, JD

Introduction

"Visual Phasing" is a process in which DNA segments are allocated to a test taker's grandparents by comparing recombination events between a set of siblings sharing those grandparents. Test results from the siblings' parents or grandparents are not required. Once segments are allocated, known matches are used to identify which segments belong to which grandparent. These labeled segments form a chromosome map, back to the grandparents, for each of the siblings.

Kathy Johnston first described this methodology and coined the terminology in 2015. She shared how-to documents in a FamilyTreeDNA forum.[1] Randy Whited also independently worked with these recombination events to generate grandparent maps. He presented his work at the 2016 Southern California Genealogical Society Jamboree.[2] Thousands of genealogists have now used visual phasing to create chromosome maps.

Although visual phasing is a complicated manual process, the chromosome maps created have many beneficial uses. First, the maps inform the genealogist which segments and how much DNA each sibling inherited from each grandparent. Although individuals inherit 50% of their DNA from their maternal grandparents, for example, it will rarely be 25% from the maternal grandmother and 25% from the maternal grandfather. Instead it will be some

All URLs were accessed 8 December 2018 unless otherwise indicated.

1. Kathy Johnston, "Segment Matching with Grandparents Using Crossover Lines," *FamilyTreeDNA* Forums (http://forums.familytreedna.com/showthread.php?t=36812), registration required.

2. Randy Whited, "Reconstructing Grandparent DNA Using Sibling Results," Southern California Genealogical Society Jamboree 2016 Session #TH023 (http://www.myconferenceresource.com/products/45-01-scgs-genetic-dna-jamboree-conference-2016.aspx).

variation that totals 50%. In one extreme example, visual phasing revealed that an individual inherited 32% from their maternal grandmother and 18% from their maternal grandfather.

Second, the chromosome maps can narrow the number of lines to be investigated when analyzing DNA matches. If a new match shares DNA at a location on chromosome 12 where an individual knows they inherited DNA only from the maternal grandfather on one copy of chromosome 12 and from the paternal grandmother on the other copy, the individual has already eliminated 50% of their tree when looking for a common ancestor shared with the match. They have eliminated the maternal grandmother and the paternal grandfather as sources of the shared DNA.

These and other benefits of visual phasing will become clear as the methodology is examined in detail.

Methodology

Visual phasing is a multi-step process. It begins by identifying segments of DNA shared by a set of siblings using a comparison tool such as the one-to-one compare tool on *GEDmatch*.[3] Recombination points are identified and assigned using shared segments of DNA. Chromosome maps are filled in based on recombination assignments. The chromosome maps are augmented using known-cousin matches to label the grandparental source of each segment.

It is essential that genealogists understand the fundamental logic of visual phasing. Although currently a manual process, it will be mostly or entirely automated in the coming years. Understanding the methodology allows the genealogist to evaluate the results of an automated process, to detect errors, and understand any assumptions made.

Step 1: providing the visual phasing input

The first step in visual phasing is identifying the input. Genealogists should begin with a set of three siblings. Although fewer or more than three siblings can be used, these scenarios complicate the analysis.

The input for visual phasing is a set of DNA comparisons for the siblings using a tool such as the one-to-one compare at *GEDmatch* (raw data could be used by automated tools in the future). The raw data of each of the siblings in the set must be downloaded from the DNA testing company and uploaded to *GEDmatch*. It is essential that each test taker provide explicit permission for their raw data to be uploaded to any third-party tool. It is preferred that all siblings in the set test at the same time with the same company; however, any test results can be used if the results can be compared to each other at *GEDmatch*.

3. *GEDmatch*® (http://www.gedmatch.com/). As of January 2019, *GEDmatch* Classic calls this the 'One-to-one' compare tool, while *GEDmatch* Genesis calls it the One-to-One Autosomal DNA Comparison tool. This refers to a tool that compares one selected test taker's genome to that of another selected test taker.

Once the raw data for each of the siblings in the set is uploaded to *GEDmatch,* the DNA will be ready for comparisons using the one-to-one compare tool.

Step 2: create a workspace for visual phasing

Before sibling comparisons are possible, a visual phasing workspace must be selected. The sibling comparisons will be inserted and the chromosome maps will be generated in the workspace.

Visual phasing workspace software must meet several criteria. It must allow insertion of sibling comparisons into the workspace, usually in the form of screenshots, and must allow chromosome mapping. The most popular options are word processing software such as Microsoft Word and its equivalents. Other options include spreadsheet software such as Microsoft Excel, presentation software such as Microsoft PowerPoint, and equivalents.

There is a third-party tool created by Steven Fox that uses Excel to create a visual phasing workspace. The Fox tool is free and is currently limited to Microsoft Windows systems. It can be accessed in the Files section of the "Visual Phasing Working Group" on *Facebook*.[4] Among other benefits, the Fox tool uses login information to access *GEDmatch* directly and import images of the siblings' one-to-one comparisons. Generating and importing these comparisons can be time-consuming and the Fox tool saves a significant amount of time.

Researchers may wish to try several different workspace options before deciding on one. Once a workspace is established, the visual phasing process can begin.

Step 3: obtain sibling comparisons

Visual phasing uses the results of sibling DNA comparisons to identify recombination points. The next step in the process is to generate sibling comparisons and import them into the workspace.

To use the one-to-one compare tool for visual phasing, enter the *GEDmatch* kit number (for example, A123456) for one of the siblings in the field labeled "Kit Number 1" and the *GEDmatch* kit number for a second sibling in the field labeled "Kit Number 2" in the fields shown in figure 2.1. Select the "Graphics and Positions" button under "Show graphic bar/ numeric positions for each Chromosome?," then select the "Submit" button to process the data. Do not change any of the other default settings of the tool. *GEDmatch* compares the DNA of the first sibling to the DNA of the second sibling to identify any segments of DNA they share.

4. "Visual Phasing Working Group," *Facebook* (https://www.facebook.com/groups/visualphasing).

GEDmatch.Com DNA one-to-one Comparison Entry Form

This utility allows you to make detailed comparisons of 2 DNA kits. Results may be based on either default thresholds, or thresholds that you provide. Estimates of 'generations' are provided as a relative means of comparison, and should not be taken too literally, especially for more than a couple of generations back.

Kit Number 1:	Sibling #1
Kit Number 2:	Sibling #2
Show graphic bar/numeric positions for each Chromosome?	(•) Graphics and Positions ( ) Position Only ( ) Graphic Only
For Full resolution graphic, check 'Full resolution' Window width in pixels:	[ ] Full resolution 1000
SNP count minimum threshold to be considered a matching segment (Leave blank for default value = 500)	
Minimum segment cM size to be included in total: (Leave blank for default value = 7)	
Size (in SNPs) of Mismatch Evaluation window. (Leave blank for default = SNP threshold)	
Size (in SNPs) of Mismatch-Bunching limit. (Leave blank for default mismatch eval window / 2)	
	Submit

Figure 2.1. The one-to-one compare tool at *GEDmatch*. The organization and inputs for this tool are likely to change over time, but the fundamental functioning will remain the same.

The output of the one-to-one compare tool identifies the DNA that the two siblings share on each chromosome. For example, figure 2.2 shows the comparison of chromosome 1 from two siblings. As with most chromosome browsers, the output displays only a single chromosome image although the siblings each have two copies of chromosome 1, one copy from their father and one copy from their mother.

For any location along a chromosome, the siblings will share DNA in one of three possible ways, each illustrated in figure 2.2:

1. No DNA is shared on either copy of the chromosome. In the one-to-one compare tool this will appear as either a gray/black or red segment, sometimes with thin yellow, red, or green lines.

2. DNA is shared on one copy of their chromosome, either maternal or paternal. In the one-to-one compare tool this will appear as a yellow segment underlined in blue, sometimes with thin green lines.

3. DNA is shared on both copies of the chromosome, the maternal and the paternal copy. In the one-to-one compare tool this will appear as a green segment underlined in blue.

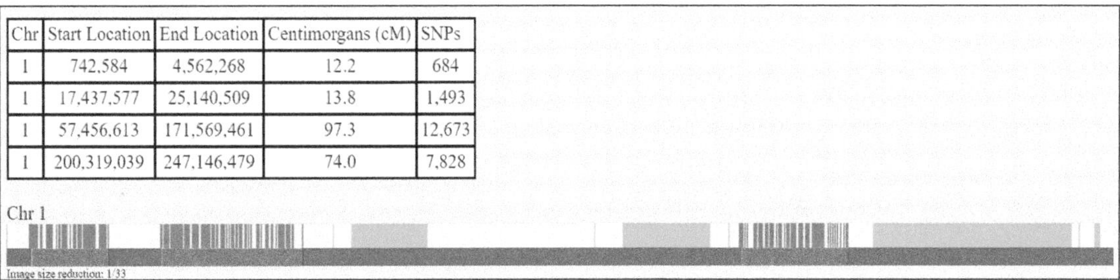

Chr	Start Location	End Location	Centimorgans (cM)	SNPs
1	742,584	4,562,268	12.2	684
1	17,437,577	25,140,509	13.8	1,493
1	57,456,613	171,569,461	97.3	12,673
1	200,319,039	247,146,479	74.0	7,828

Figure 2.2. One-to-one comparison of siblings

A comparison for each sibling pair at each chromosome must be imported into the workspace. There are several ways to do this, including taking screenshots and importing or pasting the screenshots into the workspace.

For example, figure 2.3 shows a workspace illustrating a comparison of chromosome 20 for the siblings Binh, Janya, and Saeed. This family will be used throughout this chapter. Visual phasing using three siblings will include three screenshots for each chromosome comparing: (a) sibling number 1 to sibling number 2, (b) sibling number 1 to sibling number 3, and (c) sibling number 2 to sibling number 3.

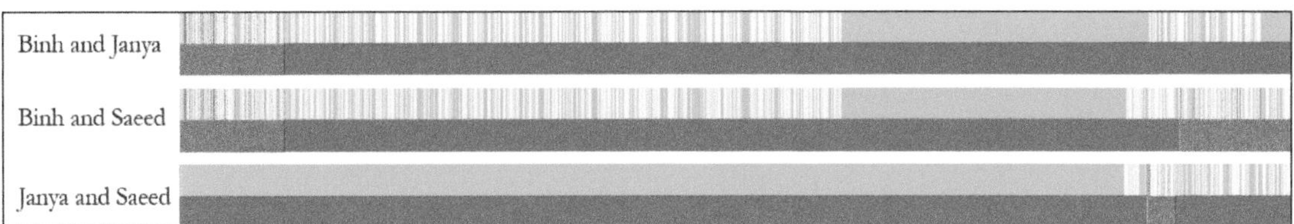

Figure 2.3. Workspace for chromosome 20

This is repeated for each chromosome, keeping each separate in the workspace. For example, each chromosome can be placed on a separate page in word processing or presentation software or on a separate tab in spreadsheet software.

Now that the sibling comparisons are within the workspace, visual phasing can commence.

Step 4: identify recombination points

The next step is to identify each location where recombination occurred for each of the three siblings. Recombination is a process that occurs during the formation of sperm and eggs in which chromosomes are broken, switched, and stitched back together. During recombination, each parent can optionally create an entirely new copy of a chromosome by stitching together segments of their maternal and paternal copies of that chromosome. This process is mostly random. There are recombination hotspots, but these are not currently factored into the visual phasing methodology.

Figure 2.4 shows recombination of chromosome 1 during formation of an egg. The mother recombined her paternal copy of chromosome 1 (inherited from her father) and her maternal copy of chromosome 1 (inherited from her mother) to generate an entirely new chromosome that she passed down to her child. This chromosome is a mixture of the DNA from the child's maternal grandmother and grandfather.

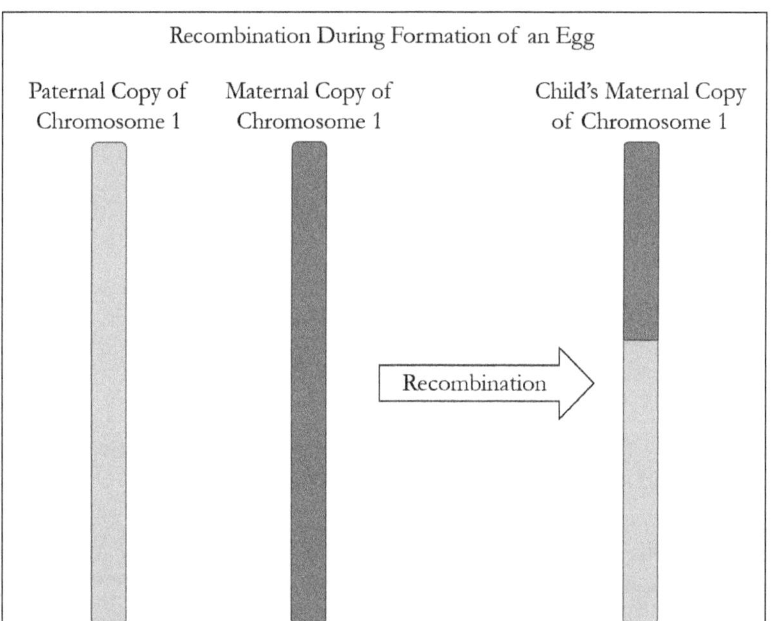

Figure 2.4. Example of recombination during formation of an egg

In the one-to-one comparisons of the three siblings, a recombination point is identified anywhere there is a change in matching. In other words, there must have been a recombination event in one or both siblings at each point where there is a change from grey/black and red to yellow or green, a change from yellow to green, or a change from green to yellow.

If two siblings inherited the same copy of a chromosome from each parent, the one-to-one comparison of the siblings would be fully green along the length of the chromosome, as shown in figure 2.5. This would most likely mean that there was no recombination on that chromosome by either parent, a somewhat rare scenario.

Chr	Start Location	End Location	Centimorgans (cM)	SNPs
1	742,584	247,164,683	281.5	35,454

Chr 1

Image size reduction: 1/36

Figure 2.5. One-to-one comparison

Instead, the siblings will mostly likely have one or more recombination events along their chromosomes, which manifests as a change in matching. In figure 2.6 there are three recombination events identified by the red arrows. Child number 1 has one recombination event, and Child number 2 has two recombination events. Each recombination event changes the matching between the two siblings.

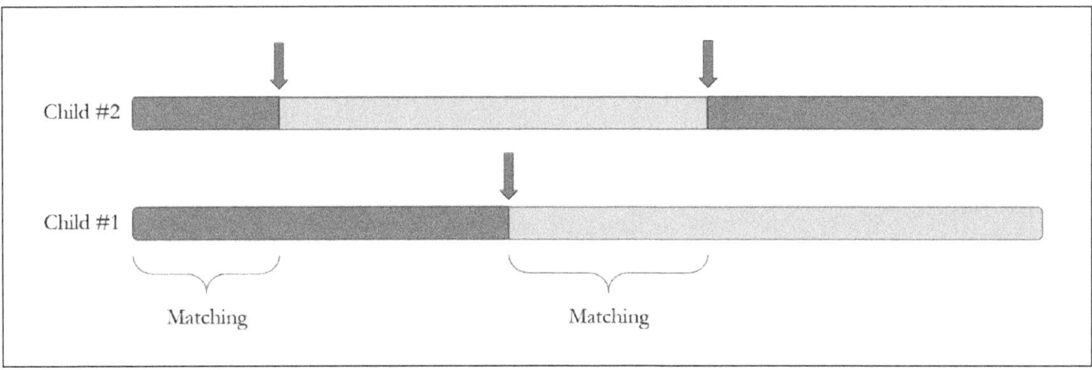

Figure 2.6. Recombination events

Matching changes in the one-to-one compare tool are used to identify the recombination events. For example, the arrows in figure 2.7 show each recombination point—characterized by every location where there is a change from one type of matching or non-matching to another type of matching.

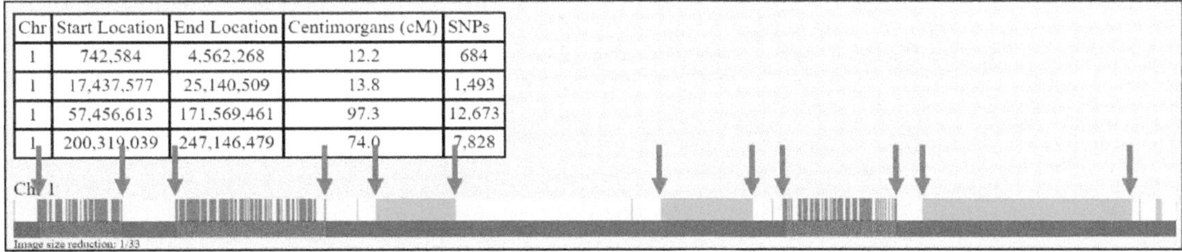

Chr	Start Location	End Location	Centimorgans (cM)	SNPs
1	742,584	4,562,268	12.2	684
1	17,437,577	25,140,509	13.8	1,493
1	57,456,613	171,569,461	97.3	12,673
1	200,319,039	247,146,479	74.0	7,828

Chr 1

Image size reduction: 1/33

Figure 2.7. One-to-one comparison of siblings with recombination points identified

What these recombination points do not reveal, however, is which sibling had the recombination that caused the change in matching and whether it is on a maternal or paternal chromosome. This information will be inferred in later steps.

Genealogists should begin visual phasing with smaller chromosomes as they will have fewer recombination points on average. Chromosomes are numbered 1 through 22 based on size.

Chromosome 22 is approximately the smallest and is a good place to begin visual phasing. In figure 2.8 each of the recombination locations from the three-sibling comparison shown in figure 2.3 has been identified. These recombination locations can be marked using any method such as the black vertical line used here.

Figure 2.8. Labeling recombination locations

The next step is to assign each recombination point to one of the three siblings.

Step 5: assign recombination points using siblings

Assigning a recombination point to one of the three siblings requires an identification of which sibling "owns" the recombination point. The owner will change sharing from one type of matching to another at the same location in both comparisons in which he or she is included. Look for the sibling found in two comparisons at the same vertical line.

In figure 2.9 Binh is the owner of the first identified recombination event. He is the sibling found in both comparisons where there is a change in matching at that location: the Binh to Janya comparison and the Binh to Saeed comparison. Binh is also the owner of the second recombination event. The third recombination event belongs to Saeed. He is the only sibling found in both comparisons where there is a change in matching at that location: the Binh to Saeed comparison and the Janya to Saeed comparison.

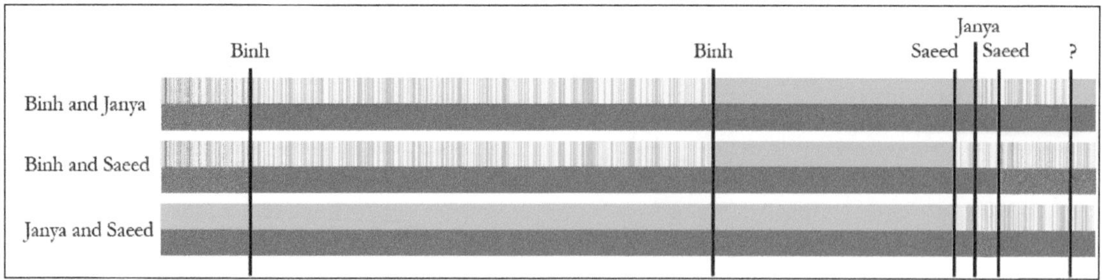

Figure 2.9. Labeling recombination-event owners

Each recombination event here is straightforward until the final one. Only the Binh to Janya comparison has a change in matching, so that location cannot be assigned. This recombination point will be examined later.

Although it is typically a simple matter to identify the owner of a recombination event, there can be complications. For example, figure 2.10 shows the recombination event identified by the vertical black line is not located at the same position in the Binh to Saeed and the Janya to Saeed comparison. Although Saeed owns this recombination event, it does not line up perfectly.

Figure 2.10. Labeling recombination owners

Recombination points that do not line up perfectly are a common occurrence and largely an artifact of the one-to-one compare tool. The tool scales chromosomes so they are roughly the same size. The scaling process can affect where the recombination point is displayed.

Another issue that results in misaligned recombination events is comparisons from different testing companies. For example, if Binh and Janya tested at Company One and Saeed tested at Company Two, the differences in tested SNPs can result in slight differences in the start and stop locations determined by the matching algorithm.

Alignment issues can be examined in greater detail using the one-to-one compare tool. By default, the tool scales the chromosomes to fit easily on a screen. It is possible to show the chromosomes with greater detail. On the comparison tool page, click the "Full resolution" option seen in figure 2.1 and re-run the comparisons.

The full resolution display spans many screen widths and contains additional information including megabase and SNP locations. Figure 2.11 shows a small portion of a full resolution display of the misaligned region from figure 2.10. The chromosome images have carets (^) that designate megabase locations, or millions of base pairs. The location in figure 2.11 is situated around 90,000,000 or 90M. The caret to the left of 90M is 89,000,000, and the caret to the right of 90M is 91,000,000. The short spaced vertical lines just under the chromosome represent every 100 tested SNPs.

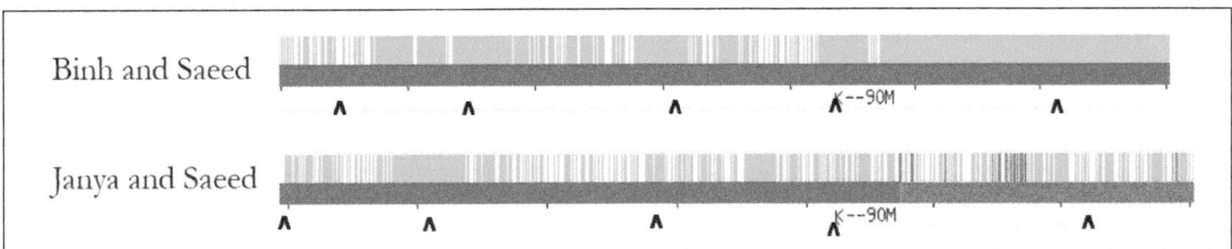

Figure 2.11. Full resolution chromosome comparisons

Based on the full resolution display in figure 2.11, the matching in the Janya to Saeed comparison terminates between 90,000,000 and 91,000,000. The matching in the Binh to Saeed comparison is harder to pinpoint, but appears to terminate in roughly the same location. The one-to-one comparison of Janya to Saeed in figure 2.12 shows that the termination point for the matching in this location is approximately at location 90,187,957.

Chr	Start Location	End Location	Centimorgans (cM)	SNPs
6	148,878	90,187,957	99.9	14,972
6	108,576,950	163,979,086	61.9	8,998

Figure 2.12. One-to-one comparison of Janya and Saeed

Figure 2.12 also demonstrates that *GEDmatch* labels start and stop points for the yellow half identical regions where the matches share DNA on one copy of their chromosomes (either paternal or maternal), but does not label the start and stop points for the green fully-identical regions where the matches share DNA on both chromosomes (paternal and maternal). These locations can be determined if one of the siblings has a start or stop point for a half identical region at the recombination point. This is shown in figure 2.12 by the arrow and the circled "End Location." Alternatively, a start or stop point can be approximated using the full resolution chromosomes as seen in figure 2.11. Later steps will make use of the megabase location of each recombination point. This can be noted in the workspace or in a spreadsheet.

A close examination of the end of chromosome 20 in figure 2.9, where there was a problematic recombination event, suggests that there might be a small segment shared by Binh and Saeed that is not being reported due to the default matching thresholds. This region can be examined by adjusting the default matching thresholds of the one-to-one compare tool. Adjusting thresholds should be done with *extreme caution*. Many small segments are false positives and lowering the matching thresholds below 7 cM raises the danger of false matching. However, valid small segments are sometimes formed by recombination that occurs at the ends of the chromosomes, so the thresholds can be lowered if there is an understanding of the risk of false matching.

Currently, the default matching thresholds for the one-to-one compare tool, as shown in figure 2.1, require that a matching segment must be at least 7 cM long and contain at least 500 SNPs to be identified as a matching segment. Incrementally lowering the "minimum segment cM size" may reveal a small shared segment at the end of a chromosome. Since there are fewer tested SNPs in smaller segments, it will sometimes be necessary to lower the SNP threshold to obtain a matching segment. For more information see the 'SNPs' section of the "Would You Like Your Data raw or Cooked?" chapter.

After lowering the thresholds to 3 cM and 300 SNPs, the small problematic segment shared by Binh and Saeed at the end of chromosome 20 is revealed. Figure 2.13 shows that the segment is 3.4 cM and contains 312 SNPs. This segment will be utilized in the visual phasing, with the caveat that it may yet be a false segment.

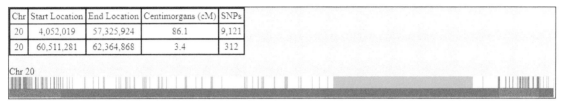

Chr	Start Location	End Location	Centimorgans (cM)	SNPs
20	4,052,019	57,325,924	86.1	9,121
20	60,511,281	62,364,868	3.4	312

Chr 20

Figure 2.13. Chromosome one-to-one comparison of Binh and Saeed

Figure 2.14 shows this chromosome with all the recombination points identified and labeled with both the owner and the approximate megabase location.

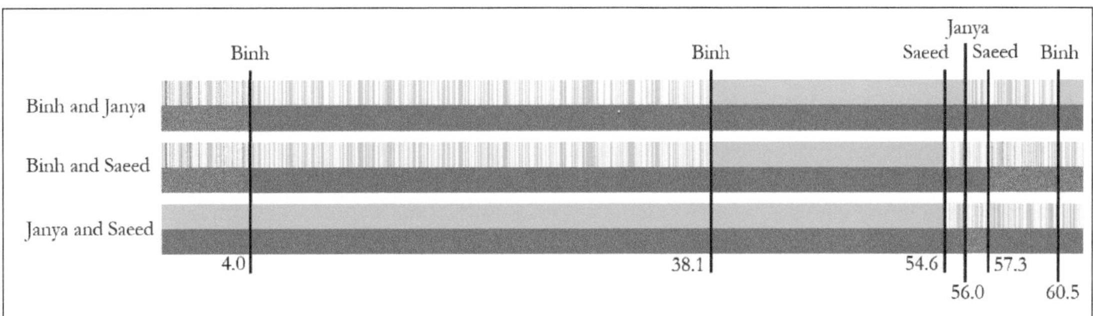

Figure 2.14. Labeled recombination points

Siblings sometimes experience recombination at the same location, or approximately the same location. Typically, this will be obvious when a recombination event appears to have occurred at the same location in all three comparisons. Sometimes, however, these recombination events can be difficult to detect.

Indeed, there will be instances where a recombination event is not detected, or where the owner of a recombination point cannot be determined. This should eventually be resolved in later steps of the visual phasing process. Therefore, if there is a location where a recombination point cannot be assigned, set it aside and move on to the next step.

Step 6: fill in the chromosome maps using siblings

With the recombination points identified and assigned, the DNA segments for the siblings can be filled in on their chromosome maps. Figure 2.15 includes the three sibling comparisons with the recombination points identified and assigned, as well as an empty set of chromosome map boxes for each sibling. Each set of chromosomes includes two copies, one copy from the mother and one copy from the father. The order of the chromosome maps is irrelevant and could be listed in any preferred order. Although any order could be used, the siblings are listed here in the order they first appear in the sibling comparisons.

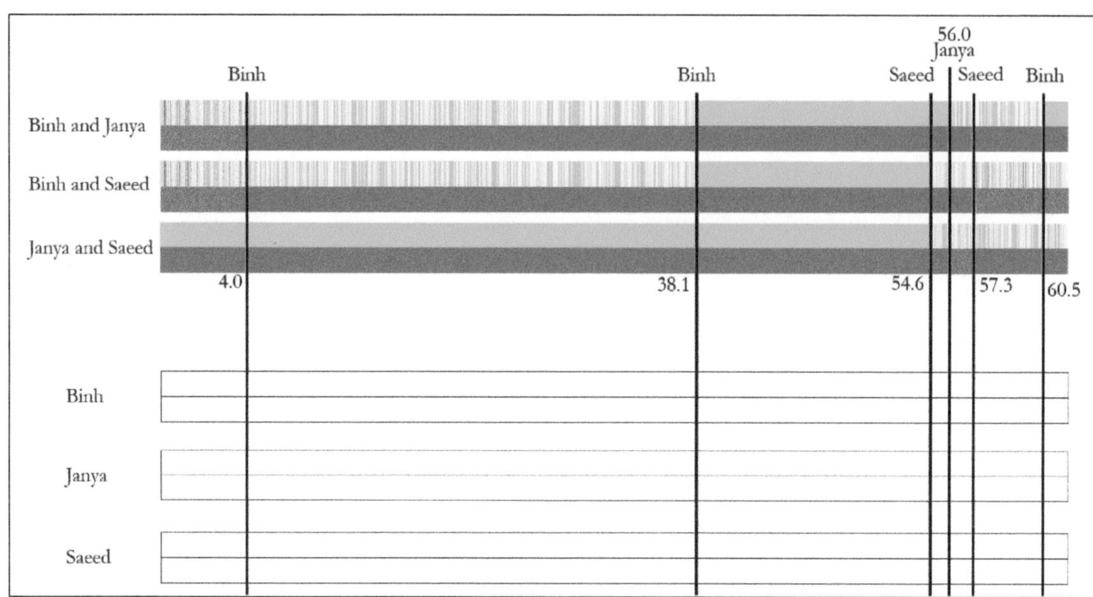

Figure 2.15. Empty chromosome map

A fully identical region (FIR) between two or more siblings should be mapped first. These are the green segments indicating that the two siblings share all DNA in those segments on both copies of their chromosomes. Thus, they inherited the same segment from their shared father on their paternal copy of this chromosome *and* the same segment from their shared mother on their maternal copy of this chromosome.

For example, between 38.1 Mb and 54.6 Mb, the three siblings all inherited the same segments of DNA from both parents. The chromosome map can be filled in with any two colors to represent this shared DNA as shown in figure 2.16.

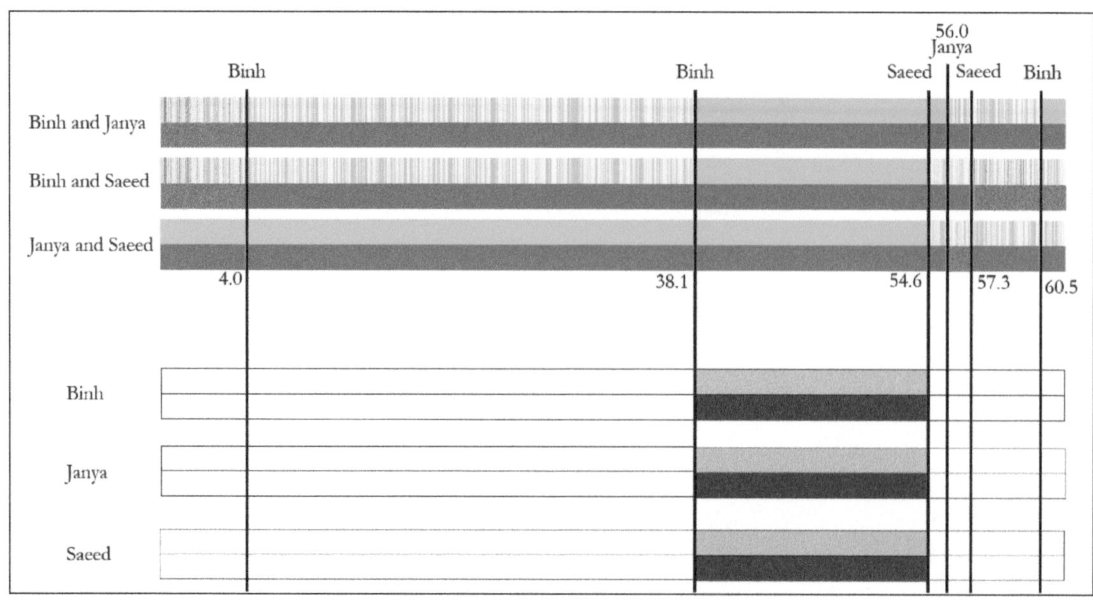

Figure 2.16. Chromosome map with filled FIR

Once a segment of DNA is mapped, the next step is to extend that mapped segment as far as it can go. A person's DNA segment extends along a chromosome until it reaches a recombination point for that person. The recombination point changes the segment and changes the color of the segment on the chromosome map. In other words, *unless there is a recombination event*, a segment of DNA will continue in both directions to the end of the chromosome. Think of a person's recombination event as being a STOP sign for segment extension.

Looking at Binh, he has a recombination event at 38.1 Mb. This means that one of his chromosomes recombines at that location, but we do not yet know which one. However, since he does not have a recombination event to the right of these mapped segments until 60.5 Mb, we can extend his two mapped segments all the way to 60.5 Mb.

Similarly, for Janya and Saeed, there is no recombination for these two siblings from 38.1 Mb to the left end of the chromosomes. Thus, their segments can be extended from 38.1 Mb to the end of the chromosome. On the right side of Janya and Saeed's segments, Saeed has a recombination point at 54.6 Mb so his segments cannot yet be extended. Janya has a recombination point at 56.0 Mb, so her segments can be extended to 56.0 Mb. The chromosome map in figure 2.17 shows the FIR segments extended as far as possible, for now.

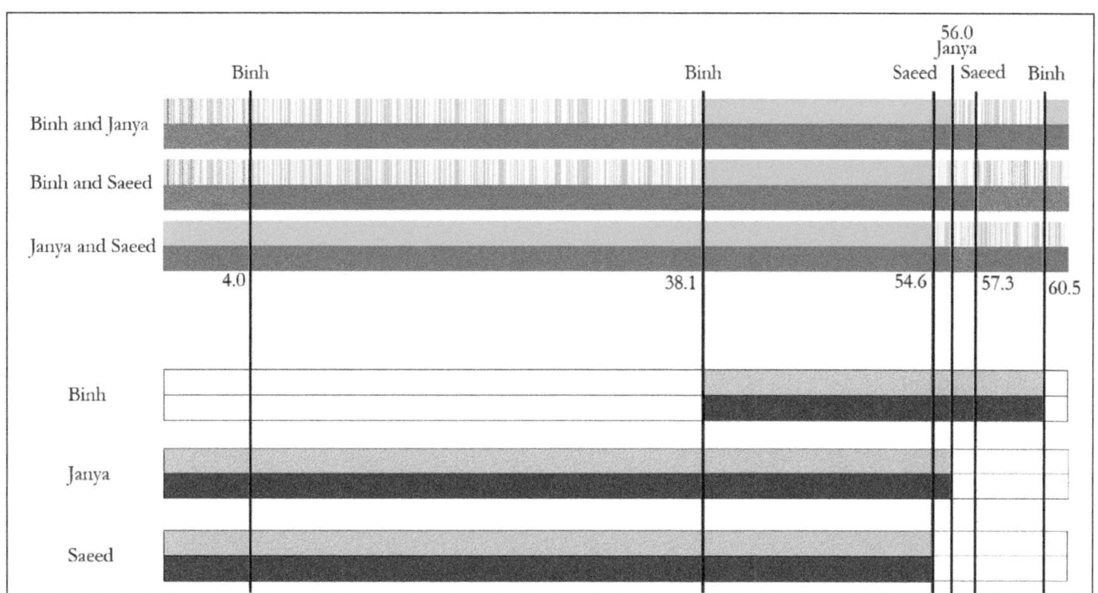

Figure 2.17. Chromosome map with extended segments

After these large FIR segments mapped, it is often possible to map segments where there is no sharing between siblings. If one sibling in a comparison is fully mapped and the second sibling shares no DNA at that location, then the second sibling's segments can be mapped.

Figure 2.18 shows that Binh and Janya share no DNA between 0 Mb and 4 Mb. Since Janya's segments fully mapped between 0 Mb and 4 Mb, Binh must have the *opposite* DNA. While

Janya and Saeed inherited certain segments from one copy of their mother and father's chromosomes at this location, Binh must have inherited the segments from the other copies of their mother and father's chromosome.

Thus, Binh can be assigned the opposite two colors between 0 Mb and 4 Mb, as shown in figure 2.18. These two colors can be any colors that are different from the previously-used colors. Visual phasing chromosome maps for a set of full siblings can only have four colors, with each of the colors representing one of the four grandparents.

Similarly, Binh and Saeed do not share any DNA between 57.3 Mb and the end of the chromosome. Since Binh's segments are filled at this region, Saeed's chromosomes will be the opposite of Binh's chromosomes at this location.

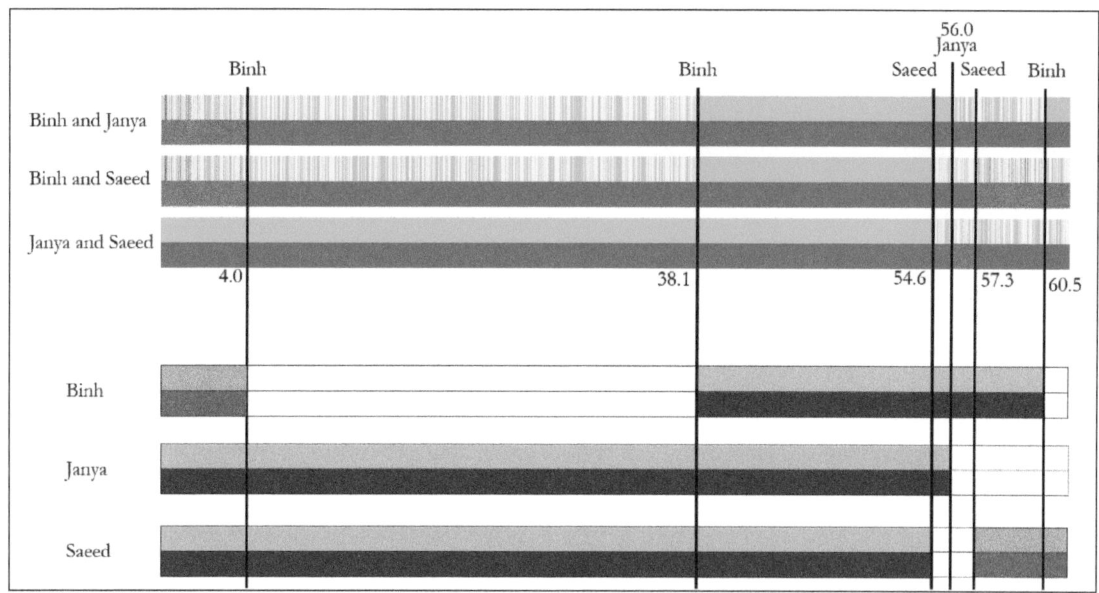

Figure 2.18. Chromosome map with non-sharing regions filled

Almost all FIR and non-matching regions now have been mapped. The remaining regions are all or mostly HIR, half identical regions, where siblings share DNA on just one copy of their chromosomes.

Mapping half identical regions is typically the most difficult aspect of visual phasing since it is not immediately clear on *which* copy the two siblings share DNA. Figure 2.18 shows Binh shares one copy of his chromosomes between 4.0 Mb and 38.1 Mb with Janya and Saeed. Thus, Binh has either the light blue or the dark blue extending from 4.0 Mb to 38.1 Mb, but which should be extended? At this point, either color can be extended because the colors are arbitrary. In other words, at this stage, any color can represent any of the four grandparents. However, the next extension at a "decision point" will effectively assign a color to a specific grandparent.

For every chromosome there can be one single decision point where a color is extended across a half identical region. Once extended at the decision point, the colors are no longer arbitrary. For example, Binh has a recombination point at 38.1 Mb. Thus, there are two options:

1. Extend the dark blue segment from 38.1 Mb to 4.0 Mb (and then fill in the yellow segment from 4.0 Mb to 38.1 Mb).
2. Extend the light blue segment from 38.1 Mb to 4.0 Mb (and then fill in the green segment from 4.0 Mb to 38.1 Mb).

At this point, either extension can be made without affecting the outcome of the mapping. However, only a single decision point can be made for each chromosome. If the dark blue segment is extended, that means that whichever grandparent has the segment inherited by Binh that extends from 4.0 Mb to 60.5 Mb is now represented by dark blue. Similarly, if the light blue segment is extended, that means that the same grandparent is now represented by light blue. The decision point effectively assigns the colors to a particular grandparent, although which grandparent is not clear until later.

In figure 2.19 circled locations indicate where an arbitrary decision point can be made and one of the two colors can be extended at a half identical region. But only a *single* decision point can be extended for each chromosome, as that will lock in the color pattern at every location.

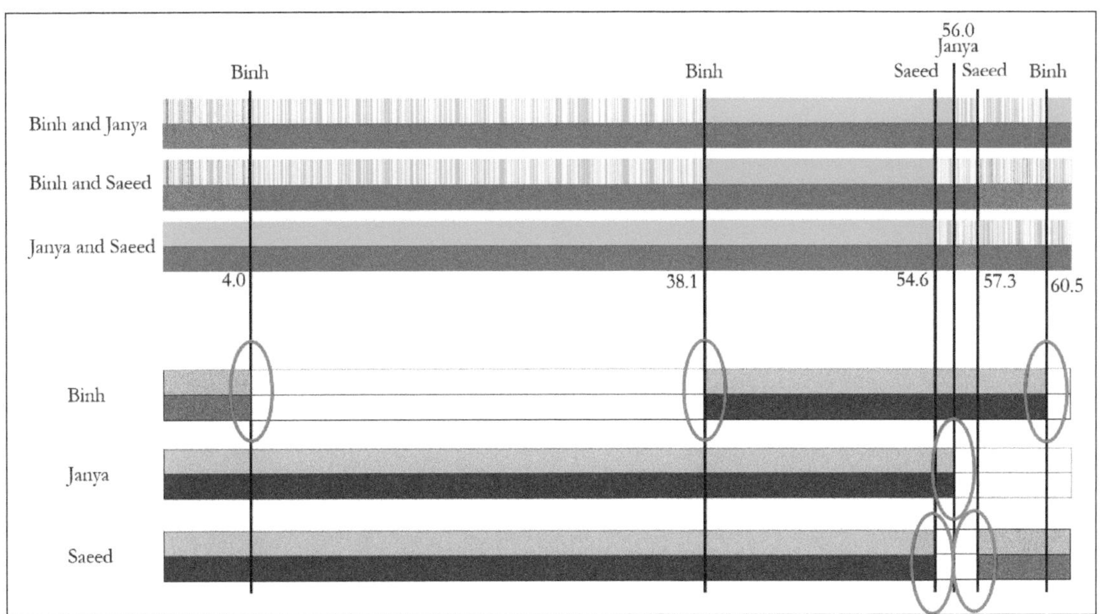

Figure 2.19. Chromosome map with non-sharing regions filled

Sometimes a decision point is made and a half identical region is filled in, but very few other segments can be filled in as a result. A possible resolution is to undo the mapping back to the decision point and extend a different segment at a different decision point.

In figure 2.20 Janya's chromosome map has been chosen for a decision point at 56.0 Mb. Either the light blue or the dark blue segment could have been extended from the recombination point at 56.0 Mb, but the dark blue was chosen for extension. Since it is now known that the light blue segment experienced the recombination at 56.0 Mb, the light blue segment switches to the opposite color, which is yellow. Janya's chromosome has now been completely mapped.

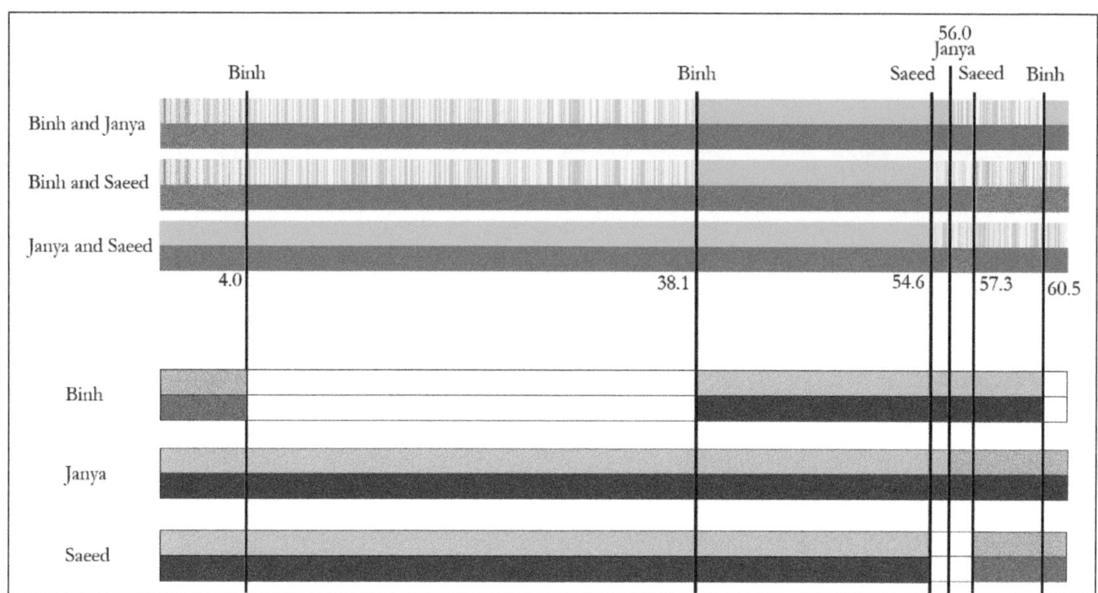

Figure 2.20. Chromosome Map with a decision-point extension

With Janya's chromosome completely mapped, additional portions of Binh and Saeed's chromosomes can be mapped. Since Binh and Janya are fully identical between 60.5 Mb and the end of the chromosome, and since Janya is fully mapped in this region, Binh will have the same two colors (yellow and dark blue) at this location. Similarly, since Janya and Saeed share no DNA between 56.0 Mb and 57.3 Mb, and since Janya is fully mapped in this region, Saeed will have colors (light blue and green) opposite to Janya in this region. Saeed's segments can also be extended from 56.0 Mb to 54.6 Mb, the adjacent recombination point for Saeed.

Thus, both Janya and Saeed's chromosomes are completely mapped, as shown in figure 2.21.

All that remains is to map Binh's chromosomes between 4.0 Mb and 38.1 Mb. However, without more information, it is not possible to extend either of Binh's segments in this region. An arbitrary decision point has already been made, and there's no way to determine which grandparent (represented by light blue or dark blue) spans the region between 4.0 Mb and 38.1 Mb.

It is possible to go back and switch the decision point from Janya at 56.0 Mb to Binh at 4.0 Mb or 38.1Mb, but this will leave similarly unresolved regions for Binh and Saeed between 54.6 Mb and the end of the chromosome as in figure 2.20.

Additional information will be required to complete the visual phasing of this chromosome. The information used to assign the mapped segments to specific grandparents will come from comparisons with other relatives.

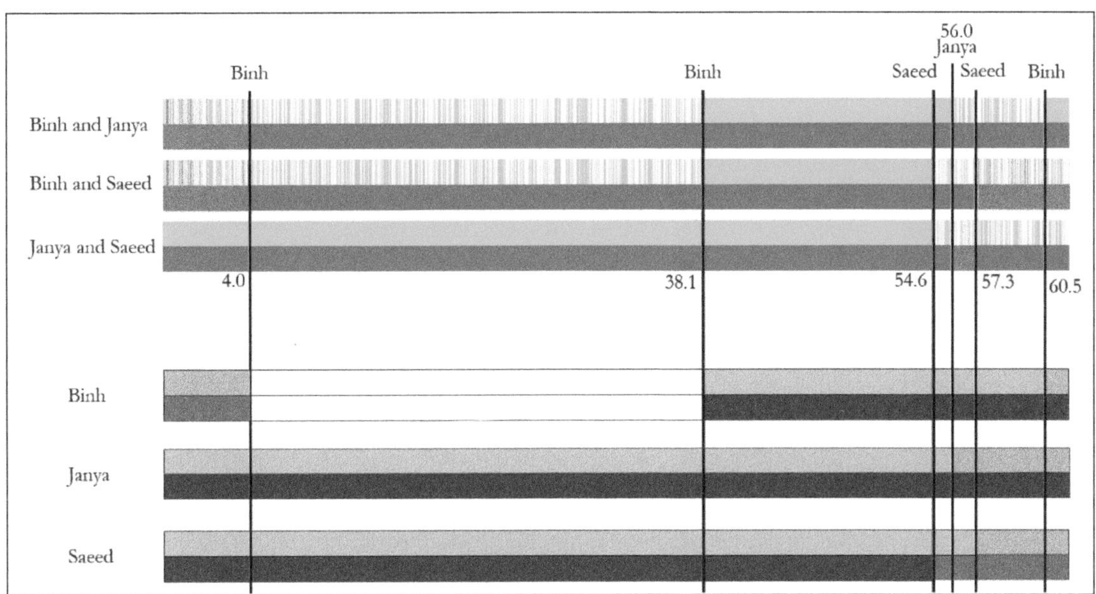

Figure 2.21. Chromosome map with Janya and Saeed's completed segments

Step 7: assign segments in the chromosome maps using other relatives

As seen in figure 2.21, the mapped chromosomes are broken down into four grandparents represented by the four colors. The identity of the light blue, dark blue, yellow, or green grandparents is unknown. A specific grandparent can be assigned to each color using cousin, aunt, and uncle matches.

Any match will work if there is some confidence that the person is only related through a single grandparent. If a match is potentially related through a maternal grandparent and a paternal grandparent, using a shared segment with that match will not help in assigning grandparents. The closer the match, the more segments that can be assigned. A third or fourth cousin may only enable assignment of a segment on a single chromosome as they typically only share one or two segments. A great-aunt or second cousin will enable assignment of many segments on many chromosomes.

To use these non-sibling matches, use the one-to-one tool comparing the relative to each of the three siblings. Then align the resulting images with the mapped chromosomes.

Using a first cousin once removed

Figure 2.22 shows the chromosome maps for the three siblings compared to a first cousin once removed (1C1R) shared with the siblings' paternal grandfather. The segments shared with this 1C1R can be assigned to the paternal grandfather. For example, the 1C1R shares segments with Janya and Saeed that appear to align with the yellow segment that each owns at this region. The matching cannot be with the dark blue owned by Janya since Saeed does not have dark blue in that region. The same conclusion applies to Saeed's green segment. The yellow segments for this chromosome can be assigned to the paternal grandfather. Note that since the yellow paternal grandfather segment on the end of Binh's chromosome is so small, it might fall below the threshold for matching. However, a quick visual inspection of the 1C1R to Binh comparison suggests that there is matching at the end of Binh's chromosome.

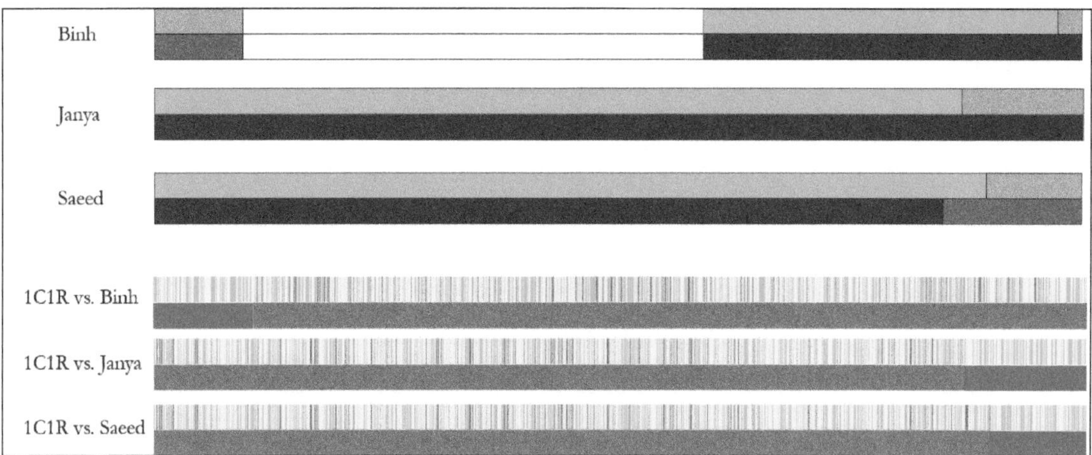

Figure 2.22. Comparing the chromosome maps to a paternal 1C1R

With the yellow segments assigned to the paternal grandfather, the corresponding light blue segments can be assigned to the paternal grandmother. The top two colors will always belong to grandparents on one side (either maternal or paternal), and the bottom two colors will always belong to the grandparents on the other side).

Using a great-aunt

Figure 2.23 shows the sibling chromosome maps compared to the DNA of a maternal great-aunt, the sister of the maternal grandfather. The pattern of matching between Saeed and the great-aunt suggests that the maternal grandfather is the green segment, since only Saeed matches the great-aunt and only Saeed has the green segment in this region. Accordingly, the green color is the maternal grandfather and the yellow color is the maternal grandmother.

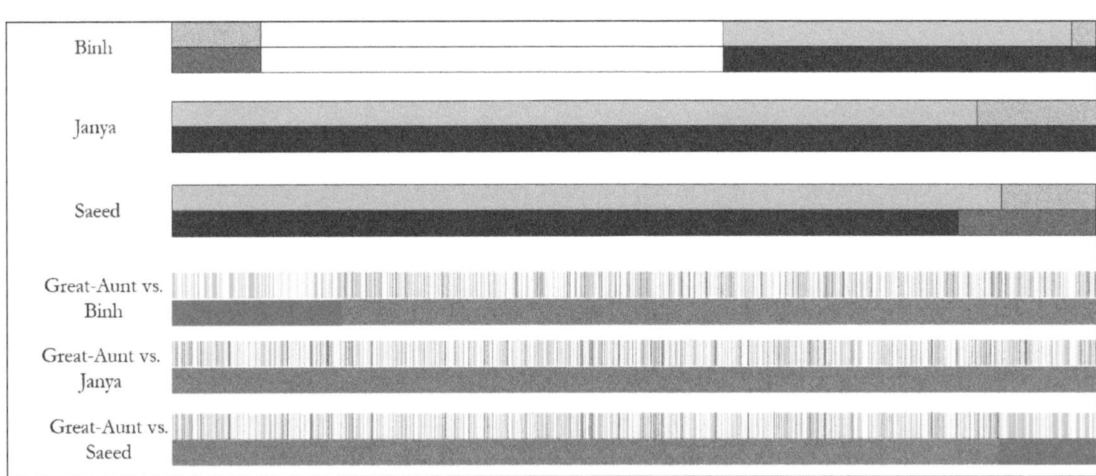

Figure 2.23. Comparing the chromosome maps to a maternal great-aunt

The matching with the great-aunt also helps resolve the missing portion of Binh's chromosome map. Since the great-aunt's matching segment is green and the matching with Binh extends into the unmapped region, the green segment must extend into the unmapped region. This means that it was the top chromosome, which has been identified as the paternal chromosome, that experienced recombination at 4.0 Mb. As a result of the matching with the great-aunt, the chromosome map is complete, and the segments have been assigned to the four grandparents as shown in figure 2.24.

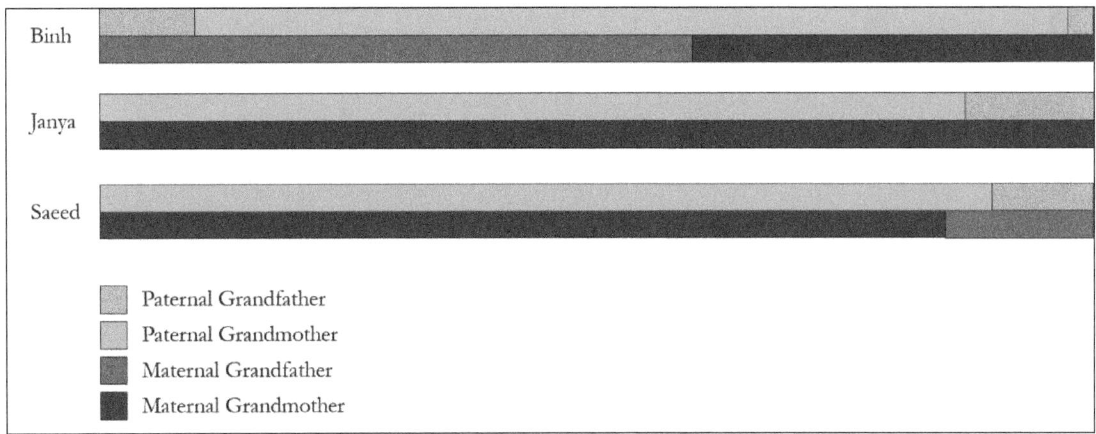

Figure 2.24. Completed chromosome maps for the three siblings

The visual phasing of chromosome 20 is tentatively complete. Any chromosome map is tentative, especially when created by visual phasing. Hidden recombination points, relationships through multiple lines, and other errors can result in changes to the map.

Continue visual phasing one chromosome at a time until all 22 autosomes and the X chromosome are mapped for all three siblings. The "X-DNA Techniques and Limitations" chapter contains information that will be helpful when phasing the X chromosome.

Upon initial completion, the color pattern can vary from chromosome to chromosome, since colors are initially assigned without knowing which grandparent belongs to which segments. Although yellow in figure 2.24 turned out to belong to the paternal grandfather on chromosome 20, it could belong to the maternal grandmother on another chromosome. As each chromosome is mapped and the colors are assigned to specific grandparents, the colors can be normalized so the same color represents the same grandparent on all chromosomes. For example, once a chromosome is mapped, the colors can be changed such that light yellow always represents the paternal grandfather, and so on.

Visual phasing with fewer than three siblings

A question frequently asked is whether visual phasing can be performed with fewer than three siblings. Not everyone has a set of three or more siblings. Most researchers have parents who are too old to generate a new sibling just for visual phasing!

The most common combinations are two siblings, two half siblings, or two full siblings and a half sibling. Although it is much more challenging, visual phasing can be performed using any of these combinations. In each case, it requires close cousin matches to help decipher the recombination points and owners. First cousins, aunts and uncles, great-aunts and great-uncles, and similar relationships work best.

Visual phasing with two siblings

Figure 2.25 shows a partially completed visual phasing of chromosome 17 for two siblings, Chris and Dana. With only two siblings, it is not possible to assign the owner of the recombination points using just the one-to-one comparison. Instead, cousin matching will identify the owners, or provide hypotheses that can be tested.

For this chromosome, the visual phasing begins with the fully identical region between 8.5 Mb and 29.5 Mb. The recombination point at 29.5 Mb was assigned to Chris' paternal chromosome based on the matching pattern with the paternal great-uncle (brother of their paternal grandmother). Chris' matching with the paternal great-uncle ended at 29.5 Mb, while Dana's matching continued (even through the next recombination point at 53.1 Mb). Thus, it was reasonable to hypothesize that it was Chris that owns the recombination point at 29.5 Mb, on the paternal chromosome. Since Dana's matching with the paternal great-uncle extended through all three recombination points, Dana must have inherited a full chromosome from the paternal grandmother.

Matching with a maternal great-aunt (sister of their maternal grandfather) allowed the tentative assignment of the recombination point at 53.1 Mb to Dana, and that portion of the chromosome could be filled out since Chris does not share any DNA with the maternal great-aunt.

The region between 0 Mb and 8.5 Mb will require comparisons to other close matches to determine who owns the recombination point at 8.5 Mb and complete the map.

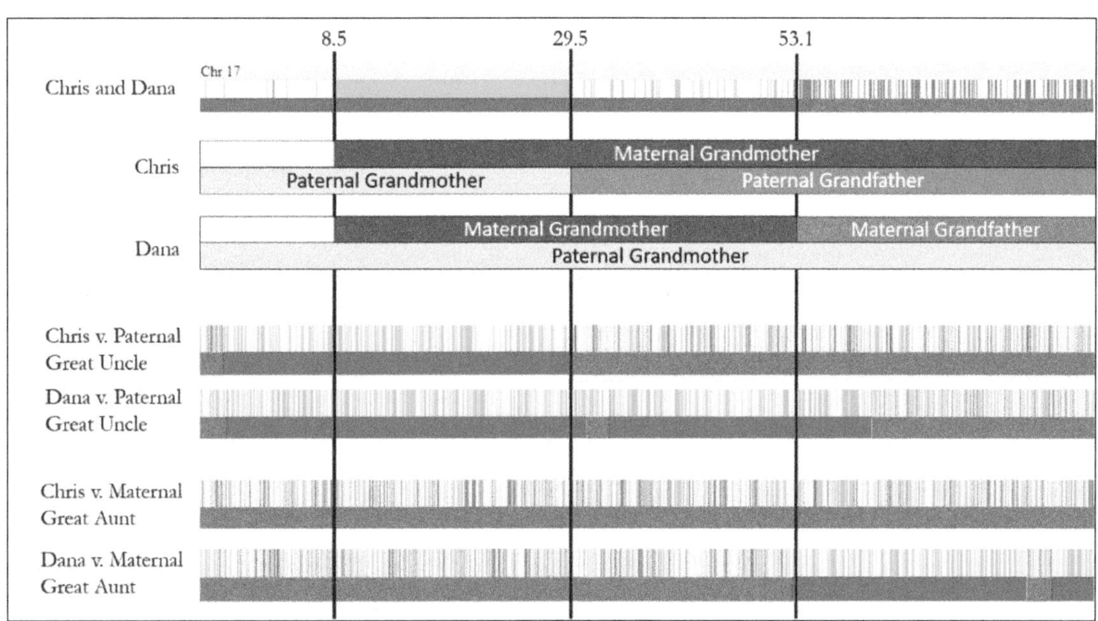

Figure 2.25. Visual phasing with two siblings

Visual phasing with half siblings

Visual phasing with a half sibling is also challenging, but not impossible, provided there are close matches to facilitate the process. In figure 2.26 half siblings Esther and Fran are compared using the one-to-one tool. They are also both compared to a maternal great-aunt (sister to their shared maternal grandmother).

We expect sharing on only one chromosome (an assumption that must always be remembered). Either the maternal or paternal chromosome is shared depending on which parent is shared. Only one chromosome is identified in the workspace. In this case, only the maternal copy of chromosome 19 can be mapped using the maternal half sibling.

Visual phasing for half siblings can begin anywhere along the chromosome, as there will not be any FIRs because they are not related on both sides of the family. In figure 2.26 the long, shared segment between 17.6 Mb and 56.8 Mb was mapped first. Based on sharing with the maternal grandmother's sister, the blue shared segment is labeled as being inherited from the maternal grandmother. The sharing with the great-aunt also suggests that Esther owns the recombination point at 17.6 Mb, and the green segment between 13.6 Mb and 17.6 Mb must belong to the maternal grandfather.

The remainder of the chromosome will require additional matches to determine who owns the recombination points at 13.6 Mb and 56.8 Mb.

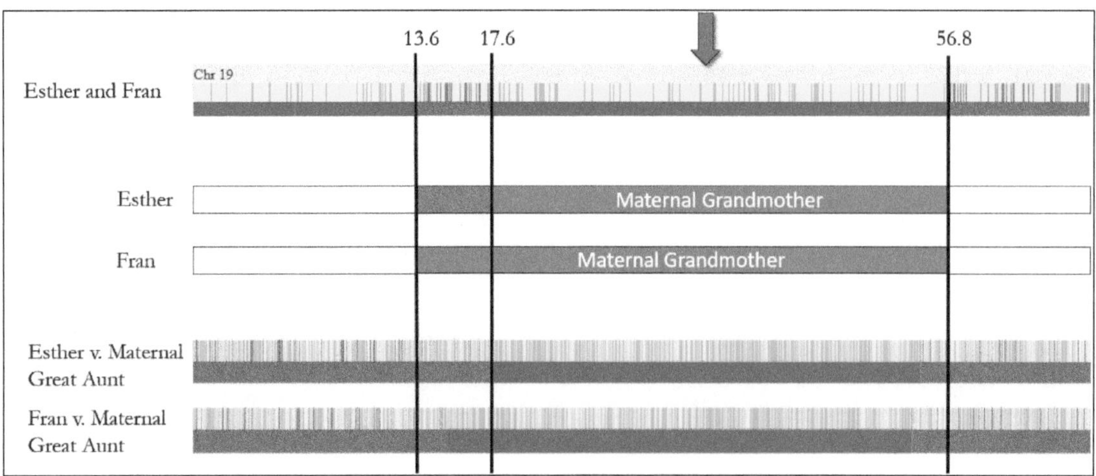

Figure 2.26. Visual phasing with a half sibling

One possible method for identifying the owner of a recombination point is to look for matching with known or unknown genetic cousins around that recombination point. The pattern of matching around a location may suggest which sibling owns the recombination point.

For example, figure 2.27 shows the recombination points at 27.6 Mb and 33.0 Mb can be examined by looking for known or unknown matches to full siblings Gill and Henry at these locations. This may reveal that Gill has several matches that end or begin at approximately 27.6 Mb, while Henry only has matches that span the region at 27.6 Mb. This would suggest, but does not prove, that Gill owns the recombination point. The conclusion would await support or confirmation from other matches. Figure 2.27 names the relationship between Gill, Henry, and the match being compared, but you may not know that relationship when doing the initial analysis.

With visual phasing, as with all chromosome mapping, an assignment of a segment is just a hypothesis. This is especially true when visual phasing with fewer than three siblings.

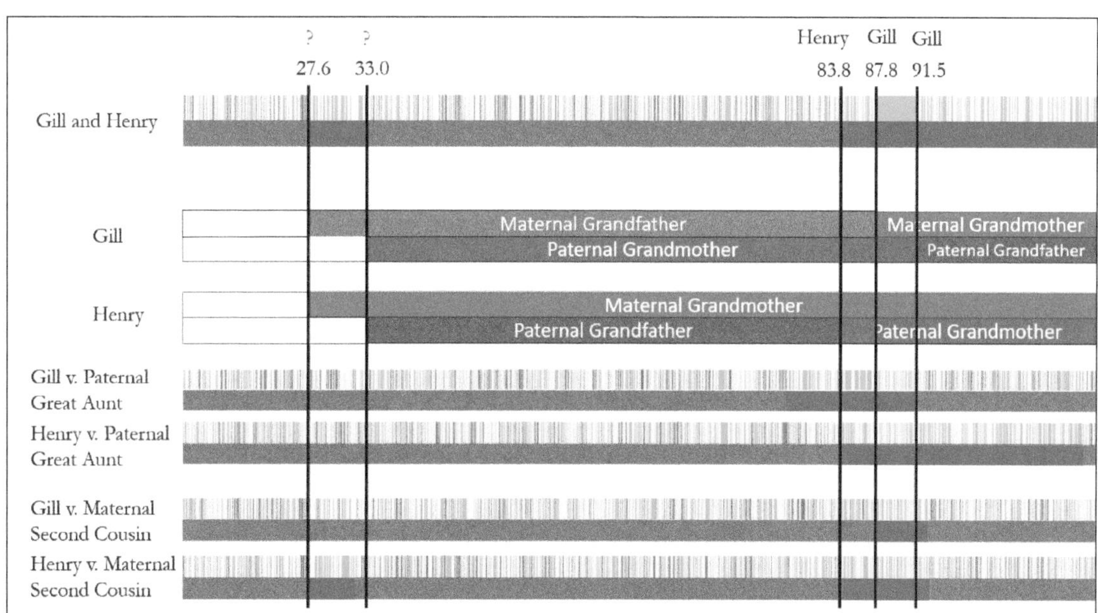

Figure 2.27. Using unknown matches to assign recombination points

Applications for Visual Phasing

Although visual phasing is currently hands-on and time-consuming, the work is both rewarding and useful. Even without any other information, the chromosome maps created by visual phasing reveal a real and physical relationship with previous generations.

When looking at the completed map for chromosome 20 in figure 2.24, for example, Binh immediately sees that he inherited DNA from all four grandparents on this chromosome. Janya sees that she inherited a full, unrecombined copy of the chromosome from her mother, the very same chromosome that her maternal grandmother passed down many decades earlier.

The chromosome maps created by visual phasing also have many genealogical applications. These maps help genealogists explore and potentially determine their relationships with new genetic cousins.

Using the results of visual phasing to examine new matches

With completed chromosome maps, the siblings always know from which grandparent they inherited DNA at any location on any chromosome. They can use this information to determine whether a match is paternal or maternal and with which grandparent they share a match, even if the parents and grandparents have not tested. This eliminates 75% of the family tree when looking for a common ancestor.

In figure 2.28 the one-to-one compare tool is used to compare each of the three siblings to a new match that recently uploaded raw data to *GEDmatch*. On this chromosome, the new match shares DNA only with Binh at locations 10.9Mb to 18.9 Mb. In this location, Binh received DNA from his paternal grandmother and his maternal grandfather, and thus he has already excluded 50% of his family tree as the source of this shared segment. In most cases, a map can further narrow down the shared line. Only Binh shares DNA with the new match at this location and only Binh has DNA from the maternal grandfather at this location. If the segment shared with the new match came from the paternal grandmother (light blue), all three siblings would match the person at this location. Thus, the chromosome map has identified the family tree of the maternal grandfather as the source of this segment of DNA.

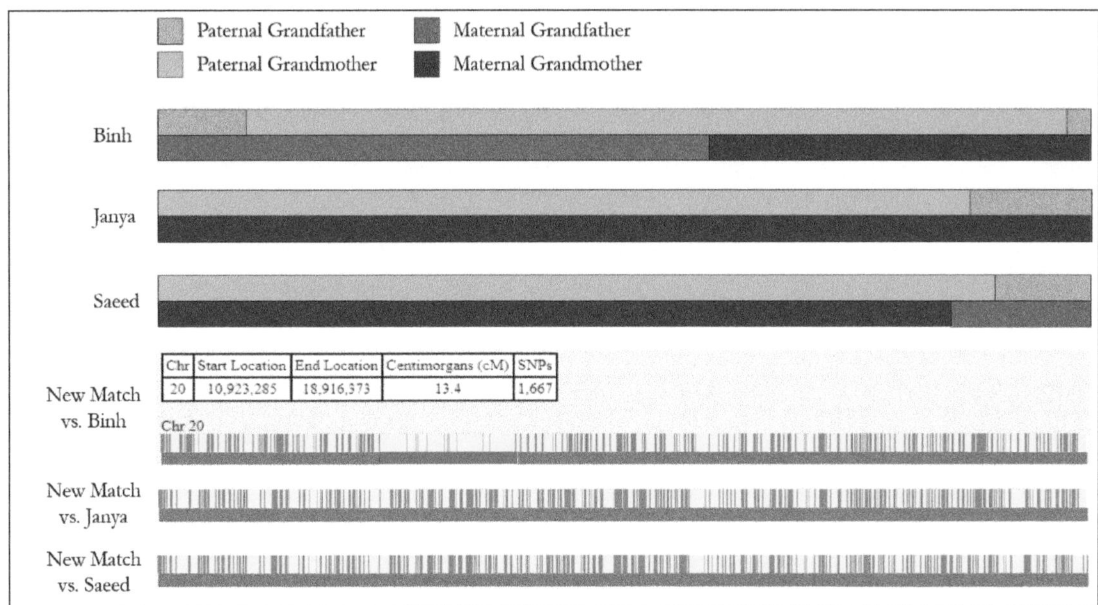

Figure 2.28. Comparing the chromosome maps to an unknown match

Take care when working with matches that share multiple segments. Although this segment is identified as coming from the maternal grandfather, another segment on another chromosome could possibly come from a different grandparent, which would suggest multiple relationships.

Using the results of visual phasing for advanced chromosome mapping

In addition to identifying the grandparental source of a segment of DNA shared with a match, the results of visual phasing can be used to construct advanced chromosome maps that combine visual phasing with traditional chromosome mapping.

In figure 2.29 the results of the author's visual phasing for the first eight chromosomes was manually imported into the chromosome mapping tool *DNA Painter*.[5] The contributions of each of the four grandparents to these chromosomes is shown in the figure and new matches can be added to this map.

5. *DNA Painter* (http://www.dnapainter.com/).

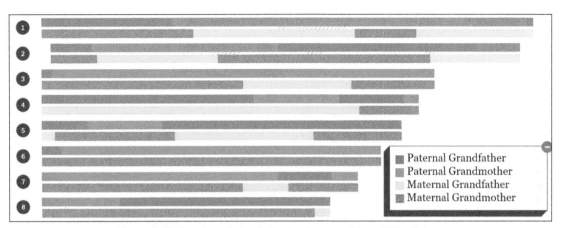

Figure 2.29. Results of visual phasing imported into DNA Painter

As each new match is added to the growing chromosome map, there is an opportunity to check the accuracy of the visual phasing as well as any assumptions made about the match. For example, if a match believed to be related via a maternal grandmother only has a segment that overlaps with the maternal grandfather, either the visual phasing is incorrect or the match is related through a different line.

In figure 2.30 the visual phasing map of chromosome 6 is expanded after adding three segments shared with a 1C1R (in black). The common ancestor is the paternal grandmother (in pink). This might be additional evidence that the visual phasing of chromosome 6 is correct and that this match is indeed related through the paternal grandmother.

As more matches are added to the combined map, more evidence is provided to support the visual phasing. Once there is confidence that the visual phasing is correct, the map can be used to explore connections with new matches to identify shared ancestors.

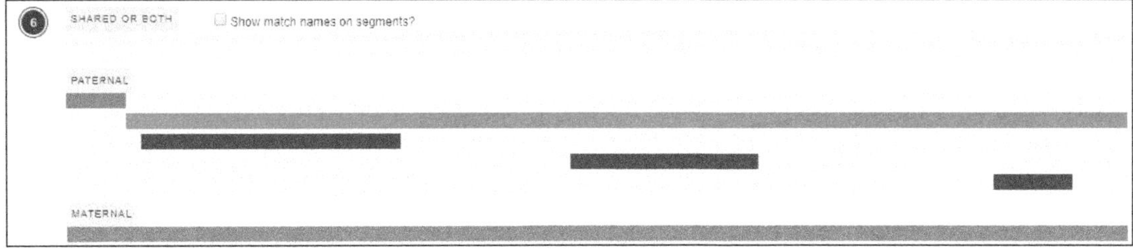

Figure 2.30. Map for chromosome 6 combining visual phasing and traditional chromosome mapping

In figure 2.31 a new match shares a segment on chromosome 5 (using the same color key for the four grandparents as in figure 2.29). This segment is aligned with the current map along the top in crosshatch. The match can be related through the paternal grandfather (blue), the paternal grandmother (pink), the maternal grandfather (green), or the maternal grandmother (gray). Based on this alignment, this segment appears to have come from the paternal grandfather. The segment would span a recombination point on the maternal chromosome where the color switches from gray (maternal grandmother) to green (maternal grandfather). Such an event is not impossible, but it is statistically unlikely. Accordingly, this match is *tentatively* assigned to the paternal grandfather.

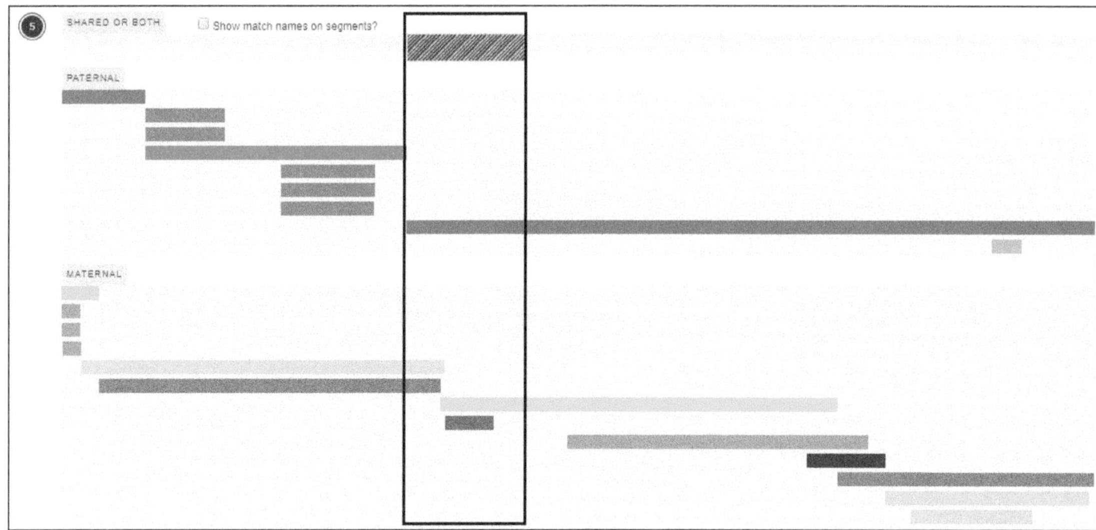

Figure 2.31. Testing hypotheses using a chromosome map

These are just a few current uses of the results of visual phasing for genealogical research.

The Future of Visual Phasing

The visual phasing methodology is just a few years old, yet there are already many genealogists using visual phasing to explore their ancestry. In the future, automated tools will help more genealogists map their chromosomes and even recreate raw data files for their grandparents.

Automated visual phasing

Although the output of visual phasing has many different applications, obtaining this output is a challenging and time-consuming process. This is a barrier for many genealogists. Tools that automate visual phasing would lower the barrier to entry.

A tool could perform most of the visual phasing steps, only requesting input from a user if there are any issues or uncertainties. However, in order to assist the tool or review the tool's output, genealogists must understand the logic behind visual phasing.

An automated tool could, for example, perform one-to-one comparisons of three siblings using either kits uploaded to *GEDmatch* or raw data. The tool could use the comparison data to create a visual representation in the workspace. A fully automated visual phasing tool may not need such a visual representation, but user intervention might require some visual analysis. The tool could then identify the recombination points and their owners, map the chromosomes, and use identified cousin matches to assign the segments. An entire genome could be mapped in a matter of minutes.

One of the first automated tools for visual phasing was the Excel spreadsheet created by Steven Fox for Windows systems. After receiving a *GEDmatch* user name and password from the user, the spreadsheet automatically logs into *GEDmatch* and copies one-to-one compare data for the siblings and identified cousins into the workspace. Manually copying the hundreds of images can take hours while the tool accomplishes this in minutes. Once the images are populated into the spreadsheet, the user completes most of the remaining steps of the visual phasing process. The Steven Fox spreadsheet is currently available for download in the Files section of the *Facebook* group.[6]

Grandparent reconstruction

As discussed throughout, visual phasing assigns segments of DNA to the four grandparents of a set of siblings. It is a logical extension of this process to take those assigned segments and generate raw data files for each of the grandparents. Although the generated files will likely not contain the grandparent's full set of markers tested by the grandchildren, having these kits to upload to *GEDmatch* or a third-party site could be beneficial.

Grandparent files created by visual phasing are likely to be more of an organizational tool than a DNA profile used for new discoveries. The siblings used to generate the files already possess the segments and have identified the matches, and the results of visual phasing will show through which grandparent a match shares DNA. Although recreating raw data files may restore some matching by stitching together segments, these files will likely not identify any close matches that were not already known. There is the potential, however, that as more people recreate ancestors, collaboration could lead to considerable new information and discoveries.

Creating grandparent kits from the results of visual phasing will require two things. First, the DNA of the three or more siblings must be visually phased, with segments assigned to the grandparents in a way that can be accessed by a "recreation" tool. The more siblings that are visually phased, the more DNA that will be populated into the grandparent kits. Second, the recreation tool must have access to raw data files so that it can retrieve the raw data for assigned segments and place it into the respective grandparent's file.

For example, if a recreation tool is creating a kit for grandfather Robin, the tool will find the chromosome number, start location, and stop location for each segment assigned to Robin via visual phasing. If the tool can determine which of the sibling's DNA belongs to that segment, either through shared matching with a sibling or with cousin matches, it will extract that raw data from the sibling's raw data file and copy it into a new file for Robin. Robin's raw data file could then be used for analysis and potentially uploaded to a third-party website.

6. "Visual Phasing Working Group," *Facebook* (https://www.facebook.com/groups/visualphasing).

These are just a few of the potential future uses of visual phasing. As more genealogists work with the methodology, new creative uses will undoubtedly be developed.

Conclusions

Visual phasing is the ultimate logic puzzle for genealogists. In addition to presenting an enjoyable challenge, the chromosome maps created by visual phasing provide information that furthers genealogical research.

Soon, automated visual phasing tools will simplify visual phasing and provide new methods of analysis, including creating raw data profiles for grandparents. The uses of visual phasing are limited only by the imagination of genealogists!

X-DNA Techniques and Limitations

Kathryn J. Johnston, MD

Introduction to the X Chromosome

The X chromosome represents one of the two sex chromosomes. The Y is the sex chromosome that creates the biological male. He gets his Y from the father and his X from the mother. The female receives two X chromosomes, one from her father and one from her mother.

The X chromosomes recombine in the same way as the autosomes, and yet the inheritance pattern is unique in some ways. Some researchers avoid X comparisons for pedigree building because of the complexities. However, such avoidance can lead to missed cousin matches. Some researchers place too much emphasis on a small X segment match. This chapter will examine the limitations and techniques for using X chromosome matching to help find family members and build family trees.

Research indicates that during recombination the X chromosomes from the maternal side behaves similarly to the autosomes. Many genealogists seem to think there is a huge difference, and that does not seem to be the case.

Males need to look at their mother's side of the family in order to identify common ancestors when they have a match on the X chromosome. However, do not confuse the X chromosome with the mitochondrial DNA (mtDNA). For many years, the mtDNA test was the only one that could be ordered by women. When autosomal testing was added, there was a misconception among beginners that the X inheritance was also passed mainly along the same mitochondrial line.

Understanding X chromosome inheritance patterns

For those involved in the search for close family, an X match can be the smoking gun that identifies relationships. However, one would not want to reach a wrong conclusion based only on a continuous X match.

An X-only match that shares no autosomal segments can come from a common ancestor within a genealogical time frame or a more distant ancestor. For some, pedigrees do not extend many generations back. Therefore, it may be difficult to identify a distant common ancestor, but that is true for the autosomes as well when there is just one segment in common on a single chromosome. For those with deep family trees, it is a nice surprise when you find an unexpected connection or when you break through a genealogical brick wall using the X chromosome. For the less experienced, it is advisable to first review some of the basic blogs about the X chromosome[1] and X recombination, tips, and warnings[2] in order to understand the advanced discussions in this chapter.

When there is a solid X segment match, you should consult the X pedigree charts by Blaine T. Bettinger[3] and Debbie Parker Wayne.[4] These charts identify the ancestors who may have contributed to your X chromosome—a small subset of your overall family tree. Just as with all chromosome matching, the lack of matching on the X chromosome does not disprove any ancestral lines except when dealing with specific close relationships. Even full siblings may not match each other on the X coming from their mother. They can match each other anywhere from zero to 100% on the X chromosome. Not matching on the X chromosome is only helpful in the identification of a relationship when dealing with specific close relationships.

The X chromosome can be useful when there is unknown parentage, particularly half sisters who may share the same father. Sperm donors need to be aware of these capabilities because their daughters who test will discover they share the father's entire X chromosome and its origins may be found by comparing their X with others.

All URLs were accessed 28 January 2019 unless otherwise indicated.

1. Louise Coakley, "X-DNA's helpful inheritance patterns," *Genie1*, 12 June 2015 (http://www.genie1.com.au/blog/63-x-dna/). Also, Roberta Estes, "X Marks the Spot," *DNAeXplained – Genetic Genealogy,* 27 September 2012 (https://dna-explained.com/2012/09/27 /x-marks-the-spot). Also, Jim Owston, "Phasing the X-Chromosome," *The Lineal Arboretum*, 21 November 2012 (http://linealarboretum.blogspot.com/2012/11/phasing-x-chromosome.html).

2. Jared Smith, "X Chromosome Recombination's Impact on DNA Genealogy," *SmithPlanet* (http://smithplanet.com/stuff/x-chromosome.htm). See also Blaine T. Bettinger and Debbie Parker Wayne, *Genetic Genealogy in Practice* (Arlington, Va.: National Genealogical Society, 2016).

3. Blaine T. Bettinger, "Unlocking the Genealogical Secrets of the X Chromosome," *The Genetic Genealogist* (https://thegeneticgenealogist.com/2008/12/21/unlocking-the-genealogical-secrets-of-the-x-chromosome/). Blaine T. Bettinger, "More X-Chromosome Charts," *The Genetic Genealogist*, 12 January 2009 (http://www.thegeneticgenealogist.com/2009/01/12/more-x-chromosome-charts/).

4. Debbie Parker Wayne, "X-DNA Inheritance Charts," *Deb's Delvings in Genealogy*, 25 October 2013 (http://debsdelvings.blogspot.com/2013/10/x-dna-inheritance-charts.html).

Biological parents can often be accurately identified based not only on whether there is an X chromosome match but also by the matching along the entire lengths of the autosomes as well. Biologically, the twenty-third pair of chromosomes, the sex chromosomes, in females is the XX pair and in males the XY pair. The testing companies can generally determine if a submitted DNA test is from a male or a female just by looking at the X results. Parents give 50% of their total autosomal DNA to each child and the matching along the entire length of each chromosome can be easily calculated in cM.

A father passes his single X chromosome essentially intact, with no recombination except at the tips where it crosses over with the Y chromosome, to all of his daughters. Therefore, sisters who share the same father will match each other and their father's X almost entirely (anywhere from about 181 to 196 cM).

Statistically, a test taker may inherit more X-DNA from some ancestors in the X inheritance line than from others. There is a chance of X chromosome recombination in every generation of a line including only females while fewer chances for recombination exist in a line alternating between male and female. The rule for transmission of the X chromosome is "never two consecutive males," that is, there is no father to son transmission of an X chromosome. Notice that the left-most shaded side of the male's fan chart in figure 3.1 shows a line of alternating male to female transmission. Bettinger's probability chart indicates the specific lines that are more likely to be involved when there is a robust X match and a small autosomal match; however, analysis of the family trees of shared matches help identify the specific ancestral line resulting in a shared X chromosome segment. Note that one of the fifth-great-grandmothers is eight times more likely to give a male descendant a segment of her X as compared to his fifth-great-grandmother on the all maternal line, 12.5% compared to 1.6%. If you extend those lines out to the eleventh generation, there is one female, a ninth-great-grandmother, who still has a 3% chance of passing on a significant segment. That is thirty-one times the probability of the female in the mitochondrial line at the same eleventh generation. Some ancestors do have a markedly higher chance of passing down an X segment. That is why for an X-only match, it is important to concentrate first on those distant ancestors with the highest likelihood of X transmission. Genealogists who are successful in X matching often have created extensive family trees along those high probability lines of descent.

Several families have reported that the size of X segments among cousins averages longer than the average size of their autosomal counterparts when there are alternating male-female-male-female lineage patterns. Segment matches are likely to be retained longer among descendants when males are involved in the transmission because the genetic code on most of the father's X chromosome does not get "mixed" by recombination prior to passing it to the next generation. The X remains essentially unchanged when passing through a male. The X chromosome a daughter inherits from her father came only from the father's mother. These segments recombine less often and retain length for more generations.

It is actually uncommon to find a distant X match along the all-female line in a pedigree chart. The probability is much higher in those lines of transmission where the males alternate with females. This disparity becomes more evident as you go farther back in time when seeking a most recent common ancestor (MRCA) in your X pedigree chart.

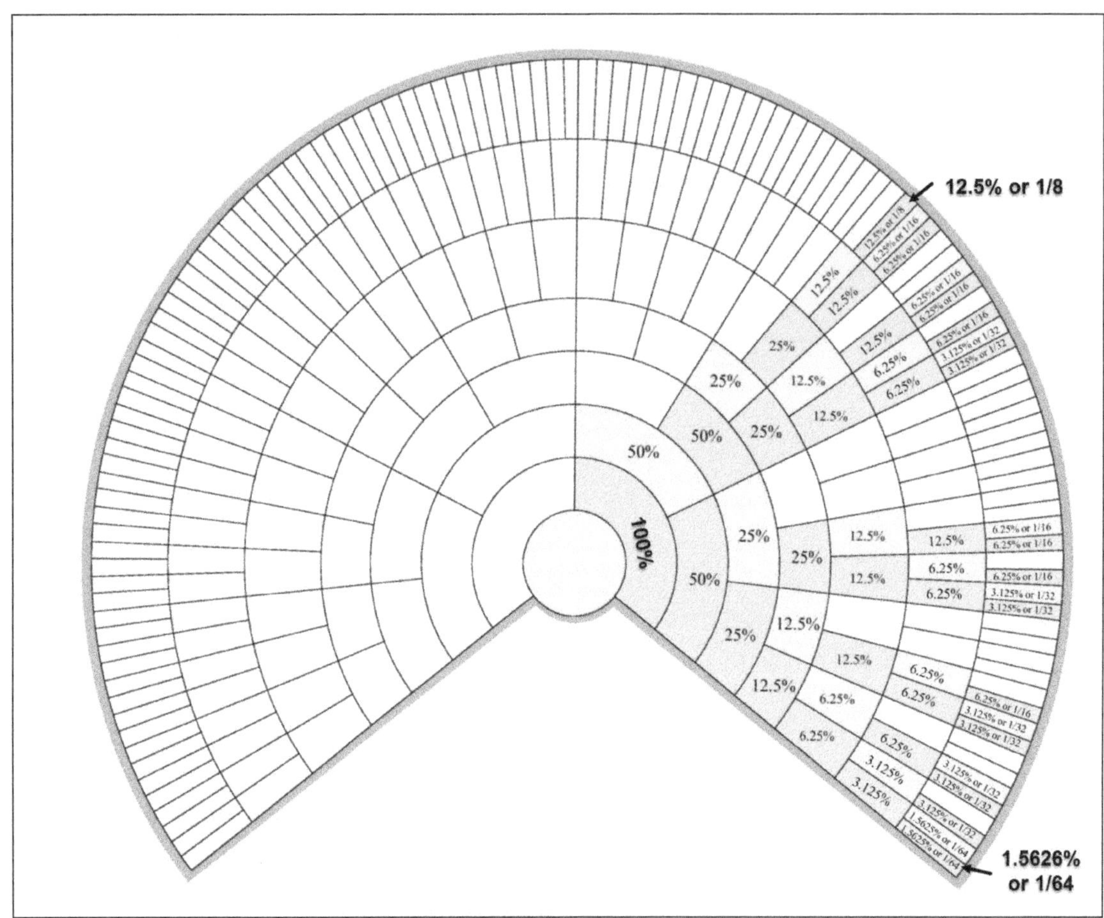

Figure 3.1. Male's X Chromosome Chart estimating the contribution percentages. These percentages are statistical averages and should not be taken literally.[5]

Theoretical probabilities are well worth studying. However, in reality, recombination is random and highly variable. An entire X, like the autosomes, can be retained or lost easily in a single generation. When there is a long X segment match but no autosomal match, the prediction of time to most recent common ancestor can range anywhere from third to very distant cousins because of the unique inheritance patterns. One must not depend on cousinship predictions based on any single chromosome, especially the X, because the common ancestor could be as close as, say, four generations back, but could also be many more generations back. People often ask where they can find a shared cM chart for the X chromosome that gives a prediction of an MRCA when they have an X-only cousin match. There simply

5. Blaine T. Bettinger, "More X Chromosome Charts," *The Genetic Genealogist*, 12 January 2009 (https://thegeneticgenealogist.com/2009/01/12/more-x-chromosome-charts/); click on thumbnail to access chart image at (https://thegeneticgenealogist.com/wp-content/uploads/NewChart5.jpg).

is *no such chart* because there is such a wide range of segment sizes coming from distant ancestors. Always look at X matches that are in combination with autosomal matches first because those could be closer, but rely on the cousinship predictions based on total amount of shared DNA as determined through the Shared cM Project.[6]

Remember that, as is true for other chromosomes, when an X match is associated with an autosomal match to the same person, it does not mean both segments came from the same ancestor. Sometimes the X segment match came from one member in a couple and the autosomal match came down from the other member of that same couple. Segment matches can also be from very divergent branches in your family tree so a thorough search for other possibilities should always be undertaken.

Direct-to-consumer test companies that offer X testing

Inclusion of the X chromosome in matching algorithms varies. Of the genealogical testing companies, 23andMe has provided X chromosome matching in a chromosome browser from the beginning. This company also paints the ancestral origins for the X in a second chromosome browser called "Ancestry Composition." 23andMe includes the X chromosome segments in the total cM count. 23andMe counts an X-only segment match in cousinship predictions; in contrast, most companies require an autosomal DNA match before checking for a match on the X chromosomes. The total shared DNA as measured in cM depends on the company algorithms.

At FamilyTreeDNA, the X chromosome was added to the chromosome browser in 2014. There, you must have an autosomal match that meets the company threshold before an X match is reported. Most of the X matches are likely too small to be significant because segments down to 1 cM in size are reported. Although only matching X segments above 10 cM should be considered, those who share above 20 cM are most likely to have common ancestors that can be identified. Researchers should increase the minimum segment size displayed by the chromosome browser to find a match likely to be Identical By Descent (IBD). Those who download to spreadsheets or use other third-party tools are able to sort the X chromosome according to shared segment sizes.

AncestryDNA does not utilize the X chromosome in its matching algorithm, but like the other companies, downloadable raw data is available. AncestryDNA does not have a chromosome browser or provide locations of shared segments, but comparisons can be made at third-party sites such as David Pike's DNA analysis tools[7] or *GEDmatch*.[8] At *GEDmatch*,

6. Blaine T. Bettinger, "Shared cM Project," *The Genetic Genealogist* (https://thegeneticgenealogist.com/). Search the blog posts for the most recent update to the project such as the August 2017 update at https://thegeneticgenealogist.com/2017/08/26/august-2017-update-to-the-shared-cm-project/. The project charts are periodically updated.

7. David Pike, Autosomal Utilities (https://www.math.mun.ca/~dapike/FF23utils).

8. *GEDmatch* (https://www.gedmatch.com/). As of January 2019, *GEDmatch* Classic and *GEDmatch* Genesis have slightly differing names for similar tools. Generic names are used in this chapter to reference the tools.

test takers often discover X matches among those who upload raw data files after testing at the other companies.

MyHeritage does not utilize the X chromosome in the matching algorithm. Like the other companies, MyHeritage permits downloading of raw data that can be included in X-DNA comparisons elsewhere.

Popular Third-Party Tools

If you have access to the raw data of your relative, you can use third-party tools to analyze the data. Pay attention to privacy considerations and obtain test taker permission before uploading raw data.[9] If you wish to compare the X chromosome coming from specific individuals as reported in megabase pairs, you can make comparisons of up to three people using David Pike's utilities.[10] You can also use these tools to phase a child's X chromosome when data is available from the child and both parents.

The displays for the X-chromosome browser can differ from the autosomal displays and may vary by company. The *GEDmatch* chromosome browser[11] provides the most X matches among test takers and shows graphics, positions, and cM size. *GEDmatch* uses different match thresholds on the X one-to-one comparison than those used in the one-to-many comparison tool.

Figure 3.2 displays a comparison between two males who tested at AncestryDNA. A male's raw data for the X chromosome from AncestryDNA and FamilyTreeDNA is listed twice even though there is only one copy of the chromosome. The raw data from 23andMe only lists the X data one time, although the current browser uses the duplicated markers. 23andMe raw data uploaded by the male test taker will appear with a single blue bar as a segment in common at *GEDmatch* as opposed to the duplicated green and blue bars shown in figure 3.2 if both males tested at the other companies. However, when comparing a male to himself, the data from 23andMe will sometimes show the green bar, perhaps depending on the date the download was obtained. The current 23andMe v5 test version is not compatible with *GEDmatch*'s traditional database. Since the new Genesis at *GEDmatch* was still in beta test at the time this was written, it is not reviewed here.[12]

The *GEDmatch* chromosome browser when used together with pedigree searches is indispensable for finding cousin matches and providing evidence.

---

9. *Genetic Genealogy Standards* (http://www.geneticgenealogystandards.com/).

10. David Pike, Autosomal Utilities (https://www.math.mun.ca/~dapike/FF23utils).

11. *GEDmatch* (https://www.gedmatch.com/).

12. *GEDmatch* Genesis (https://www.gedmatch.com/). Also see the "Would You Like Your Data Raw or Cooked?" chapter for more on Genesis.

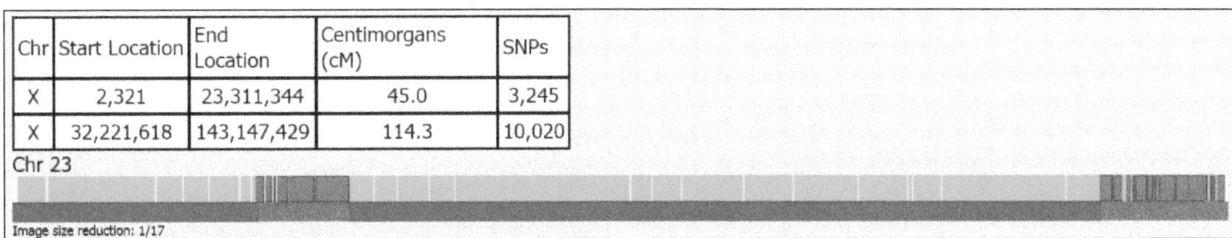

Chr	Start Location	End Location	Centimorgans (cM)	SNPs
X	2,321	23,311,344	45.0	3,245
X	32,221,618	143,147,429	114.3	10,020

Figure 3.2. Comparison between X chromosomes in brothers who tested at AncestryDNA seems to show duplication of the X (indicated by the green) even though males only have one X chromosome.

Limitations of the X and the Differences Between Autosomal and X-DNA

Meiosis and recombination in the X chromosome

Empirical evidence indicates that the female's pair of X chromosomes engage in crossover recombination[13] similarly to autosomes that are of equivalent size. During recombination, some chromosomes get "mixed" or recombined fairly well, but some do not seem to "mix" at all. This "mixing" is uneven and the crossover points are infrequent in general. The X may be inherited intact with no recombination in about 14% of cases when transmitted by the mother. Figure 3.3 shows how the X recombination events compare to chromosome 7 which is of similar size. There is not enough evidence so far to conclude that the X behaves differently than the autosomes when recombination takes place in the oocyte that forms the mother's egg. More research like this is needed in the field of recombination where test takers can make informative contributions to science.[14]

Test takers often ask, "why do so many X chromosomes seem to be passed down unrecombined?" The male passes his entire X chromosome to all of his daughters without "mixing" or recombination except in what is called the pseudoautosomal regions (PAR) at the very ends of the chromosome.[15]

The total SNP counts change as new chips are introduced. Instead of matching a paternal half sister for the full 196 cM size of the X chromosome, the matching size may fall below 190 cM and still be considered a full X match. The start locations often differ by company. MyHeritage and FamilyTreeDNA report a start location that is farther along the chromosome than 23andMe and AncestryDNA.[16] The start point can range anywhere from about

13. "Recombination," *International Society of Genetic Genealogists (ISOGG) Wiki* (https://isogg.org/wiki/Recombination).

14. Blaine T. Bettinger, "Preliminary Results from The X-DNA Inheritance Project," *The Genetic Genealogist* (https://thegeneticgenealogist.com/2017/02/12/preliminary-results-from-the-x-dna-inheritance-project/).

15. The X and the Y do exchange information, but it rarely is a problem for the genealogist because the number of reported alleles is so small at the tips.

16. *GEDmatch* aligns the uploaded raw data to a specific build number. Other tools and techniques analyzing the raw data files may report differences that are because of the compared files representing different "build" versions. Also see "Builds" in the "Would You Like Your Data Raw or Cooked?" chapter for more information.

2,321 to 2,701,157. The mismatches between two otherwise matching X chromosomes can be expected to occur more often at the start of the X chromosome than at the end. Figure 3.4 gives an indication of where these PAR segments are located at each end of the X by comparing a male test taker to his own data from a different testing company. This test taker tested at both AncestryDNA and 23andMe. The small green bars at each end illustrate where to find the PARs.

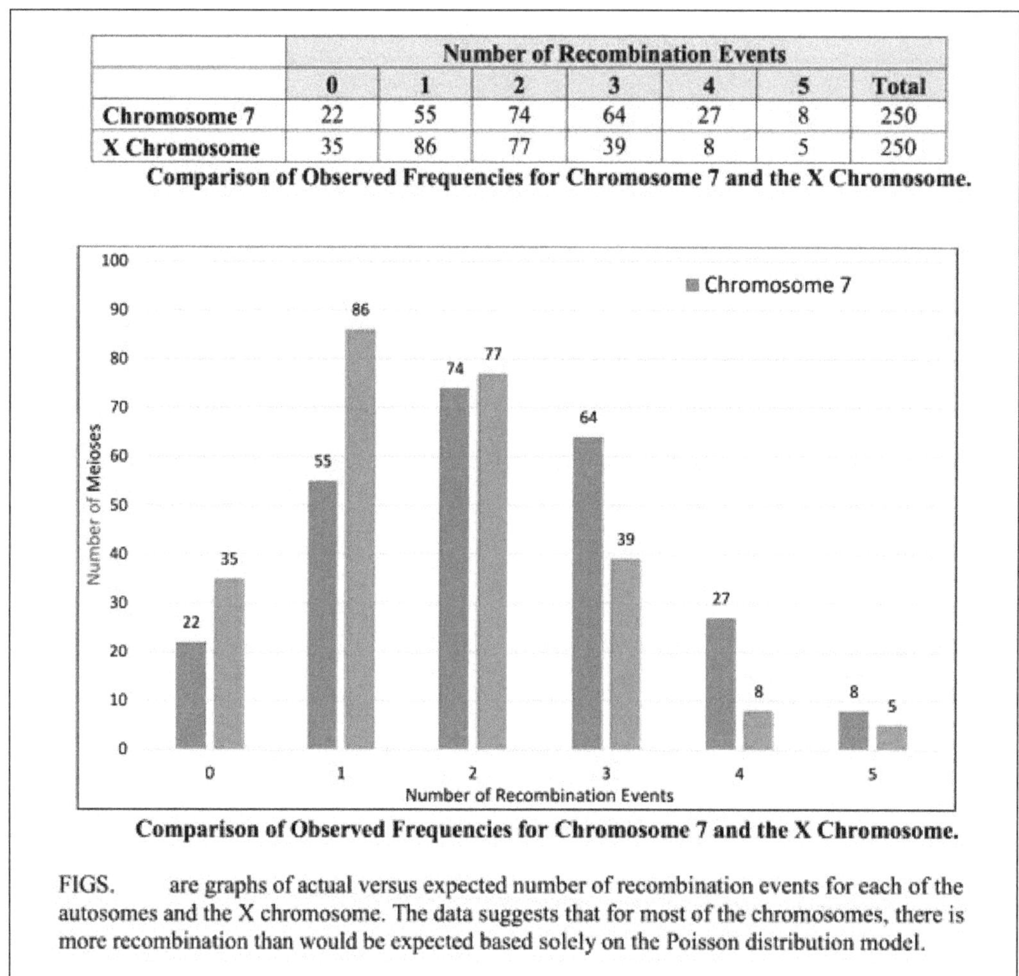

	Number of Recombination Events						
	0	**1**	**2**	**3**	**4**	**5**	**Total**
Chromosome 7	22	55	74	64	27	8	250
X Chromosome	35	86	77	39	8	5	250

Comparison of Observed Frequencies for Chromosome 7 and the X Chromosome.

Comparison of Observed Frequencies for Chromosome 7 and the X Chromosome.

FIGS. are graphs of actual versus expected number of recombination events for each of the autosomes and the X chromosome. The data suggests that for most of the chromosomes, there is more recombination than would be expected based solely on the Poisson distribution model.

Figure 3.3. The maternal X recombination resembles autosomal crossing over[17]

Chr	Start Location	End Location	Centimorgans (cM)	SNPs
X	2,321	154,886,292	196.1	16,584

Chr 23

Image size reduction: 1/17

Figure 3.4. A male's X chromosome from two testing companies is compared

17. Blaine T. Bettinger, "The Recombination Project: Analyzing Recombination Frequencies Using Crowdsourced Data," *The Genetic Genealogist*, 20 February 2017 (https://thegeneticgenealogist.com/wp-content/uploads/2017/02/Recombination_Preprint.pdf).

"Meiosis is a process where a single cell divides twice to produce four cells containing half the original amount of genetic information. These cells are our sex cells—sperm in males, eggs in females."[18] During meiosis the autosomes line up with their homologous counterparts. These are equal in size. These pairs of autosomes have corresponding alleles. The first stage of meiosis occurs in females (the future mothers) primarily while they are still in the womb, but it occurs in males after they reach puberty and just prior to fertilization.

Mixing occurs through recombination, which resembles the slicing and splicing together of blocks of DNA code. The father's X and Y recognize each other prior to mixing but these two do not have enough in common to be able to engage in much recombination except at the very tips. The Y is quite small, around 59 million base pairs, but the X is around 155 million base pairs, so size alone makes it difficult for the X and Y to pair up.

In order to understand the reason for an unmixed chromosome though, it is necessary to study what happens in the mother's oocyte (egg-creating cells) during recombination. Each of the pairs of autosomes and X chromosomes lines up with its homologue and joins at the centromere to form a fused chromosome pair. A homologue is a chromosome that is similar in physical attributes and genetic information to another chromosome with which it pairs during meiosis.

When a mother's pair of X chromosomes mix, each chromosome in a pair copies itself during meiosis first to become two "sister chromatids" just like the autosomes. One pair of identical sister X chromatids came from your maternal grandmother and the other pair of identical sister X chromatids came from your maternal grandfather. Non-sister chromatid exchanges create new segments and make you a combination of your grandparents.

During recombination, each of the pairs first reproduce themselves so that four chromatids are able to mix. Because there are now *four* chromatids instead of just the pair of chromosomes, you can imagine why some copies of chromosomes can now mix better with each other than others. Only one out of the four chromatids can result in a new chromosome to be included in the egg. The X chromosomes mix differently for each egg.

Electron microscopy shows that sister chromatids remain mostly connected to each other during the time recombination takes place, but at least one of them must seek to recombine with one or two of the homologous non-sister chromatids through crossing over. Out of the four chromatids, all can undergo mixing, but in many instances one or two of the chromatids seem to be able to escape the mixing process altogether. This a hot topic of conversation among test takers who are able to identify these parental, non-recombinant chromosomes that seem to match just one grandparent.

---

18. "What is meiosis?," *Your Genome*, Wellcome Genome Campus (https://www.yourgenome.org/facts/what-is-meiosis).

One of the current favored models of homologous recombination in college textbooks is the double strand break and repair model. Two sister chromatids are broken, then seek to repair themselves by forming junctions with the non-sisters. Even in the most complicated examples in which the chromatids go through contortions to mix with each other more than once, one of the resulting four chromosomes can remain essentially intact and appears to be unbroken. It appears that one sister chromatid can sometimes do all the work by recombining with her non-sisters while her former twin appears to lend stability to the structure by remaining essentially unchanged but in close contact and intertwined with her sister chromatid.[19] There could be some sister recombinations that cannot be detected and perhaps these are more common on the X. In any event, this lack of visible recombination is more often noticed by test takers when inspecting their isolated X chromosomes coming from the mother, even though the same phenomenon is common among the autosomes as well, especially the shorter ones.

As with autosomes, some resulting chromosomes are well mixed, but in many cases the X does not appear to mix well at all. For example, my uncle got five or six crossover points from my grandmother, but I got none from the X that I received from my mother. There is a large range in the number of crossovers, but the longer the chromosome, the more crossovers there are in general. The X is about average in size compared to the autosomes.

A word of caution for adoptees and those with unknown parentage

Two female adoptees who suddenly discover they share an average of about 1800 cM on the autosomes should check their X chromosome match at *GEDmatch*. If they share a continuous length of around 196 cM on the X they are often told, "congratulations, you are half sisters and share the same father." However, that is *not always the case*.

The chance that two siblings share an entire X from the mother should be around 1%. The probabilities are multiplied in this case. There is about a 14% chance that no recombination will be observed in a single recombination event for the X chromosome. There is about a 2% (0.14 times 0.14) chance that no recombination takes place for *both* siblings. Therefore, 1% is the expected probability of fully matching on the X and 1% is the expected probability of not matching at all on the X. Adoptees should not make the mistake of assigning parentage to the wrong parent based solely on an X match. Half sisters who share the same father always share an X chromosome inherited from him. However, they could uncommonly, perhaps in less than 1% of cases, share an entire X coming from a mother or they could even more likely share with another relative such as a grandmother or maternal aunt.

There are exceptions and it is important to investigate all possibilities before jumping to such conclusions. If parentage and age are unknown, females who match an entire X could easily have a paternal grandmother-granddaughter relationship. Another common mistake

19. Daniel L. Hartl and Elizabeth W. Jones, *Genetics: Analysis of Genes and Genomes* (Sudbury, Mass: Jones and Bartlett Publ., 2005), 248–251.

occurs when a maternal aunt just happens to share an entire X with her niece. The closeness of the ages of the two individuals is not always a reason to assume sisters. A mother and her full sister will share the paternal X from their father, but they also are likely to share on their maternal X as well. Even with crossover recombination, an aunt can easily match the entire X of her sister's child. In that case, it could be impossible to tell a half sister from an aunt-niece relationship without further investigation. This is a very common mistake made among adoptees. Figure 3.5 shows an aunt and her full sister who share extensively on both X chromosomes (top comparison). After mixing is carried out by the mother through recombination, her daughter may match the aunt along the entire length of her X chromosome. The resulting X match (bottom comparison) can be indistinguishable in a browser when comparing two paternal half sisters.

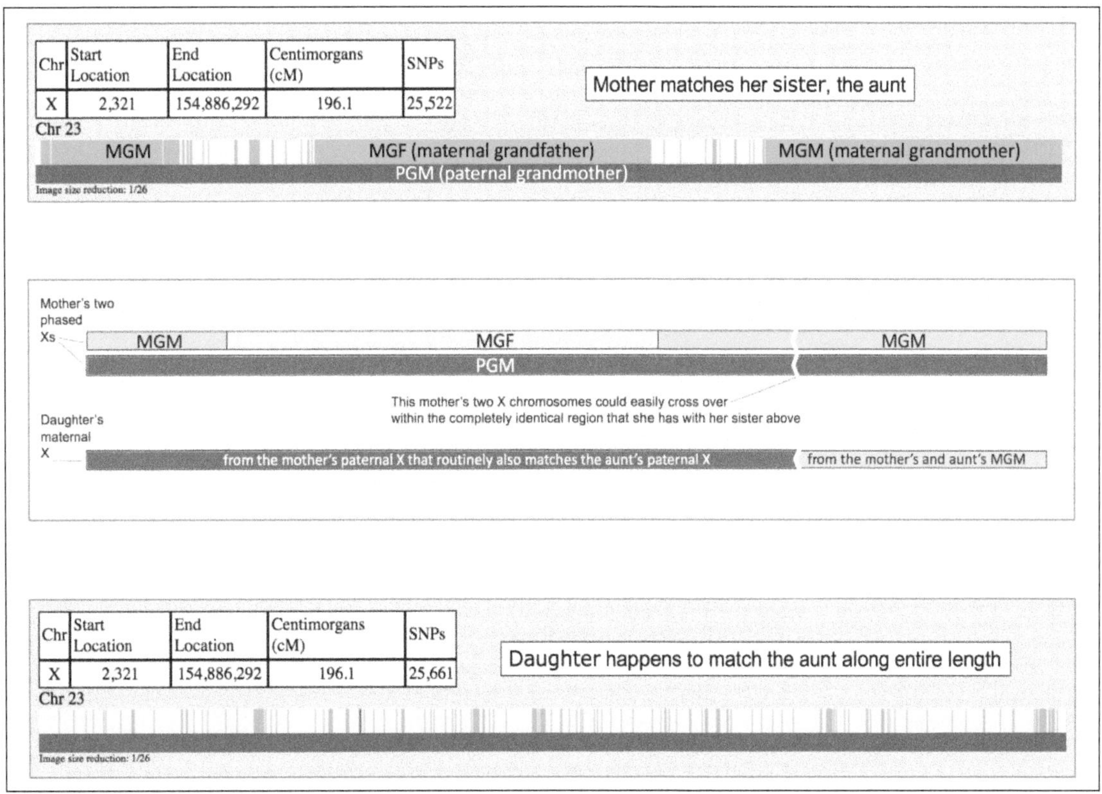

Figure 3.5. Mother, daughter, and aunt X chromosome comparisons

The green in the top image in figure 3.5 indicates fully identical regions (FIRs) among these two full siblings. The X chromosome is the only chromosome in the browser in which the segment borders can be visually phased to the maternal side without further investigation. These particular segments have been phased and mapped to the maternal grandmother (MGM) and maternal grandfather (MGF) because one of these grandparents was tested. The chromosome browsers at 23andMe and *GEDmatch* are often the most useful because of the visible FIRs. Any crossover borders made by the mother should be visible unless she makes crossovers for two of her children at the same or nearby locations.

X chromosome sibling matches at *GEDmatch* show FIRs representing maternal DNA colored in green. A full sister match to the paternal side corresponds to the blue bar along the entire length. Fortunately, this paternal blue X match in the top comparison will map only to their paternal grandmother. In this case, the aunt's and mother's paternal grandfather can be ruled out as a match altogether because he is on a non-X inheritance line. Each maternal match (as is demonstrated by one of the green segments) can only map back to one maternal grandparent and not both. These particular solid green segments in the first GEDmatch comparison can alternate between MGM and MGF or could map only to one grandparent depending on which sibling received the crossover borders.

Limitations and Gender Differences in the Matching Process

Male and female comparisons differ on the X. The X chromosome is essentially already phased in the male since he only has one X and it came from the mother. There is usually no confusion as to which of his alleles are in common on this single X when two males are compared. The matches among males are more likely to be IBD. In general, males will have far fewer cousin matches on their X chromosomes than females do and usually do not match many other males. Males are understandably surprised by this difference. It is not just the fact that females have twice as many X chromosomes. There are other differences as well between males and females.

False positive matches are particularly problematic for females because of the lack of phasing. There may also be excess IBD segments and SNP-poor regions tested on current chips. Therefore, care must be taken when making assumptions for the X chromosome. The X also needs to be included more in scientific research studies to increase our knowledge of the characteristics of this chromosome.

Companies were not originally concerned with the density of X SNPs tested as they never intended to utilize the X chromosome data for cousin matching. The SNP density was probably not considered as important for the X chromosome when the chips were designed for genealogy purposes. There may not have been as many variants as could have been included. Gaps in the matching have been a known problem on the X.[20]

23andMe paid attention to the X chromosome from the beginning for both cousin matching and biogeographical assignments.

FamilyTreeDNA improved its SNP orientation and matching display for the X chromosome in 2016. However, as new DNA chips are replacing old ones, it remains to be seen how well we will be able to continue to use third-party tools for X matches.

20. Blaine T. Bettinger, *The Family Tree Guide to DNA Testing and Genetic Genealogy* (Blue Ash, Ohio: Family Tree Books, 2016), 130–1.

The International Society of Genetic Genealogy (ISOGG) Wiki reports that, through simulations, 23andMe determined that X chromosome segment matches between males could be IBD at a much lower threshold than the standard recommendations for autosomes. Our criteria for X chromosome matches appears to have changed in recent years because the current thinking is that one should *not* use 1 cM matches to attempt to find an MRCA even among males. However, at one time the reports showed the following recommendations for matching and 23andMe could still be using these particular criteria in its own comparisons. 23andMe also advises using higher than average SNP concentrations between two females, shown here at 1200 SNPs.

For half identical regions at 23andMe, the suggested matching thresholds are listed below:

- Male to male X: 200 SNPs, 1 cM minimum length
- Male to female X: 600 SNPs, 6 cM minimum length
- Female to female X: 1200 SNPs, 6 cM minimum length[21]

FamilyTreeDNA still reports an X match segment down to 1 cM for both males and females, so the match list generated there for the X is likely to be full of false positive matches. In a comparison done in the large FamilyTreeDNA project of Southern California genealogists, most X matches were determined to be false.[22]

There are many nuances that need to be considered when dealing with X matches. To date, there is no minimum segment size agreed upon for matching on the X. Analyzing matches between males first is recommended because the index of suspicion seems lower for phased males. Even for males, segments below 10 cM should rarely be used unless there is some compelling reason to do so. Such reasons might be comparing siblings during the phasing and mapping process or when there are rare variants. At *GEDmatch,* for the more distant matches, users should examine the one-to-one X matches prior to contacting the match through email. Many of the reported X matches in one-to-many lists are false positives that can be screened out with the higher thresholds used in the one-to-one comparison. Alternatively, the SNP count may actually have to be lowered at *GEDmatch* to find a cousin that matches above a 10 cM threshold cutoff because the SNP density can fall below the default cut off.

Despite its limitations, success stories are often reported with X matching as long as the size in cM is high enough. Even a single chromosome match can be successful in breaking through brick walls or may point you in the direction of a common geographical origin.

21. Blaine T. Bettinger and Debbie Parker Wayne, *Genetic Genealogy in Practice* (Arlington, Va.: National Genealogical Society, 2016), 76.

22. "Kathy" Johnston, "What Can the X Chromosome Tell Us About the Importance of Small Segments?" *Kitty Cooper's Blog*, 19 December 2014 guest post (http://blog.kittycooper.com/2014/12/small-segments-on-the-x-by-kathy-johnston).

Case Studies

Some names are anonymized in the following examples, but the DNA analysis is from cases of real test takers.

Case 1: Varley family X-DNA breaks through a brick wall

This case study demonstrates how DNA can break through a brick-wall. Figure 3.6 shows an X match between my mother-in-law Donna, who is now deceased, and a test taker who tested at a different company. This is a robust match and would likely indicate an ancestor in common within a genealogical timeframe, especially if there also is an autosomal match. There were three smaller shared autosomal segment matches as well, but we will focus on the shared X match.

Chr	Start Location	End Location	Centimorgans (cM)	SNPs
X	2,707,868	41,854,099	67.5	5,426

Figure 3.6. Long X segment match between third cousins once removed

Some *GEDmatch* users have also shared GEDCOM files on the site. Comparing the trees of the two test takers, focusing on the X chromosome inheritance paths, provides clues to potentially break through a brick-wall.

The alternating male to female zigzag pathway is striking as we trace this X chromosome back in time. This cousin's father's mother's father's mother's parents were probably the same couple as Donna's mother's father's mother's parents in Ireland. Donna identified a Bridget Vorley or Varley born mid-1800s in County Mayo, Ireland, as her great-grandmother. Bridget Varley is thought to be the daughter of Thomas Varley who died in 1860 in Creevagh Mayo, Ireland, and his wife, Barbara Cunningham, but we need more evidence.

In the cousin's GEDCOM at *GEDmatch*, the candidates for common ancestors can be identified as this same couple. There appears to be an unnamed child of Thomas Varley and Barbara Cunningham. Only Kate, John, Edward, Patrick, Maggie, and Maria Varley are named. The pedigree chart is conveniently colored red for females and blue for males so that you can easily see father to son inheritance that would eliminate a shared X chromosome. Figure 3.7 illustrates *GEDmatch*'s Individual Detail Display and pedigree display for this family. The likely most recent common ancestral couple is Thomas Varley and Barbara Cunningham based on documentary research and the X pattern of inheritance. That documentary research is not described here because of space considerations. The test takers are potential third cousins once removed. Figure 3.8 illustrates the likely links between Donna and her Varley cousin based on analyzing the trees and the X chromosome match.

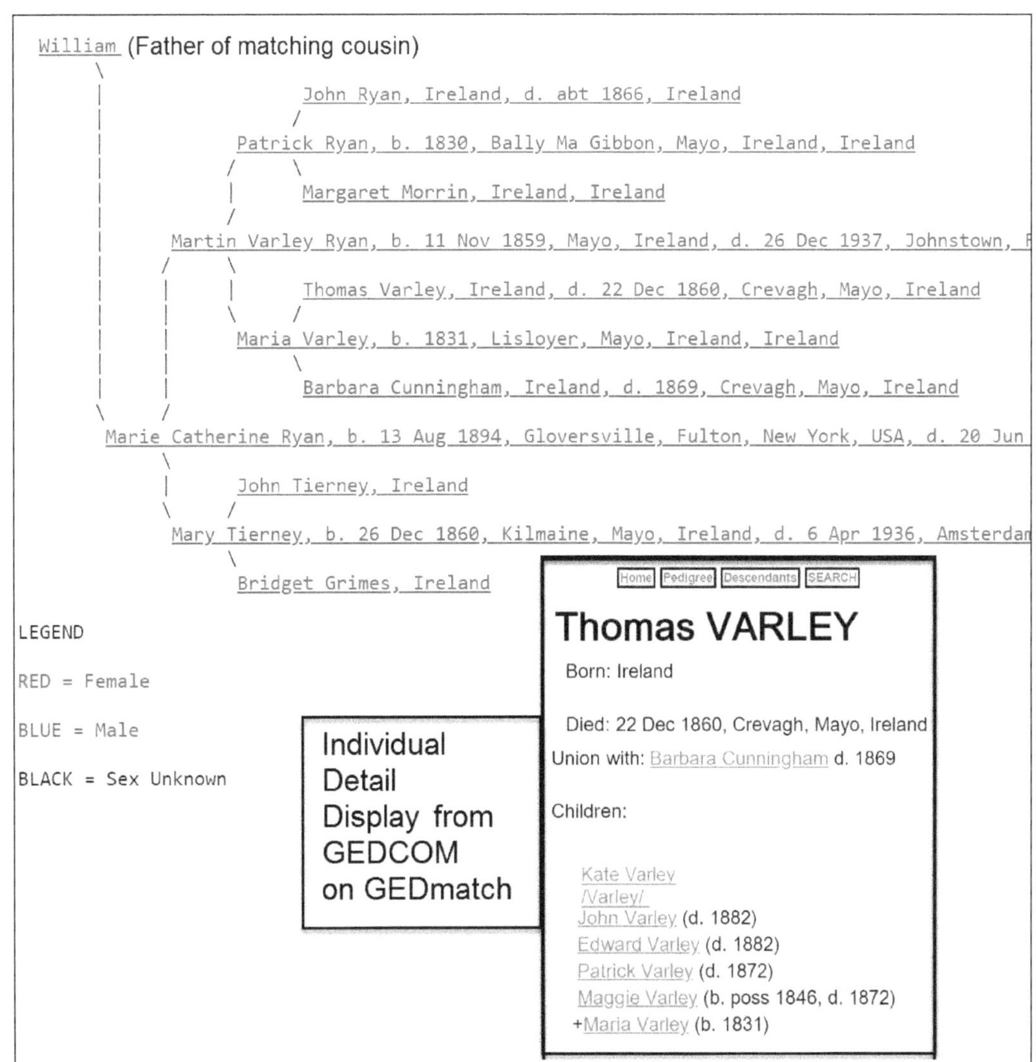

Figure 3.7. Thomas Varley and Barbara Cunningham family

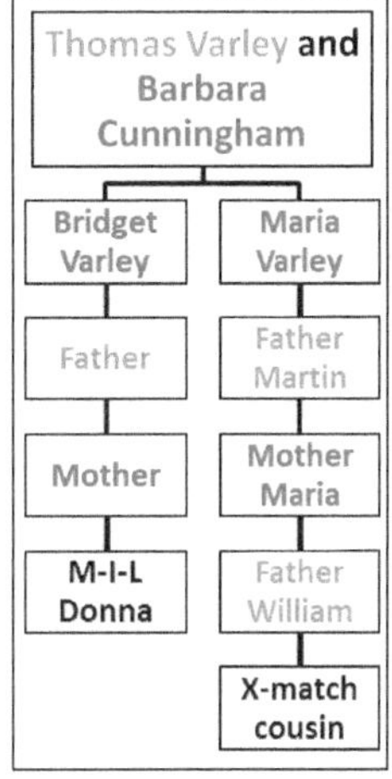

Figure 3.8. Likely relationship between Donna and her Varley Cousin with X match

Case 2: X-DNA mapped to a specific ancestral line

This second case study involves an X chromosome comparison in which the MRCA can be identified and the exact inheritance path of the segment can be mapped to a likely individual. This third cousin twice removed relationship between "Female" family member and "Male Cousin" provides evidence to identify the single ancestral owner of the original X chromosome segment, a female born in 1780 in County Armagh, Ireland. Irish and Scots-Irish ancestry is difficult to research because so many records were destroyed. Convincing evidence of these distant matches through DNA is very satisfying. These cousins both tested at AncestryDNA[23] but the X segment match was discovered at *GEDmatch*.

Figure 3.9 shows that a 21 cM segment is shared at AncestryDNA. An even longer segment appears as an X chromosome match between these two cousins at *GEDmatch*, but is not shown on AncestryDNA. This same autosomal segment is measured as 25.1 cM at *GEDmatch* (using a different algorithm) and there is an additional 33.4 cM X chromosome segment match.

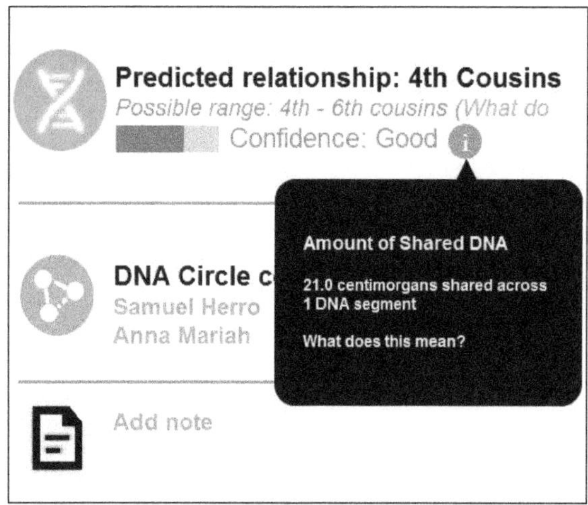

Figure 3.9. This 21 cM segment match represents autosomal DNA at *AncestryDNA*. X chromosome segments are not included

The default display at *GEDmatch* must be modified to pull up the most promising X chromosome matches that potentially could break through brick walls. An autosomal match of at least 20 cM along with an X match of at least 30 cM should always be investigated. Figure 3.10 shows how *GEDmatch* one-to-many sorts based on the amount of shared autosomal DNA. This cousin who has a 30.2 cM total shared autosomal match with 25.1 cM as the largest segment size, plus a shared 33.4 cM X-DNA match is an excellent candidate for a pedigree search, but the X match could easily be missed as it is far down in the list. It is better to sort by the amount of shared X when focusing on the X chromosome. Clicking on the triangle under X-DNA re-sorts the matches listing largest X matches first as shown in figure 3.11.

23. *AncestryDNA* (https://www.ancestry.com/).

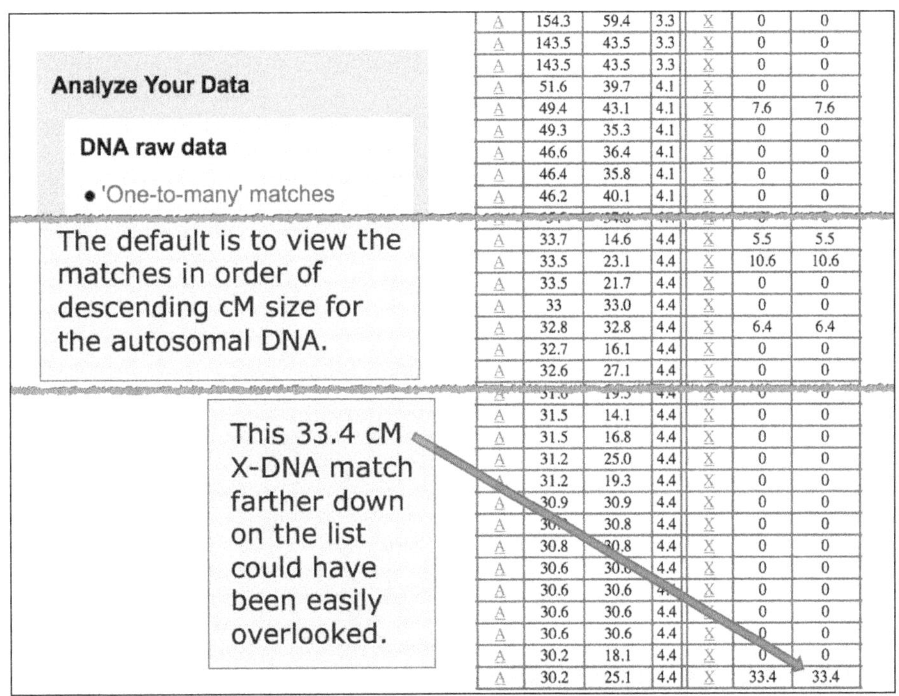

A	154.3	59.4	3.3	X	0	0
A	143.5	43.5	3.3	X	0	0
A	143.5	43.5	3.3	X	0	0
A	51.6	39.7	4.1	X	0	0
A	49.4	43.1	4.1	X	7.6	7.6
A	49.3	35.3	4.1	X	0	0
A	46.6	36.4	4.1	X	0	0
A	46.4	35.8	4.1	X	0	0
A	46.2	40.1	4.1	X	0	0
A	33.7	14.6	4.4	X	5.5	5.5
A	33.5	23.1	4.4	X	10.6	10.6
A	33.5	21.7	4.4	X	0	0
A	33	33.0	4.4	X	0	0
A	32.8	32.8	4.4	X	6.4	6.4
A	32.7	16.1	4.4	X	0	0
A	32.6	27.1	4.4	X	0	0
A	31.5	14.1	4.4	X	0	0
A	31.5	16.8	4.4	X	0	0
A	31.2	25.0	4.4	X	0	0
A	31.2	19.3	4.4	X	0	0
A	30.9	30.9	4.4	X	0	0
A	[illegible]	30.8	4.4	X	0	0
A	30.8	30.8	4.4	X	0	0
A	30.6	[illegible]	4.4	X	0	0
A	30.6	30.6	[illegible]	X	0	0
A	30.6	30.6	4.4	[illegible]	0	0
A	30.6	30.6	4.4	X	0	0
A	30.2	18.1	4.4	X	0	0
A	30.2	25.1	4.4	X	33.4	33.4

Figure 3.10. *GEDmatch* one-to-many sorts by shared autosomal DNA by default

Autosomal				X-DNA		
Details	Total cM	largest cM	Gen	Details	Total cM	largest cM
A	5.4	5.4	7.7	X	45.4	38.6
A	0	0		X	45.4	38.6
A	5.4	5.4	7.7	X	38.6	38.6
A	0	0		X	37.9	37.9
A	0	0		X	37.9	37.9
A	0	0		X	37.9	37.9
A	0	0		X	37.9	37.9
A	1648.1	137.3	1.6	X	93.6	36.2
A	1694.1	118.5	1.5	X	93.3	36
A	2568.8	196.7	1.2	X	91.9	35.3
A	0	0		X	34.5	34.5
A	849.8	60.8	2	X	63.5	34.2
A	0	0		X	34.2	34.2
A	0	0		X	33.8	33.8
A	30.2	25.1	4.4	X	33.4	33.4

Click on the blue triangle to create a red triangle.

The matches will now appear in descending order of X segments.

Figure 3.11. Sorting by shared X segments lists largest X matches first

The X match between these two cousins is indicated by the arrow in the stacked comparisons in figure 3.12, which should be used for the following discussion. Cutting and pasting comparisons in a stacked manner is helpful when using the process of elimination to determine pedigree lines. DNA of siblings can sometimes be used to determine which side of the family should be searched for a specific match.

This "Sister" and her "Male Cousin" appear as an X match at *GEDmatch*. She matches her "Brother 3" on her second X chromosome at the same location. The match with "Brother 3" must be from their mother's side of the family as the brother inherited X-DNA only from their shared mother. Because the cousin is male his X chromosome came from his mother. None of the brothers match this cousin at the same segment site. Her maternal X chromosome could, therefore, reasonably be ruled out as a match with this cousin through the process of elimination. The MRCA is likely to be on the paternal side of her family; therefore, the segment traveled through her paternal grandmother.

Three different companies are represented in these comparisons. Any solid yellow color between this female and her brothers is because two of them tested at 23andMe. We are indeed fortunate to have a third-party tool like *GEDmatch* because all of these brothers are now deceased and this is one of the few ways to make comparisons among all family members. Their DNA lives on.

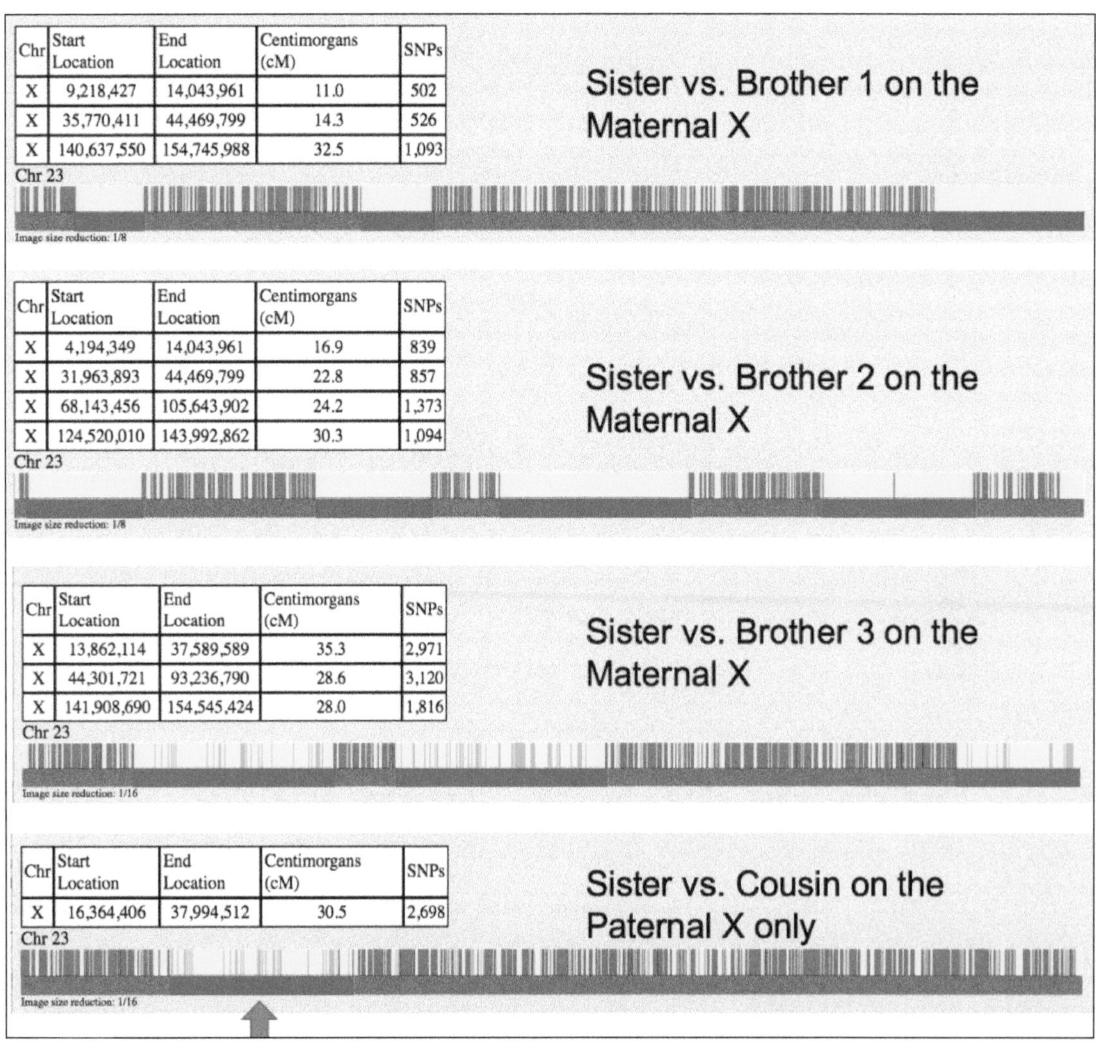

Chr	Start Location	End Location	Centimorgans (cM)	SNPs
X	9,218,427	14,043,961	11.0	502
X	35,770,411	44,469,799	14.3	526
X	140,637,550	154,745,988	32.5	1,093

Chr	Start Location	End Location	Centimorgans (cM)	SNPs
X	4,194,349	14,043,961	16.9	839
X	31,963,893	44,469,799	22.8	857
X	68,143,456	105,643,902	24.2	1,373
X	124,520,010	143,992,862	30.3	1,094

Chr	Start Location	End Location	Centimorgans (cM)	SNPs
X	13,862,114	37,589,589	35.3	2,971
X	44,301,721	93,236,790	28.6	3,120
X	141,908,690	154,545,424	28.0	1,816

Chr	Start Location	End Location	Centimorgans (cM)	SNPs
X	16,364,406	37,994,512	30.5	2,698

Figure 3.12. X chromosome comparisons between siblings and a cousin

Case 3: Anna Mariah Piper and siblings X-DNA

This case includes a male cousin in the right-side lineage column in figure 3.13 who tested at AncestryDNA. He matches a female's X chromosome at *GEDmatch* but does not match the X of any of her three brothers there. This X chromosome pattern of inheritance shown in figure 3.13 (alternating female to male) is likely to be correct; both of the female's maternal grandparents and many other relatives could be ruled out. That research is not described here because of space considerations. For autosomal DNA, we often do not know which ancestor of a couple is the source of a specific DNA segment. However, in this case, the X segment could only be inherited from Anna Mariah Piper born 1780 in County Armagh, Ireland. Her son, James Herron, born in 1802 must have been one of the carriers and he did not get an X from his father.

Similar to the Varley success story, the X here also exhibits an alternating inheritance path between males and females. This alternating pattern minimizes the recombination that occurs because of fathers passing an X chromosome to daughters without recombination. The male cousin represented in the right-side lineage traces back six generations to this MRCA couple, Samuel Herron and Anna Mariah Piper.

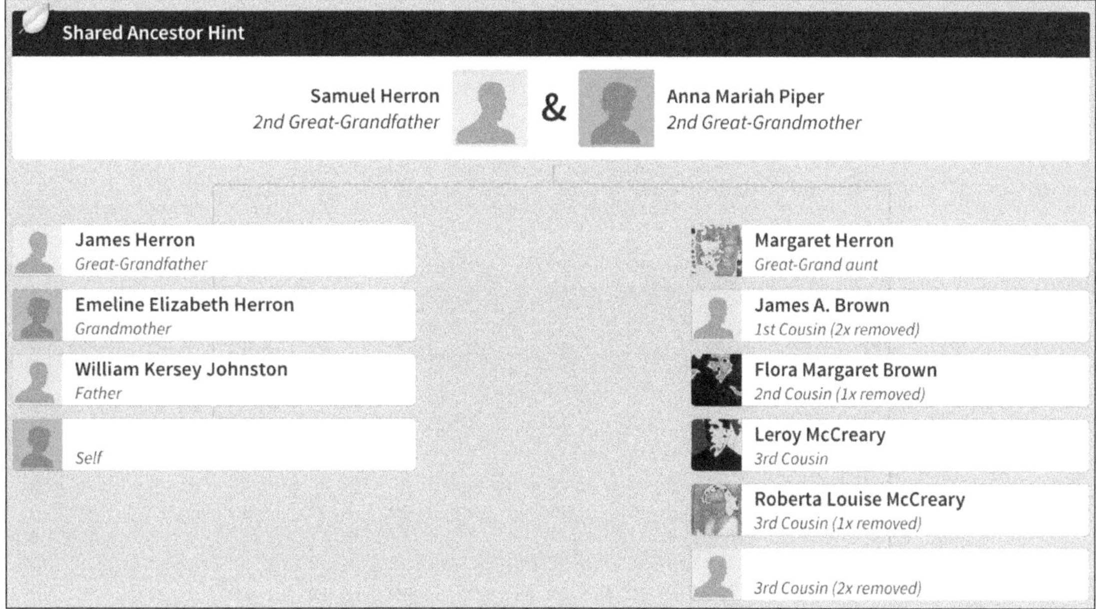

Figure 3.13. Shared Ancestor Hint at AncestryDNA

We must perform reasonably exhaustive research on each of the families to rule out other potential shared ancestors. There can never be absolute certainty that we have the only explanation for this "Scots Irish" segment origin, but the result of this analysis is highly suggestive because so many other ancestors can be ruled out as the source of the X chromosome. Oral family history and extensive documentary research can be used to build each of these pedigrees, but it may be the DNA evidence that leads to networking among descendants and the mapping of specific DNA segments back to a single ancestor. Ruling in and also ruling out other lines in the respective pedigrees, along with the rest of the body of evidence, can confirm and credibly support the conclusions.[24]

Case 4: three-quarter siblings

A very unusual case is one in which sisters are expected to share their father's X but do not share with him *at all* on the X chromosome. This is based on the *GEDmatch* one-to-one comparison between these sisters. They share close to 2,300 cM on the autosomes with most autosomes showing at least some FIRs. The matching demonstrates that they must be related on both sides of the family. This can be puzzling, but it was suggested by one of the sisters that the deceased paternal grandfather might be the father of one of the sisters. The mother had children with both father and son which resulted in what we call "three-quarter siblings." Technically, they would be half sisters with an additional half-aunt-to-niece relationship.

The more usual type of three-quarter siblings are often related as half siblings and first cousins at the same time. This more common type of three-quarter relationship occurs when either the mother produced offspring with two brothers or the father produced offspring with two sisters. A common father for sisters would be suggested by the matching along the entire length of the X, however. It is entirely possible, though uncommon, for two full brothers to share no segments of the X chromosome with each other. In this particular case, the family history suggests that the father and his son both fathered children with the same female. The key to the diagnosis was not only the family history, but the fact that the sisters did *not* share the X chromosome. Full sisters always share their father's X.

Why the X is so much Fun for DNA Enthusiasts

Visual phasing began with X comparisons and the X is still recommended as a starting point

Genealogists are passionate about their detective work and problem-solving skills. When we see a pattern developing through simple observation, we like to let the rest of the community know about it. We call ourselves citizen scientists. We share innovations even before there is any kind of peer review. Sometimes simple observation can lead to more advanced puzzle-solving techniques useful for pedigree building. That was certainly the case with a new methodology that we now call "visual phasing." It actually began as a result of X chromosome comparisons between my siblings with nothing more than pattern recognition. I certainly was not looking to prove any sort of scientific hypothesis at the time. I introduced it first to a handful of genealogists to see if it had utility.

The phasing of chromosomes is not a new idea. I remember hearing lectures at genetic genealogy conferences presented by PhDs like T. Whit Athey and David A. Pike about phas-

24. Board for Certification of Genealogists, *Genealogy Standards*, 2nd ed. (Nashville, Tenn.: Ancestry.com, 2019).

ing prior to the visual phasing idea. I was not even thinking about phasing at all when this application was discovered. It was the patterns of lines that caught my eye when I looked at my sibling comparisons. I wondered if these segment borders and matches could be used to map back to close relatives. As a dermatologist, I am accustomed to detecting patterns and color transitions in the skin. If you do not see all the repeated patterns of lines and circles in a skin rash, you can easily miss an important diagnosis. Some of us are visual learners and can apply logic even when we have trouble understanding advanced math and science. Visual skills can be very useful when observing matches in chromosome browsers. As the saying goes, you do not have to be a rocket scientist to make these kinds of discoveries.

Bennett Greenspan demonstrated the "missing tooth" pattern when he introduced the FamilyTreeDNA Family Finder chromosome browser at a conference held in Houston in 2010. He pointed out the fact that you sometimes flip away from matching a relative on a chromosome, then as you read the chromosome from left to right, you flip back to matching that relative again and vice versa. I was fascinated by these missing teeth patterns but I did not know how useful these patterns were at the time. It did not become obvious until I realized that you need to identify all of your own crossover borders, not just some of them in order to make the next leap in the discovery process. Figure 3.14 shows the missing teeth pattern between a test taker and three close relatives on one chromosome.

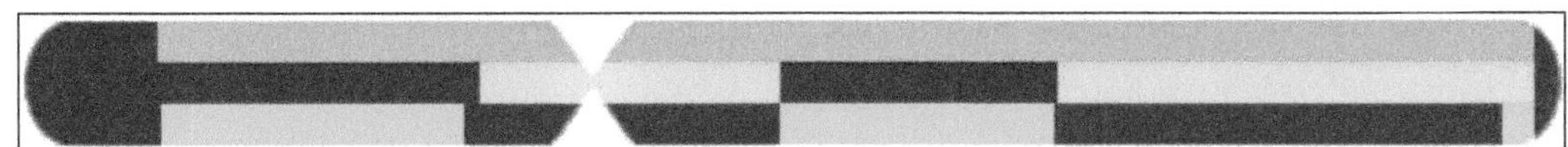

Figure 3.14. Missing teeth pattern between matches on this chromosome

I began by identifying borders on the X chromosome because it is the easiest one to determine crossover points. Understanding crossover recombination is an important advanced genetic-genealogy skill. Visual phasing is a method that is currently being used to map identical by descent (IBD) segments back to the grandparents through sibling matches when those grandparents are not available for testing. It usually requires the DNA from at least three siblings and a good chromosome browser. The first requirement is the identification of each sibling's crossover recombination points in a browser that displays both half identical regions (HIRs) and fully identical regions (FIRs).

For the X, the mother creates these crossovers points, but the child "owns" them. For me, this mapping to the maternal grandparents became apparent through a simple observation. There is a pattern when X comparisons (copied from 23andMe) between all my siblings were stacked one on top of the other as seen in figure 3.15.

The X chromosome provides segments from just one side of the family and all the crossover points seem to line up like little soldiers. The in-common-with names for each border can determine the ownership. These DNA fingerprints are unique to an individual and each segment border can be assigned to a specific sibling as shown. Instructions on this process can be found in the "Visual Phasing Methodology and Techniques" chapter.

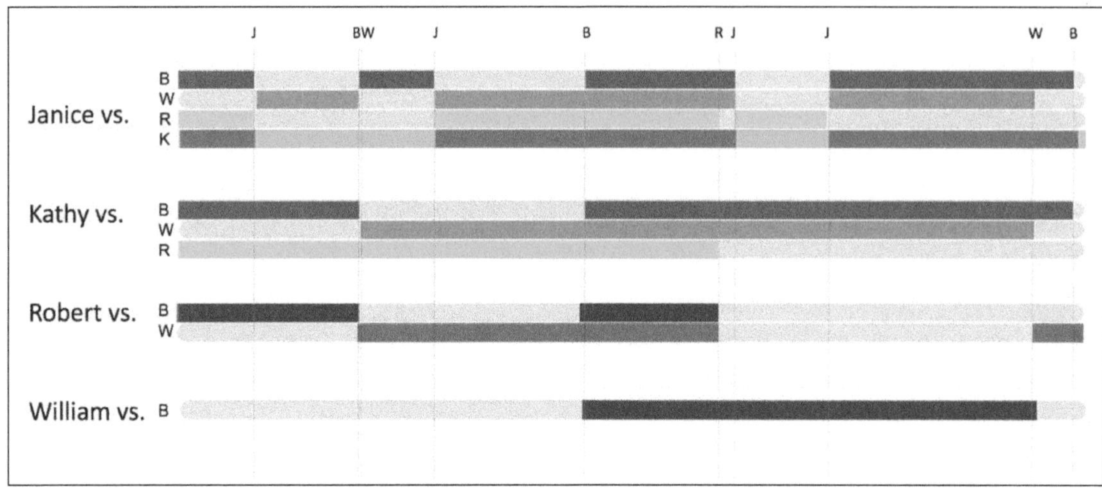

Figure 3.15. Pattern recognition paves the way to the practice of visual phasing

The Fibonacci sequence is what makes the X unique

One of the most remarkable observations involving the X chromosome is the appearance of the Fibonacci sequence of numbers as demonstrated in the X pattern of inheritance charts. The Fibonacci sequence is found throughout nature and, for those of us studying our family trees, the X chromosome is probably the best example of it. The Fibonacci code found in our X chromosome inheritance charts was first brought to the attention of genealogists by Luke Hutchison while he was working for the Sorensen Molecular Genealogy Foundation.[25]

The number of X ancestors in each generation follows the Fibonacci sequence. The Fibonacci sequence starts with number 1, and each subsequent number is the sum of the previous two numbers in the sequence. The numbers are 1, 1, 2, 3, 5, 8, 13, and so on. for a male and 1, 2, 3, 5, 8, 13 etc. for a female in the X chromosome pedigree charts. See figure 3.16 for the male version of the chart. In a five-generation pedigree chart, a female will count herself as number 1 because she is the only one who carries the X chromosomes in her generation for her female-focused chart. Her parents count as two ancestors in the second generation because both mother and father carry at least one X that could be passed on to a child. There are only three X ancestors (who together carry 100% of your X segments) in the third generation, five X ancestors in the fourth generation, and so on. Keep in mind that it is easy to lose an entire X from an ancestor in just one generation, so do not expect to carry segments from all of the more distant ancestors. Only a few X ancestors are likely to contribute to your X or portions thereof.

25. Luke A. D. Hutchison, Natalie M. Myres, and Scott R. Woodward, "Growing the Family Tree: The Power of DNA in Reconstructing Family Relationships," *Proceedings of the First Symposium on Bioinformatics and Biotechnology (BIOT-04, Colorado Springs)* (September 2004): 42–49. See also, "X Chromosome," *Wikipedia* (https://en.wikipedia.org/wiki/X_chromosome).

Why is this Fibonacci number significant? It illustrates that as you go back in time, there is an even smaller percentage of ancestors who are potential X contributors for any given generation. That means you can rule out a lot of ancestors as the source of an X match with a cousin. The utility of genetics in genealogy is not simply the knowledge of where to find an ancestor in common, but also the ability to rule out a large number of individuals as potential candidates in order to narrow the playing field.

Ann Turner also pointed out the Fibonacci code among our X ancestors. Any time there is a male to male inheritance, that ancestor can be ruled out along with all of his forebears. The number of ancestors left per generation adds up to a Fibonacci number.[26] Ahnentafel numbers along with helpful probabilities are included in the X chromosome pedigree charts provided by Sue Griffith.[27] Genome Mate Pro and *Wikitree* are two programs that will give you a list of your X ancestors.

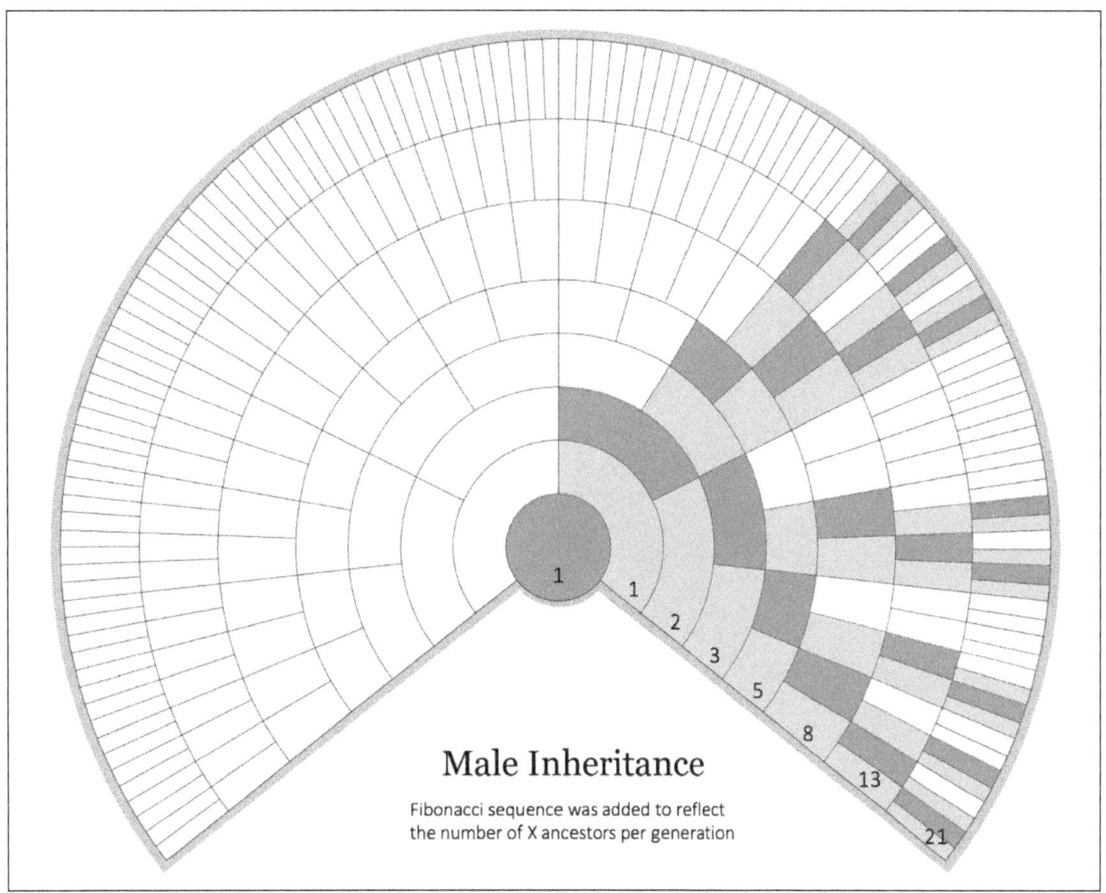

Figure 3.16. Male Inheritance with Fibonacci numbers added

26. Ann Turner, "Ahnentafel numbers of ancestors who could contribute a segment on the X chromosome," archived at *Internet Archive* (https://web.archive.org/web/20170607171833/http://dnacousins.com/AHN_X.TXT).

27. Sue Griffith, "Templates for a Pedigree Chart and X-DNA Inheritance Charts," *Genealogy Junkie* (http://www.genealogyjunkie.net/downloads.html/).

If a male counts twenty generations back in time, starting with himself as generation 0, only about 1% of potential ancestors in that generation could have possibly given him a piece of his X chromosome. Among those X ancestors, some will have a huge advantage over others because the X will not be broken up as much over time. Naturally, there will likely be varying degrees of pedigree collapse when cousins marry cousins, making any such calculation impractical in those cases.

The X chromosome and the future

In addition to the X chromosome uses in genealogy, many people have discovered unexpected origins from their biogeographical paintings of the X at 23andMe. Sometimes these origins differ from the autosomes, the Y-DNA, and the mtDNA. Figure 3.17 illustrates a person who might be surprised to find Yakut or a similar population painted orange on the X chromosome extending through the centromere. It would be interesting to see if this segment survived through many generations in the family tree along the lines of highest probability.

Figure 3.17. X chromosome with unexpected geographical origin

As of this writing, 23andMe is the only test company that provides tools on the X chromosome for both genealogical and ancestral painting. *GEDmatch* lacks biogeographical displays for the X chromosome in the admixture tools. The future uses of the X chromosome in genealogy may depend on the continuing work of geneticists, population biologists, test companies, and citizen scientists like us.

Conclusion

While full of limitations, the X chromosome also has great potential for breaking through brick-walls in the family tree. Many lines in the X chromosome charts can be eliminated when seeking an MRCA in cases where there is a solid X match.

Y-DNA Analysis for a Family Study

James M. Owston, EdD

Recently, a participant in one of the many genetic genealogy forums revealed that she considered testing her brother's Y-DNA to gather more information about their shared patrilineal line. One by one, others began to advise her against doing this. One even went as far as to say, "Why would you ever want to do this, when autosomal DNA is all you'll ever need?"

I was concerned by the overwhelming number of genealogists who attempted to dissuade her from pursuing Y-DNA testing. Unfortunately, this was not an isolated incident, as the trend to denigrate Y-DNA has continued in threads across several different forums. Has Y-DNA become an antiquated method of gathering familial data? I do not think so and neither do others.

Calafell and Larmuseau championed the interdisciplinary usefulness of Y-DNA: "the Y chromosome and the availability of markers with divergent mutation rates makes it possible to answer questions on relatedness levels which differ in time depth; from the individual and familial level to the surnames, clan and population level."[1] Genetic genealogist Diahann Southard considers Y-DNA as her favorite type of test "as it focuses on one single line of inheritance. It's often a quick way to see how individuals with the same or similar surname are or are not related to each other. In the context of a group project you can sometimes even discern between related lineages, helping further shape your paternal family history."[2]

While I could counter the arguments against Y-DNA testing *ad nauseum*, this chapter will show the possibilities that Y-DNA provides. Like autosomal DNA, Y-DNA is another tool that has a place in genealogy. Two case studies show how Y-DNA alone answered questions in my own family that documentation, autosomal DNA, or mitochondrial DNA could not.

All URLs were accessed 14 March 2018 or 2 December 2018 unless otherwise specified.

1. Francesc Calafell and Maarten H. D. Larmuseau, "The Y Chromosome as the Most Popular Marker in Genetic Genealogy Benefits Interdisciplinary Research," *Human Genetics* (2016) (https://doi.org/10.1007/s00439-016-1740-0), especially abstract.

2. Diahann Southard, *Facebook* Messenger correspondence to James Owston, 17 June 2017, privately held.

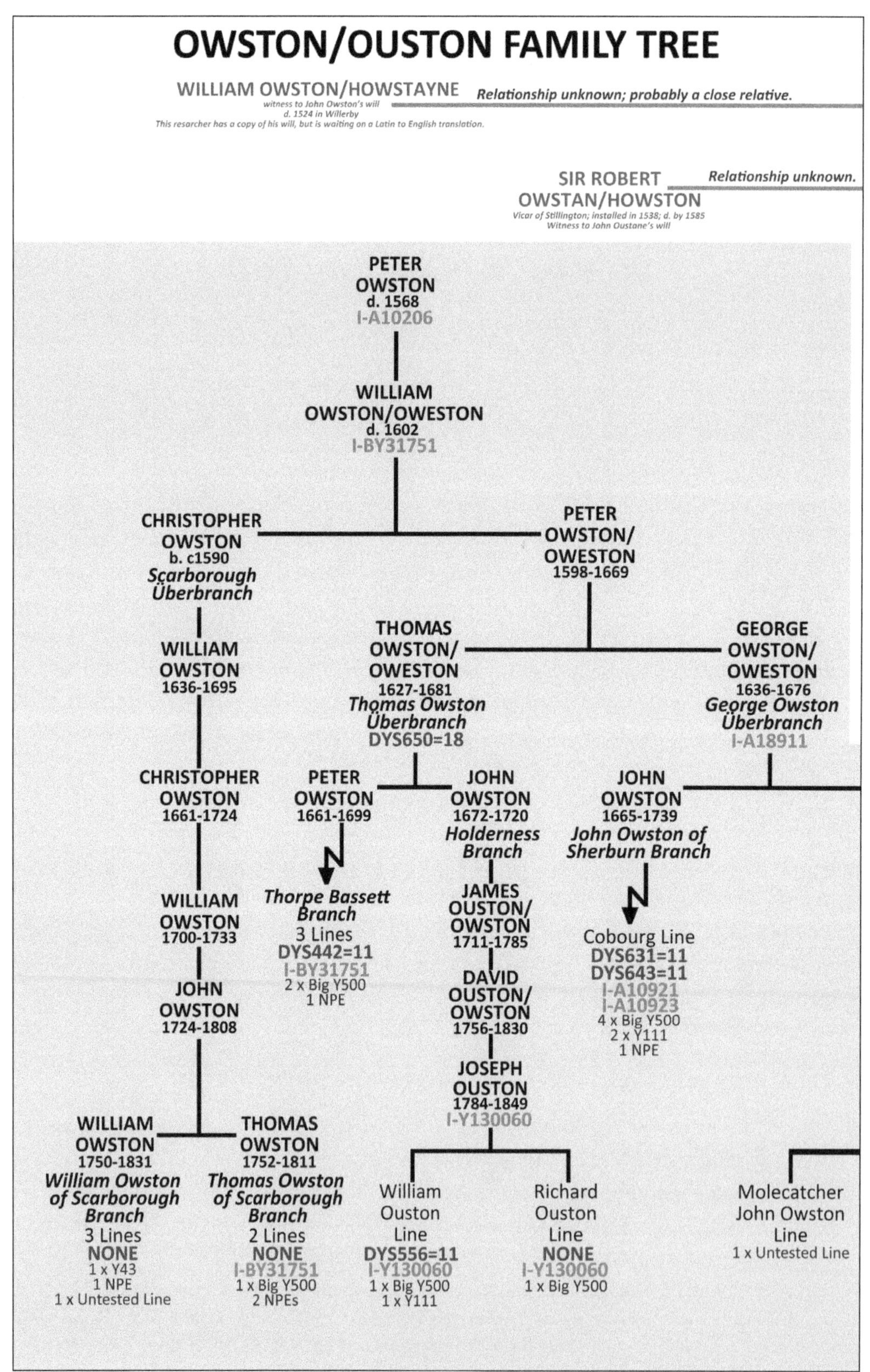

Figure 4.1a. Owston/Ouston families with Y-DNA signatures (part 1)

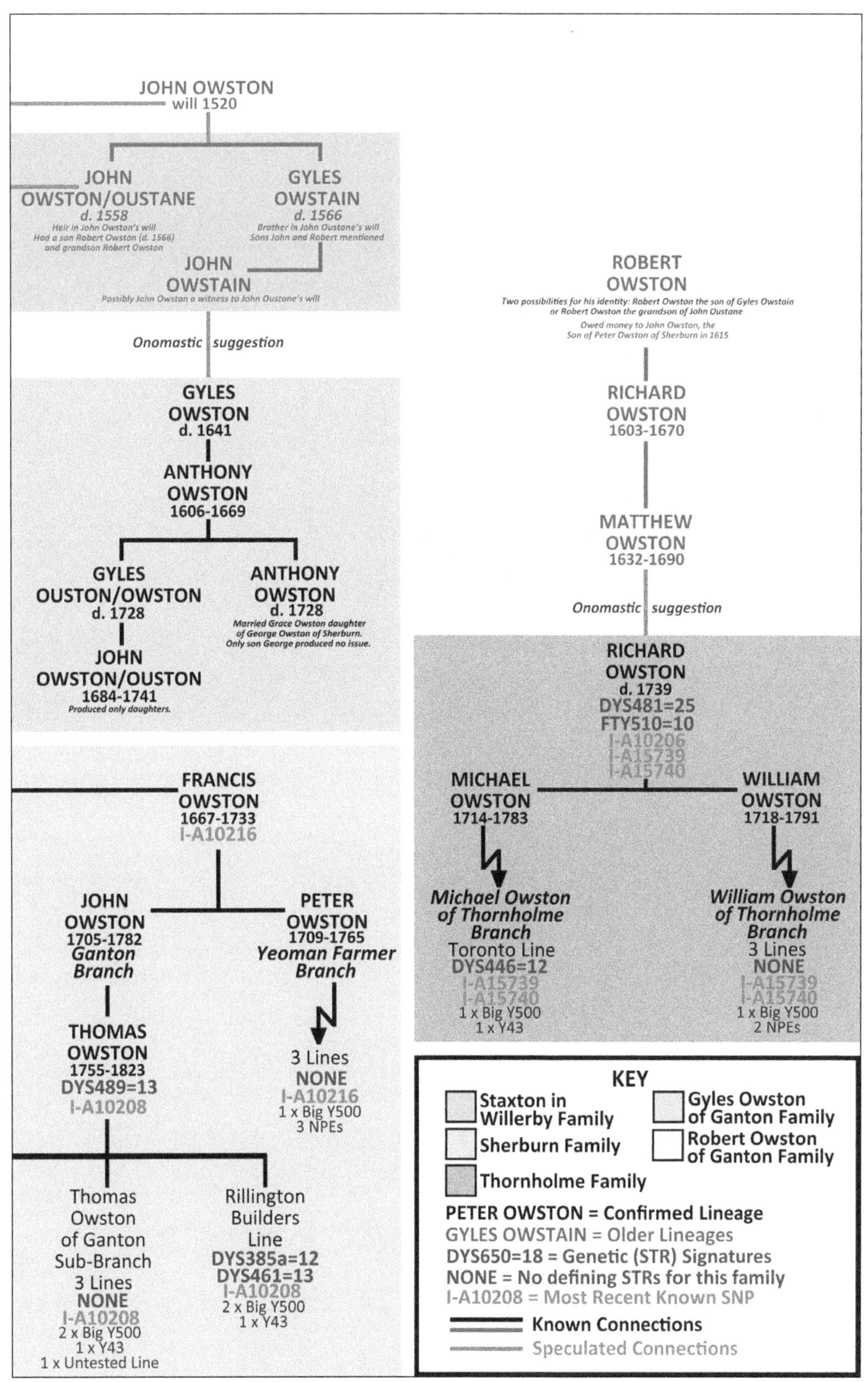

Figure 4.1b. Owston/Ouston families with Y-DNA signatures (part 2)

The Owston/Ouston Surname Project

For illustration purposes, I will be using my own surname project as an example throughout this chapter. It began in the 1970s when three independent researchers began gathering evidence on our low-frequency surname. In addition to my efforts in North America and elsewhere, Timothy J. Owston and Roger J. Ouston in the U.K. were researching our shared surname. There are three variations that appear with the following percentages of males bearing the surname: Owston (72%), Ouston (26%), and Owston-Doyle (2%).[3]

In a serendipitous fashion, we found each other in 1990 and set about sharing information from the families that we had traced. By 1993, we had determined that there were three extant and at least two extinct families bearing our surname that originated in what is now the Ryedale District of North Yorkshire. Analysis of original records and correspondence with others in the U.K., U.S., Canada, Australia, and New Zealand allowed us to place individuals into families. Roger Ouston cataloged the research into a volume, which was periodically updated.[4] Figure 4.1 summarizes our documentary and DNA research and is described in the following paragraphs. Charts for each branch and a copy of figure 4.1 can be accessed online for details and enlargement while reading this chapter.[5]

Three extant families are descended from Peter Owston (d. 1568) of Sherburn in Harford Lythe, Gyles Owston (d. 1641) of Ganton, and Richard Owston (d. 1739) of Thornholme in the parish of Burton Agnes—all in the original East Riding of Yorkshire, England.[6] At present, we believe that only the Sherburn and Thornholme families have living Y-DNA descendants and that the older Gyles Owston family of Ganton had daughtered out in the mid-eighteenth century. The later Ganton Owstons appear to be members of the Sherburn family; this will be discussed in Case Two.

Sherburn and Ganton were less than four miles apart, while Thornholme was fifteen miles to the south.[7] Due to the proximity of these families, I theorized that they all were related. Tim Owston, however, remained skeptical: "Yes, it is possible that the families do connect, but on the other hand why should they . . . apart from reasons of geography? There could have been other connections between them, perhaps untraced female ones? Name patterns are not present."[8]

3. James M. Owston, "It's Raining Men—or is it," *Owston/Ouston One-Name Study*, 9 June 2016 (https://owston.wordpress.com/2016/06/09/its-raining-men-or-is-it/).

4. Roger J. Ouston, *2003 Directory of Ouston/Owston Families*. (Highbridge, U.K.: Roger J. Ouston, 2004); in the possession of James Owston, Beckley, West Virginia; limited edition sent to family members only.

5. James M. Owston, "Owston/Ouston Family Lineage Charts," (http://www.owston.com/family/owston/Owston_Family_Charts.pdf).

6. Ouston, *2003 Directory of Ouston/Owston Families*.

7. *Google Maps* (https://www.google.com/maps/).

8. Timothy J. Owston (Preston, Lancashire) to James M. Owston, letter, 8 April 1991; privately held by James M. Owston, Beckley, West Virginia, 1991.

While I could not confirm a connection in 1991, Tim could not refute it either. This would have to wait twenty years for the Owston/Ouston DNA project—when matching Y-DNA concluded that all three families, known at the time, descended from a common male ancestor. Currently the project has sixty-four members with thirty-three Owston/Ouston surnamed males having participated in Y-DNA testing; six additional Owston men have tested atDNA but are close relatives of Y-DNA tested males. For privacy reasons, we identified each of the participants by the family of origin and a number, such as Sherburn01, Ganton03, and Thornholme05. The Cobourg line of the Sherburn family, due to the number of participants, was given its own designation.

Y-DNA STRs

Y-DNA testing comes in two forms: short tandem repeats (STRs) and single nucleotide polymorphisms (SNPs). While STRs, like SNPs, are also found on the autosomes, they have been used more extensively as markers in Y-DNA testing. These markers have a series of repeated strings of distinct nucleotides. The number of repeats is reported as it is counted in the string.

Modal haplotype

The most-common observed value for a line, family, or haplogroup is termed as the modal haplotype. According to Maurice Gleeson, "On the balance of probabilities, this is the unique genetic signature of our most recent common ancestor."[9] To aid the researcher, the modal values are automatically generated in FamilyTreeDNA's "Y-DNA Results Colorized" report.

Mutations

The differences in marker values from the modal are caused by mutations. These occur when repeats are added or lost in the string. Mutations may signify a family signature, or they may just occur indiscriminately at random. Unless you test numerous people from a family, branch, or line, you will never know if the mutations hold significance. I like the way Rebekah Canada describes STR mutations: "To put it in non-technical terms, mutations are like lightning strikes. They follow patterns, but [they] are not predictable on an individual basis."[10] Certain mutations will also provide false positives of a relationship's distance. These may occur as back mutations, parallel mutations, STR convergence, or a combination of these.

A back mutation occurs when a specific marker in a family mutates back to its modal value. The surname modal may be 9 on a specific marker, but a mutation of 10 repeats occurs

9. Maurice Gleeson, "How to Group Project Members—FTDNA Conference 2017," *DNA and Family Tree Research* (November 2017); *YouTube* video (https://youtu.be/A9JcvbFcgUI).

10. Rebekah Canada, "Response to Facebook post by Daphne Jasinski," Y-DNA - Applied Genealogy & Paternal Origins, *Facebook* group, February 2018 (https://www.facebook.com/groups/YDNA.applied/permalink/210300616189928/).

eight generations in the past within a specific line. A participant, who is descended from that line, has another mutation that returns the marker to 9 repeats. Without confirming records of descent, we might misattribute that man to another line of the family. The only way to determine if a back mutation is present is if others with a confirmed descent in that specific line have results different from the modal.[11]

Parallel mutations occur when two individuals exhibit the same mutation independently. This might be illustrated by the surname modal being 12 on a marker and two individuals who are tenth cousins share 14 repeats on the same marker. The two similar results may falsely indicate that the two men are descended from the same line when they are not.[12]

Convergence occurs when a person from a related, but different haplotype, has values that have mutated to the same value. Convergence describes "the process whereby two different genetic signatures (usually Y-STR-based haplotypes) have mutated over time to become identical or near identical resulting in an accidental or coincidental match."[13]

These converging marker values are generally found among matches of 37 markers or less, and in the Owston/Ouston study, they have genetic distances of 2, 3, and 4; however, several also exist at 67 markers. They are, however, plentiful at 12 and 25 marker resolutions—in fact most matches at these levels will be related thousands of years in the past. Both parallel mutations and back mutations influence convergence. Gleeson provides several excellent blog posts on the phenomenon of convergence and back mutations.[14]

The Owston/Ouston project can illustrate several examples of the STR match types. Being a low-frequency and significantly-researched surname, it can be a useful model in determining whether matching STRs are signatures of descent, back mutations, parallel mutations, or convergence.[15]

Figure 4.2 shows all the STR mutations among our twenty-three matching individuals in our project up through 111 markers. The four participants with missing STRs are those who have only tested at 43 markers at the now defunct GeneTree; two of these participants, Ganton01 and Ganton02, are deceased.[16] Markers, such as DYS358a and DYS458, identified in the heading with marker names in white text on a brown-shaded background, are

11. Maurice Gleeson, "Convergence: Quantifying Parallel & Back Mutations (Part 1)," *The Gleason / Gleeson DNA Project*, 27 May 2017 (http://gleesondna.blogspot.com/2017/05/convergence-quantifying-parallel-back.html).

12. Gleeson, "Convergence: Quantifying Parallel & Back Mutations (Part 1)."

13. "Convergence," *International Society of Genetic Genealogy (ISOGG) Wiki* (https://isogg.org/wiki/Convergence).

14. Maurice Gleeson, "Convergence: What is it?" *The Gleason / Gleeson DNA Project*, 27 May 2017 (http://dnaandfamilytreeresearch.blogspot.com/2017/05/convergence-what-is-it.html).

15. James Owston, *Owston/Ouston DNA Project* (https://www.familytreedna.com/public/owston?iframe=ycolorized).

16. Owston, *Owston/Ouston DNA Project.*

fast-mutating markers as noted by Ballantyne et al.[17] Markers that have identical values for all participants have been omitted.

As noted in figure 4.2, there appear to be many individual mutations (shaded black) among participants. These may be line specific, but until more subjects are tested, they appear as singletons. Complete charts of these families can be found online.[18] The two participants from the Scarborough überbranch (a term that I coined) that traces back to Christopher Owston (c. 1590) share no STRs that appear to be unique to their specific family group.[19] Further testing may indicate signature markers for its lines and/or branches.

Y-STR signatures

Figures 4.1 and 4.2 illustrate the family lines described in the following sections. Some descendants not shown in the figures are also discussed to provide more context for the study group and marker analysis details. Those descendants not shown here can be found in the online charts.[20] Family signature markers are shaded blue in figure 4.2.

The Rillington Builders line of the Ganton branch has two such signature markers: 12 repeats on DYS385a and 13 repeats on DYS461.[21] The mutations can be narrowed down to occurring in one of three generations between the years of 1793 to 1855. It can be no earlier than Francis Owston (1793–1898), as the descendants of his brother Thomas (1782–1869) do not carry either repeat. These two mutations, however, could have occurred in tandem with one ancestor or individually through different ancestors.

The Thorpe Bassett branch has an identifying mutation of 11 repeats on DYS442; this mutation can be attributed to a single ancestor, Peter Owston (1661–1699), as he is the common ancestor of Sherburn02 and Sherburn04. Descendants of Peter Owston's brother John (1672–1720) in the Holderness branch do not share this mutation.[22]

---

17. Kaye N. Ballantyne et al., "Mutability of Y-Chromosomal Microsatellites: Rates, Characteristics, Molecular Bases, and Forensic Implications," *American Journal of Human Genetics* 87 (September 2010) (https://doi.org/10.1016/j.ajhg.2010.08.006), supplemental data.
18. James M. Owston, "Owston/Ouston Family Lineage Charts," (https://www.owston.com/family/owston/Owston_Family_Charts.pdf).
19. Owston, *Owston/Ouston DNA Project.*
20. James M. Owston, "Owston/Ouston Family Lineage Charts," (http://www.owston.com/family/owston/Owston_Family_Charts.pdf).
21. Owston, *Owston/Ouston DNA Project.* James M. Owston, *His Name is my Name Too* (http://www.academia.edu/605010/HIS_NAME_IS_MY_NAME_TOO).
22. Owston, *Owston/Ouston DNA Project.* Ouston, *2003 Directory of Ouston/Owston Families.*

		DYS391	DYS385a	DYS458	DYS447	DYS449	DYS464a	DYS464b	GATA H4.1	DYS570	CDY a	CDY b	DYS442	DYS413a	DYS534	DYS444	DYS481	DYS446	DYS710	DYS714	DYS556	DYS549	DYS650	DYS504	DYS635	DYS643	DYS461
	MODAL	10	13	16	22	29	12	14	10	19	36	37	12	22	16	13	26	13	28	24	12	13	19	17	21	12	12
Scarborough Überbranch	Sherburn05	10	13	16	21	29	12	14	10				12					13							21		12
	Sherburn12	10	13	16	22	29	12	14	9	18	36	37	13	22	16	13	26	13	28	24	12	13	19	17	21	12	12
Thorpe Bassett Branch - Thomas Owston Überbranch	Sherburn04	10	13	16	22	29	12	14	10	19	36	37	11	22	16	13	26	13	28	24	12	13	18	17	21	12	12
	Sherburn02	10	13	18	22	29	12	14	10	19	36	38	11	22	16	13	26	13	28	24	12	13	18	17	21	12	12
Holderness Branch - Thomas Owston Überbranch	Sherburn09	10	13	16	22	29	12	14	10	20	36	37	12	22	16	13	26	13	28	24	11	13	18	17	21	12	12
	Sherburn14	10	13	16	22	29	12	14	10	19	36	37	12	22	16	13	26	13	28	25	11	13	18	17	21	12	12
	Sherburn01	11	13	16	22	30	12	14	10	19	37	37	12	22	16	13	26	13	28	24	12	13	18	17	21	12	12
Cobourg Line - John Owston Branch - George Owston Überbranch	Cobourg01	10	13	16	22	29	12	14	10	19	36	37	12	22	16	13	26	13	28	24	12	13	19	17	21	11	12
	Cobourg02	10	13	16	22	29	12	14	10	19	36	37	12	22	16	13	26	13	28	24	12	13	19	17	21	11	12
	Cobourg09	10	13	16	22	29	12	14	10	19	36	37	12	22	16	13	26	13	28	24	12	13	19	17	21	11	12
	Cobourg06	10	13	16	22	29	12	14	10	20	36	37	12	22	16	13	26	13	28	24	12	13	17	17	21	11	12
	Cobourg08	10	13	16	22	29	12	14	10	19	36	37	12	22	16	13	26	13	28	24	12	13	19	17	21	12	12
	Cobourg05	10	13	16	22	29	12	14	10	19	36	37	12	22	16	13	26	13	28	24	12	13	19	17	21	11	12
Thomas Owston Sub Branch - Ganton Branch - George Owston Überbranch	Ganton01	10	13	16	22	29	12	14	10				12					13							21		12
	Ganton03	10	13	16	22	29	12	14	10	19	36	37	12	22	16	13	26	13	28	24	12	13	19	17	21	12	12
	Ganton04	10	13	16	22	29	12	14	10	19	36	37	12	22	16	13	26	13	28	24	12	13	19	17	21	12	12
Rillington Builders Line - Ganton Branch - George Owston Überbranch	Ganton02	10	12	16	22	29	12	14	10				12					13							21		13
	Ganton06	10	12	16	22	29	12	14	10	18	36	37	12	25	16	13	26	13	28	24	12	13	19	17	21	12	13
	Ganton05	10	12	16	22	29	12	14	10	19	36	38	12	22	16	14	26	13	28	24	12	13	20	17	21	12	13
Yeoman Farmers Branch	Sherburn08	10	13	15	22	29	12	14	10	19	36	36	12	22	16	13	26	13	28	24	11	12	19	17	22	12	12
Toronto Line - Michael Owston Branch	Thornholme01	10	13	16	22	29	14	15	10				12					12							21		12
	Thornholme05	10	13	16	22	29	12	14	10	19	36	37	12	22	16	13	25	12	29	24	11	13	19	17	21	12	12
William Owston Branch	Thornholme04	10	13	16	22	29	12	14	10	19	36	37	12	22	15	13	25	13	28	24	12	13	19	18	21	12	12

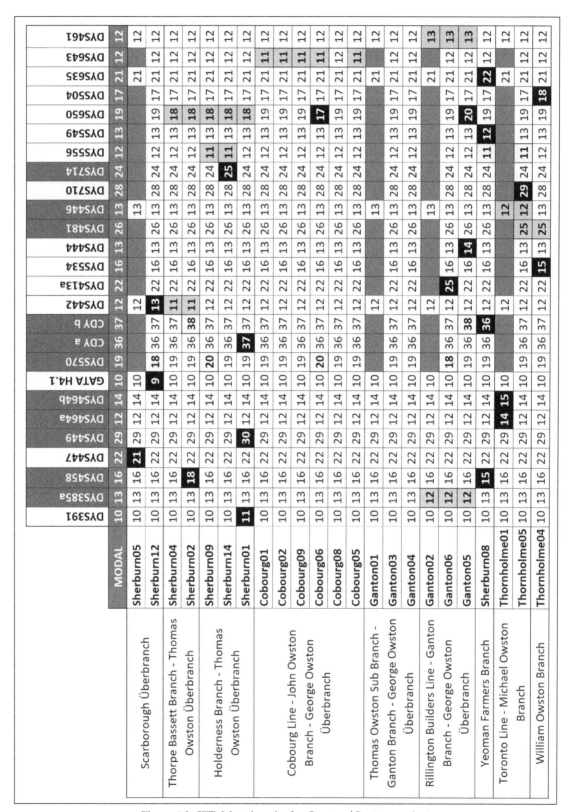

Figure 4.2. STR Mutations in the Owston/Ouston project

The Holderness branch does not exhibit a signature STR mutation like the Thorpe Bassett branch; however, one Holderness group, the William Ouston (1807–1888) line, may have a signature STR mutation. There are 11 repeats on DYS556 found in descendants of two sons of William Ouston, but not in the descendants of his brother Richard Ouston (1811–1895). To be conclusive in this seeming signature, we would need to test others descended from Richard, as our tested subject (Sherburn01) may have had a back mutation. Furthermore, because the same value is seen among two other distantly related participants (Sherburn08 and Thornholme05) as parallel mutations, these two closely related participants (third cousins once removed) may also be an example of this same phenomenon.[23]

A mutation also can be ascribed to the common ancestor of both the Thorpe Bassett and Holderness branches: Thomas Owston (1627–1681), the progenitor of the überbranch that bears his name. All Thomas's descendants who have tested to 111 markers share 18 repeats on DYS650. Descendants of Thomas's brother George (1636–1676) do not exhibit this mutation.[24]

In looking at the Thornholme family, two markers exhibit characteristics of signature STRs. Two members of the Toronto line share 12 repeats on DYS446. While this is a fast-mutating marker and the matching members of the family are closely related, the mutation could have occurred anywhere from 1714 to 1877—a six-generation spread. The Toronto line contains ten Owston males; however, three are products of adoptions and are not genetic Owstons. Another male took the surname of his mother. The most distant relationship among the remaining six males is that of first cousins twice removed. Thornholme01 and Thornholme05 are first cousins once removed.[25]

Participants from both branches of the Thornholme family share 25 repeats on DYS481 indicating that mutation occurred no later than the birth of Richard Owston who died in Thornholme in 1739—the most recent common ancestor of the Michael Owston and William Owston branches. The mutation could have occurred at any time from the common ancestor of both the Sherburn and Thornholme families to the birth of Richard (d. 1739).[26] Although Richard's ancestry cannot be confirmed, several forenames in his family mirror those found in the Robert Owston of Ganton's family. The names Richard, Michael, and Matthew that are found in both are absent from the early Sherburn family and the extinct Gyles Owston family.

Finally, another signature STR is seen in the project. DYS643 exhibits 11 repeats in five of the six members of the Cobourg line. Four men (Cobourg01, 02, 06, and 09) descend

23. Owston, *Owston/Ouston DNA Project.* Owston, *His Name is my Name Too.* Ouston, *2003 Directory of Ouston/Owston Families.*

24. Owston, *Owston/Ouston DNA Project.* Ouston, *2003 Directory of Ouston/Owston Families.*

25. Owston, *Owston/Ouston DNA Project.* Ouston, *2003 Directory of Ouston/Owston Families.*

26. Owston, *Owston/Ouston DNA Project.* Ouston, *2003 Directory of Ouston/Owston Families.* Roger J. Ouston, *An examination of the early Owston data of the parish of Willerby and adjacent parishes* (Swanland, Humberside: Roger J. Ouston, 1997).

from William Owston's (1778–1857) youngest child, John Gillon Owston (1826–1901). Cobourg05's ancestor is William's fifth child, Charles Paget Herbert Owston (1817–1858). Since the descendants of these brothers are fourth cousins or fourth cousins once removed, it appears that the mutation occurred at some point in the four generations culminating with the birth of William. The oldest possible place it entered the family was with John Owston (1665–1739), because descendants of John's brother Francis (1667–1733) do not carry this mutation.[27]

An interesting development is that Cobourg08, descendant of William Owston's third child James Wilson Owston (1809–1858), does not carry the mutation. This appears to be a back mutation to the modal value of 12. Further testing of other men descended from James may confirm this scenario.[28]

The Big Y-500 added three lineage-specific STRs in the additional 450 STRs included in FamilyTreeDNA's panel six (not shown in figure 4.2). DYS631 at 11 repeats rather than the modal of 10 is a signature of the Cobourg line of the Sherburn family. All three participants who descend from William Owston (1778–1857) carry this result.[29] DYS489 at 13 as opposed to the modal of 12 is shared by all four members of the Ganton branch of the Sherburn family who have tested with the Big Y-500. This is the only STR marker that is indicative of this branch that descends from Thomas Owston (1755–1823) of Ganton. Finally, FTY510 with 10 repeats as opposed to 9 are shared by two seventh cousins once removed who descend from Richard Owston (d. 1739) from Thornholme.[30]

Parallel mutations

While Y-STR signatures can group families, there is danger in accepting every equal marker value as a signature value; this may create a false indication of a close relationship when one is not present. Parallel mutations often mimic signature STRs. Returning to figure 4.2, there are several examples of parallel STR mutations that are shaded in yellow.

For example, the fast-mutating marker DYS570 shows two examples of parallel mutations. Cobourg06 and Sherburn09 both share a value of 20 as opposed to the modal value of 19. These ninth cousins once removed also have a genetic distance of zero at both 37 and 67 marker resolutions. Because of this, FamilyTreeDNA considers them to be "tightly related." The men's closest tested relatives do not share this mutation indicating that it was a very recent occurrence in each lineage.[31]

At the same marker, Sherburn12 and Ganton06 exhibit a loss from the modal value with 18 repeats. These eleventh cousins once removed also appear to have parallel mutations.

27. Owston, *Owston/Ouston DNA Project*. Ouston, *2003 Directory of Ouston/Owston Families*.
28. Owston, *Owston/Ouston DNA Project*. Ouston, *2003 Directory of Ouston/Owston Families*.
29. Owston, *Owston/Ouston DNA Project*. Ouston, *2003 Directory of Ouston/Owston Families*.
30. Owston, *Owston/Ouston DNA Project*. Ouston, *2003 Directory of Ouston/Owston Families*.
31. Owston, *Owston/Ouston DNA Project*. Ouston, *2003 Directory of Ouston/Owston Families*.

Although there are no close relatives of Sherburn12 who have tested at 111 markers, Ganton06's second cousin once removed (Ganton05) and two sixth cousins (Ganton03 and 04) do not share the mutation.[32]

CDYb, a fast-mutating marker, shows the shared value of 38 as opposed to the modal value of 37 in two participants: Sheburn02 and Ganton05. Like in the previous scenarios, these eighth cousins once removed have no close relatives sharing this mutation.[33]

Genetic distance

With convergence as a possibility, it is helpful to address some of the issues with genetic distance. One of the ways FamilyTreeDNA intimates relationships is through genetic distance or GD. Because of a greater likelihood of convergence, Y-12 and Y-25 testing are useless in identifying relatives within a genealogical timeframe. The number of matches at both resolutions indicates that many matches at these two levels are related thousands of years in the past. In the Owston/Ouston project, as I would assume in other projects, genetic distance varies greatly. Because STR mutations occur randomly and indiscriminately, it is a poor predictor of relationship. This is especially true at 37 markers or less.[34]

While the FamilyTreeDNA TiP (Time Predictor) Calculator takes into consideration individual marker differences, it also cannot predict exact relationships. It provides probabilities that two men are related within twenty-four generations. TiP was never updated for mutation rates beyond the markers in the 37-marker test and is known to underestimate the time to a common ancestor.[35] Where TiP can be helpful is in grouping unknown participants sharing a surname into genetic families. Gleeson advises, "It is important to note that the use of the TiP24 Score is not an attempt to date when two people are related, merely to ascertain if two people are likely to be related."[36] This concept has escaped many, as some use it to determine exact relationships.

SNP Testing

Single-nucleotide polymorphisms (SNPs) occur when there is "a variation at a single position in a DNA sequence among individuals."[37] When a nucleotide base changes from what is found in the general population, such as A to C, this is a SNP. As more people test, new

32. Owston, *Owston/Ouston DNA Project.* Ouston, *2003 Directory of Ouston/Owston Families.* .

33. Owston, *Owston/Ouston DNA Project.* Ouston, *2003 Directory of Ouston/Owston Families.*

34. James M. Owston, "Is Genetic Distance an Adequate Predictor of Relationships?" *The Lineal Arboretum,* 19 June 2014 (http://linealarboretum.blogspot.com/2014/06/is-genetic-distance-adequate-predictor.html). Owston, *Owston/Ouston DNA Project.*

35. Rebekah A. Canada, "FTDNA - Y DNA 25 matches that only tested at the YDNA 25 level," DNA-NEWBIE Yahoo Group, 7 September 2018 (online archive available to group members).

36. Maurice Gleeson, "Criteria for Grouping People into Y-DNA Genetic Families," *DNA and Family Tree Research,* 30 June 2017 (http://dnaandfamilytreeresearch.blogspot.com/2017/06/criteria-for-grouping-people-into-y-dna.html). TiP24 references 24 generations as the scope using FamilyTreeDNA's TiP tool.

37. "SNP," *World Library of Science: A Global Community of Science Education,* online dictionary (https://www.nature.com/wls/definition/single-nucleotide-polymorphism-snp-295).

SNPs are discovered. These are named by a letter or letter sequence representing the discoverer, followed by a number representing the order in which the SNP was found. The number has no relationship to upstream (earlier) or downstream (later) SNPs in the same sequence.

Prior to 2012, SNPs were primarily used for discerning haplogroups and deep ancestry. For genealogical purposes, they were not relevant other than grouping individuals into large families whose common ancestor lived thousands of years in the past. Additionally, STR repeat values were found among specific haplogroups, and these were often used to identify the haplogroup rather than testing for a specific SNP.[38]

This changed in 2012 with the introduction of National Geographic's Geno 2.0 test and FamilyTreeDNA's Walking the Y test. They ushered in what is colloquially called the "SNP Tsunami." New SNPs were identified and these narrowed time-frames for origin of specific clades or sub-branches of a haplogroup by several thousand years. As more SNPs were discovered, it was possible to do target SNP testing. This helped identify further downstream clades. Both FamilyTreeDNA and YSEQ began to offer special SNP packs to aid in determining to which of several downstream haplogroups a test taker might belong.[39]

In 2013, two companies began offering next-generation sequencing (NGS) for the Y chromosome and the discovery of new SNPs increased exponentially. The current two products are Full Genomes Corporation's (FGC) Y-Elite 2.1 and FamilyTreeDNA's Big Y-700, which replaced the Big Y-500 in January 2019.

Big Y-700

Although the Big Y test has been available since 2013, FamilyTreeDNA made some significant upgrades to the product between 2017 and 2019. The first enhancement was to move existing tests from Genome Reference Consortium Human Build 37 (hg19) to Build 38 (hg38). Build 38 allowed "[b]etter mapping of NGS [Next Generation Sequencing] data to the proper location" and the "[c]onsideration of alternative haplotypes across the genome."[40]

The initial retooling of the Big Y in April 2018 included up to 500 usable STRs and branding as the Big Y-500. By January 2019, FamilyTreeDNA included 700 usable STRs and rebranded as the Big Y-700. Previous Big Y-500 customers who received their results earlier than January 2019 are required to upgrade to get the additional 200 STRs.[41]

38. T. Whit Athey, "Haplogroup Prediction from Y-STR Values Using an Allele-Frequency Approach," *Journal of Genetic Genealogy* 1 (Spring 2005) (http://www.jogg.info/pages/11/athey.htm).

39. "DNA Tests," *FamilyTreeDNA* (https://www.familytreedna.com/products). "Welcome to the YSEQ DNA Shop," *YSEQ: DNA Origins Project* (http://yseq.net/).

40. *FamilyTreeDNA*, email newsletter to group administrators, 17 October 2017. For definitions of build numbers and hg numbers see "Reference genome," *Wikipedia, the free encyclopedia* (https://en.wikipedia.org/wiki/Reference_genome).

41. "Group Project Administrators Announcement: Product Update: Big Y-700 & Big Y Block Tree," *FamilyTreeDNA* (https://mailchi.mp/familytreedna/gap_announcement_big_y700_release).

The new version also eliminated irrelevant matches found in the previous version of the Big Y-500 and only included matches to a test taker's five most recent SNPs. Not all test takers at these five levels will show, as FamilyTreeDNA requires a match to have thirty or fewer differences in SNPs.[42]

An added issue found in the Owston/Ouston project is that there are numerous downstream SNPs from the earliest surname specific SNP and, therefore, some individuals only have matches to others with the surname. This is illustrated by the combined haplogroup tree for the Owston/Ouston project identifying the most recent ancestors for each SNP in figure 4.3.[43]

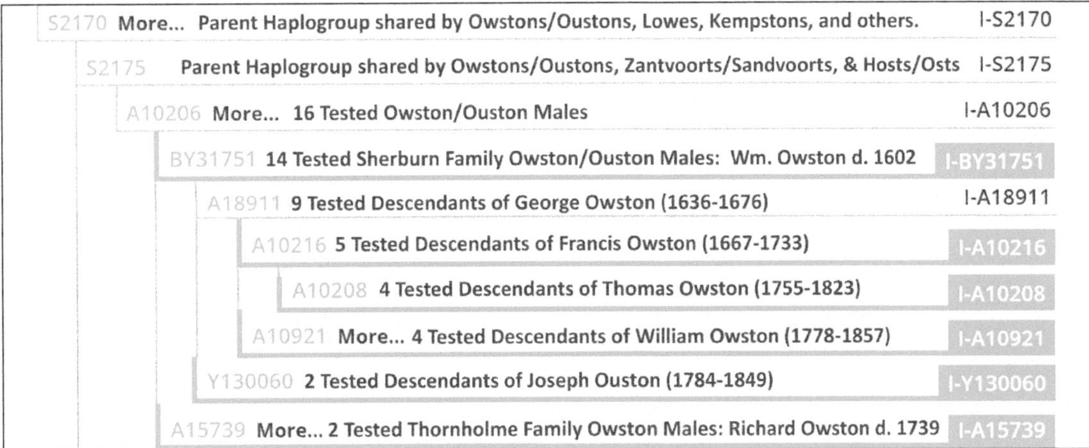

Figure 4.3. Combined haplogroup tree for the Owston/Ouston project.

Several project members have numerous downstream SNPs from the surname modal SNP A10206, and this limits the number of matches that are included in the results. For example, the four Ganton branch participants with the terminal SNP of A10208 only see members from our project as shown in figure 4.4.[44]

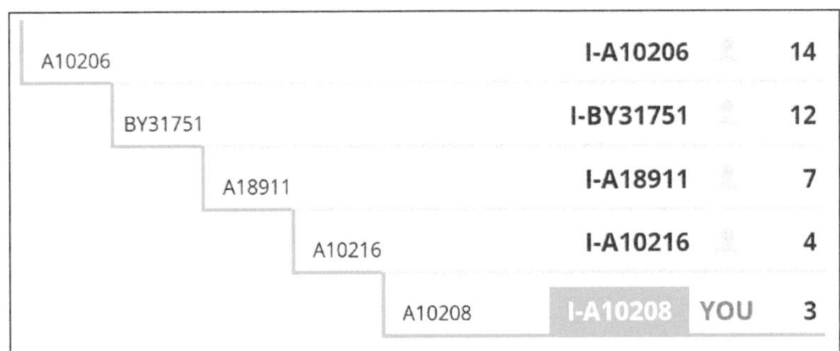

Figure 4.4. FamilyTreeDNA stair-step chart for A10208 participants

42. "Big Y-500," *FamilyTreeDNA* (https://www.familytreedna.com/learn/y-dna-testing/big-y/big-y/).
43. Owston, *Owston/Ouston DNA Project.*
44. Owston, *Owston/Ouston DNA Project.*

Those sharing A10921 or A10216 as terminals have the upstream S2175 SNP in their list. The two Holderness branch participants see both S2175 and S2170. The remaining five participants have three upstream SNPs: S2175, S2170, and FGC2491. I-S2170 as seen in figure 4.5 includes two non-Owstons. These individuals are members of the Derbyshire Lowes of Maryland project. YFull estimates that the Owstons and Lowes share a common ancestor who lived 2300 years before the present. The Owstons and Lowes have been appearing in each other's Y-STR match lists at 37 and 67 markers for years. In addition, two men who have tested at S2175, which is downstream from S2170 and upstream from the surname modal A10206, do not show as matching any member of the Owston or the Lowe projects at any level at FamilyTreeDNA.[45]

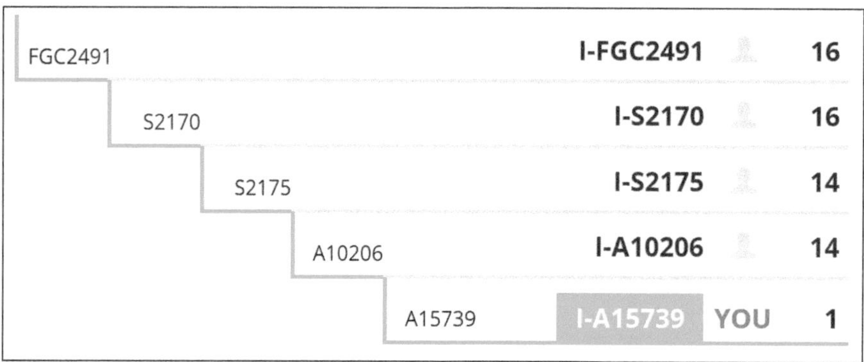

Figure 4.5. FamilyTreeDNA stair-step chart for A15739 participants

One benefit to the earlier Big Y-500 and current Big Y-700 products is that the participant need not order an STR test beforehand. All STRs found in FamilyTreeDNA Panels One (1-12), Two (13-25), Three (26-37), Four (38-67), and Five (68-111) are now included in Big Y-700. Panel Six (112-561) added up to an additional 450 STRs and the Big Y-700 adds 200 more STRs. While many of these could be extracted by YFull, the number of no calls[46] in the FamilyTreeDNA data is substantially lower than seen at YFull. Because of no calls, FamilyTreeDNA only guaranteed 389 additional STRs with the Big Y-500.[47]

On first review, I did not find these additional markers genealogically relevant. A detailed analysis of the STRs included with the Big Y-500 is available online.[48] As this volume goes to press, the Big Y-700 results are emerging and the usefulness of the additional 200 STRs remains to be seen.

45. Owston, *Owston/Ouston DNA Project.* Michael Lowe, *Derbyshire Lowe's of Maryland* (https://www.familytreedna.com/public/DerbyshireLowesofMaryland). YFull YTree v6.05.08, *YFull* (https://yfull.com/tree/I1).

46. No calls result when the DNA at a particular location cannot be accurately read by the lab process.

47. "Big Y-500," *Family Tree DNA.* "Group Project Administrators Announcement," *FamilyTreeDNA.*

48. James M. Owston, "Does the Big Y500 Provide Value?" *The Lineal Arboretum,* 21 April 2018 (http://linealarboretum.blogspot.com/2018/04/does-big-y500-provide-value.html).

In advance of the newer Big Y-700 results, FamilyTreeDNA added a new feature—the Big Y Block Tree Matching Tool. Based on the block tree used in the Big Tree website (for haplogroup R-P312), this new tool allows Big Y participants to view matches and upstream and downstream SNP blocks, as well as sibling branch blocks. Figure 4.6 provides an annotated version of a block tree. By clicking on the arrow at the top or by directly clicking on a SNP at the top of the page, a Big Y participant can move to the next higher SNP. As a user goes further back in time, downstream SNPs are truncated and combined.[49]

Known countries of origin are represented by the respective flags. All Owston/Ouston participants display the English flag. England is the surname's country of origin despite test taker residencies being in four countries today. If the flag is not familiar, the country name is displayed by hovering over the flag image.

In addition, a doughnut chart presents autosomal averages for the participants in each block. By either hovering or clicking, the average admixture (ethnicity) percentages of those within the block are revealed. In figure 4.6, significant percentages of Northeast and Southeast Asia origins display for those sharing A10208. Only one individual sharing A10208 has Asian markers, as his mother is Chinese. The usefulness of this merging of autosomal information (inherited from all ancestors) and Y-DNA information (inherited only through patrilineal ancestry) is unknown.

It is not clear how FamilyTreeDNA determines which SNP blocks have autosomal results. Blocks A10208 and A10921 each have four members tested autosomally and averages are displayed. Block S2175 has only two autosomally-tested members, and also displays averages; however, blocks BY31751 and A15739 have, respectively, three and two tested individuals and display no averages. Overall, the Big Y Block Matching Tree has received positive comments from the genealogy community. Had it been available earlier, conclusions in the Owston study might have been reached sooner.

Y-DNA can be very helpful in answering specific genealogical questions as shown in the following case studies. Case One confirms a conclusion based on traditional genealogical documentation and deduction. Case Two refutes a conclusion made thirty years ago and required us to revisit our documentary evidence. While this produced a bit of cognitive dissonance at the time, the newly found line of descent was a welcome addition to our research.

49. "Group Project Administrators Announcement," *FamilyTreeDNA*. Owston, *Owston/Ouston DNA Project*. Alex Williamson, *The Big Tree* (https://www.ytree.net/).

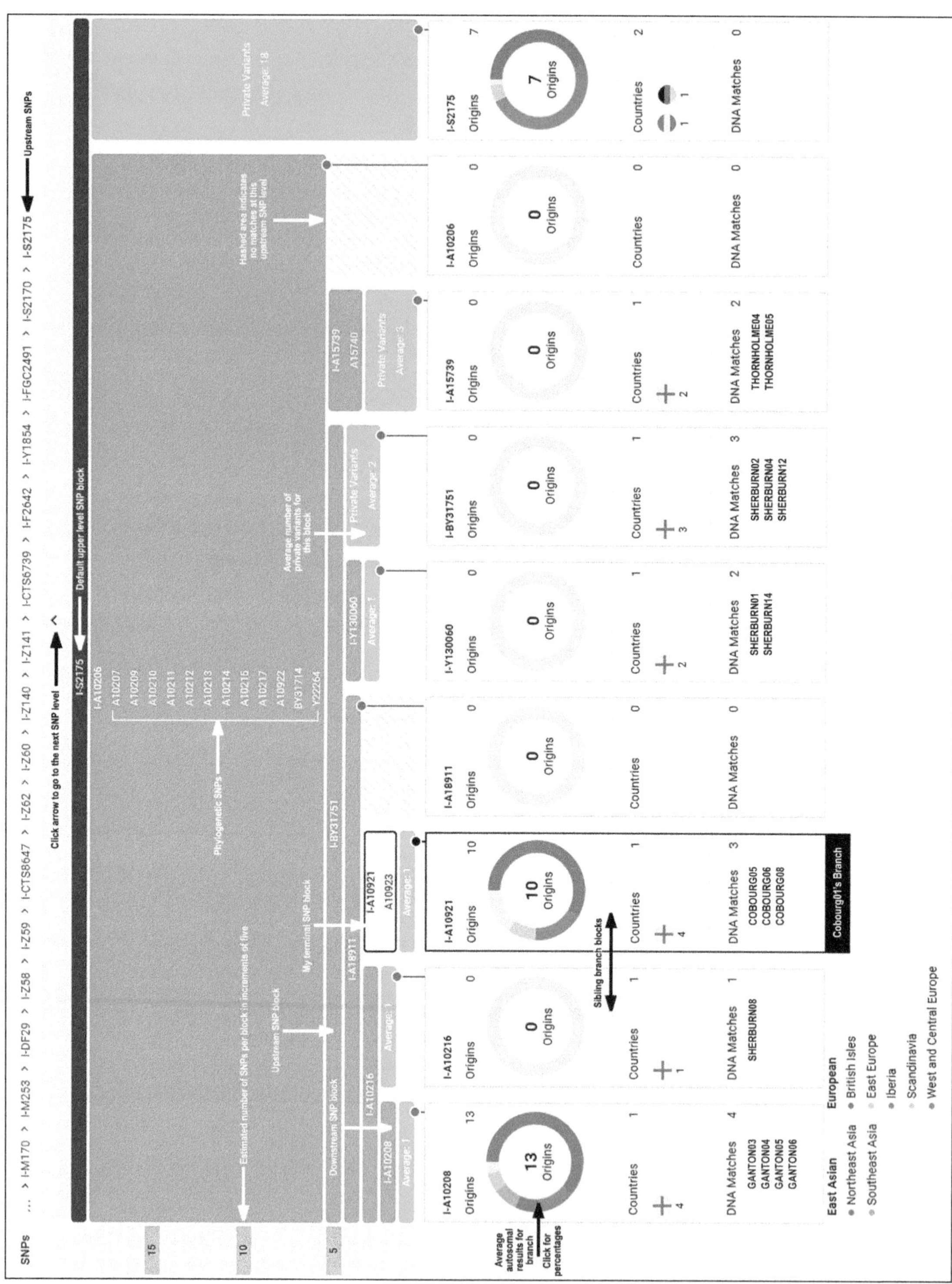

Figure 4.6. Big Y Block Tree Matching Tool for Cobourg01 (the author) with annotations

Case Studies

Case 1: strengthening a conclusion

In 1990, Tim Owston and I addressed a conundrum in the Owston families who were living in Ganton in the original East Riding of Yorkshire. In the latter part of the eighteenth century, two men named Thomas Owston were simultaneously fathering children; however, the mothers' names were absent from all fourteen baptismal records. To complicate matters, there were five shared names: Anne, Mary, John, Thomas, and William. Mary was used thrice. In addition, the unique names of Elizabeth, Francis, and Jane appear once each.[50]

In 1772, Thomas Owston (1733–1819) of Sherburn married Elizabeth Walker of Potter Brompton.[51] Potter Brompton is situated midway between Sherburn and Ganton and is within the confines of Ganton parish.[52] In November 1777, Thomas Owston (1755–1823) of Ganton married Mary Vickerman of Sherburn. It appears that this couple was situated in Potter Brompton during the years of 1788 to 1781 and from 1804 until their deaths in 1818 (Mary) and 1823 (Thomas).[53]

We identified the Thomas and Elizabeth Owston family as part of the Sherburn family. Thomas and Mary Owston were identified as being from the original Ganton family, as it appeared that the younger Thomas was descended from Gyles Owston who died in Ganton in 1641. This connection is explained more in Case Study Two.[54]

All fourteen births where Thomas Owston is the named father occurred within the twenty-year span of 1773 to 1793. While the wives were not mentioned in the baptismal records, they were identified in two burial records; this aided in sorting the children.[55] In the bishop's

50. St. Nicholas's Church (Ganton, Yorkshire, England), "Ganton Parish Register, 1553–1794," unnumbered pp. 139–154, Owston baptisms 1773–1793; digital images, *FindMyPast* (https://www.findmypast.com). St. Nicholas's Church (Ganton, Yorkshire, England), "Ganton Parish Records (Bishop's) Transcripts. 1601–1867," unnumbered pp. 164–197, Owston baptisms 1773–1793; digital images, *FindMyPast* (https://www.findmypast.com).

51. St. Nicholas's Church (Ganton, Yorkshire, England), "Ganton Parish Marriages, 1758–1793," unnumbered p. 7, Thomas Owston and Elizabeth Walker, 14 May 1772; digital images, *FindMyPast* (https://www.findmypast.com). St. Nicholas's Church (Ganton, Yorkshire, England), "Ganton Parish Records (Bishop's) Transcripts. 1601–1867," unnumbered p. 162, Thos. Owston and Eliz. Walker marriage, 14 May 1772; digital images, *FindMyPast* (https://www.findmypast.com).

52. William White, *History, Gazetteer, and Directory of the East and North Ridings of Yorkshire* (Sheffield: Robert Leader, 1840), 388.

53. St. Nicholas's Church (Ganton, Yorks), "Ganton Parish Marriages, 1758–1793," unnumbered p. 9, Thos. Owston and Mary Vickerman, 24 November 1777. St. Nicholas's Church (Ganton, Yorks), "Ganton Parish Records (Bishop's) Transcripts. 1601–1867," unnumbered p. 174, Thos. Owston and Mary Vickerman marriage, 24 November 1777. St. Nicholas's Church (Ganton, Yorks), "Ganton Parish Burials, 1813–1982," pp. 4-7, Owston burials; digital images, *FindMyPast* (https://www.findmypast.com).

54. Ouston, *2003 Directory of Ouston/Owston Families.*

55. St. Nicholas's Church (Ganton, Yorks), "Ganton Parish Register, 1553–1794," unnumbered pp. 143–154, Owston baptisms 1778–1780.

transcripts, John Owston's death in 1797 identified him as the son of Thomas and Elizabeth, while the parish register provides his grandparents' names as John and Eleanor Owston of Sherburn.[56] Jane Owston is listed in both the parish register and the bishops' transcripts as the daughter of Thomas and Mary Owston.[57] Sorting, however, was difficult, as there is a dearth of information concerning both families. Both men farmed common land and were not property owners. Further, neither left a will.

To begin our separation of the families, we first used the bishop's transcripts, which are readily available on microfilm; however, less information was contained in these records. This required determination of who belonged to whom based on the parents' marriage dates, the distance between births (correcting for premature births) to rule out overlapping pregnancies, the use of the mother's maiden name as a middle name in subsequent generations, and the age of the mother at the approximate date of birth.

In some cases, no descriptor of the father was listed—simply son or daughter of Thomas Owston. In other listings, a location is given such as Potter Brompton, Windlebeck (a farm on the northeastern edge of Ganton parish), and Ganton Walk (an unknown location within the parish). Both Thomases lived at Windlebeck at different times, but this could be sorted out by years.[58] John Owston, the father of Thomas, husband of Mary, died at Windlebeck in 1782.[59] While Elizabeth Walker's family was from Potter Brompton,[60] the location appeared to be used only for Thomas of the Ganton family. The location of Ganton Walk was used twice for the same Thomas, the husband of Mary.[61]

The identification of "farmer" nearly always was applied to Thomas, the husband of Elizabeth. It appears that the designation was used to distinguish him from the other Thomas

56. St. Nicholas's Church (Ganton, Yorks), "Ganton Parish Register, 1793–1812," unnumbered p. 36, John Owston death and burial record, 3 and 5 February 1797. St. Nicholas's Church (Ganton, Yorks), "Ganton Parish Records (Bishop's) Transcripts. 1601–1867," unnumbered p. 201, John Owston burial, 5 February 1797.

57. St. Nicholas's Church (Ganton, Yorks), "Ganton Parish Register, 1793–1812," unnumbered p. 42, Jane Owston death, 25 June 1804. St. Nicholas's Church (Ganton, Yorks), "Ganton Parish Records (Bishop's) Transcripts. 1601–1867," unnumbered p. 216, Jane Owston burial, 27 June 1804.

58. St. Nicholas's Church (Ganton, Yorks), "Ganton Parish Register, 1553–1794," unnumbered pp. 139–154, Owston baptisms 1773–1793. "Yorkshire Wolds Way: Windlebeck Farm" *National Trails* (https://www.nationaltrail.co.uk/yorkshire-wolds-way/accommodation/windlebeck-farm : accessed 8 June 2018).

59. St. Nicholas's Church (Ganton, Yorks, England), "Ganton Parish Register, 1553–1794," unnumbered p. 89, John Owston burial, 26 June 1782. St. Nicholas's Church (Ganton, Yorks), "Ganton Parish Records (Bishop's) Transcripts. 1601–1867," unnumbered p. 184, John Owston burial, 26 June 1782.

60. St. Nicholas's Church (Ganton, Yorks), "Ganton Parish Marriages, 1758-1793," unnumbered p. 7, Thomas Owston and Elizabeth Walker.

61. St. Nicholas's Church (Ganton, Yorks), "Ganton Parish Register, 1553–1794," unnumbered pp. 139–154, Owston baptisms 1773–1793. St. Nicholas's Church (Ganton, Yorks), "Ganton Parish Records (Bishop's) Transcripts. 1601–1867," unnumbered pp. 164–197, Owston baptisms 1773–1793.

Owston; however, Thomas, the husband of Mary, is also called farmer in the bishop's transcripts for his daughter Jane's baptism.[62]

Only one of the seven females born during this period was known to produce issue; since females have no Y chromosome, we concentrate on the males listed in the Ganton baptismal records.

Table 4.1. Male Children ascribed to Thomas and Elizabeth (Walker) Owston of the Sherburn family		
Name	**Baptismal Date**	**Reasoning**
Thomas[a]	18 April 1773	Prior to the marriage of Thomas and Mary; father listed as Thomas Owston, farmer.
William[b]	27 February 1778	Listed as Thomas Owston, farmer.
John[c]	22 August 1779	Listed as Thomas Owston, farmer; 1797 burial listed grandparents in PR and mother in bishop's transcripts.

a. St. Nicholas's Church (Ganton, Yorks), "Ganton Parish Register, 1553–1794," unnumbered p. 139, Thomas Owston baptism, 18 April 1773.

b. St. Nicholas's Church (Ganton, Yorks), "Ganton Parish Register, 1553–1794," unnumbered p. 142, William Owston baptism, 27 February 1778.

c. c. St. Nicholas's Church (Ganton, Yorks), "Ganton Parish Register, 1553–1794," unnumbered p. 144, John Owston baptism, 22 August 1779; p. 89, John Owston burial, 26 June 1782. St. Nicholas's Church (Ganton, Yorks), "Ganton Parish Records (Bishop's) Transcripts. 1601–1867," unnumbered p. 184, John Owston burial, 26 June 1782.

Likely sons of Thomas and Elizabeth are shown in table 4.1. By our best estimation, they had three sons: Thomas (1773–????), William (1778–1857), and John (1779–1797). There is no further record of Thomas. John died just prior to his eighteenth birthday. William, the progenitor of the Cobourg line, appeared to be the only son of this family who produced issue. The Cobourg line has twenty living Owston males with thirteen participating in DNA testing—seven of whom participated in Y-DNA testing. Of these seven, one failed to have a matching Y-DNA haplotype or matching autosomal DNA to any Cobourg participants outside of his second cousin who also failed to match any others (possibly due to an unofficial adoption of the subject's great-grandfather), four participated in Big Y-500 testing, and the remaining two have 111 Y-STR marker results.[63]

Likely sons of Thomas and Mary are shown in table 4.2. They had four sons: John (1780–1862), Thomas (1782–1869), William (1784–1792), and Francis (1793–1870). The progeny

62. St. Nicholas's Church (Ganton, Yorks), "Ganton Parish Register, 1553–1794," unnumbered pp. 139–154, Owston baptisms 1773–1793. St. Nicholas's Church (Ganton, Yorks), "Ganton Parish Records (Bishop's) Transcripts. 1601–1867," unnumbered pp. 164–197, Owston baptisms 1773–1793.

63. James M. Owston, "Pruning the Family Tree with DNA Evidence" *The Lineal Arboretum,* 13 October 2010 (http://linealarboretum.blogspot.com/2010/10/pruning-family-tree-with-dna-evidence.html). Owston, *Owston/Ouston DNA Project.*

of the Ganton branch is quite large with an estimated sixty-three living males; the majority descend from their second son, Thomas. To date, six Ganton Owston men have tested; three descended from Thomas and three from Francis. John has two living male Owston descendants; one is a minor and the other bears the surname of his mother.[64]

Table 4.2. Male Children ascribed to Thomas and Mary (Vickerman) Owston of the Ganton family		
Name	**Baptismal Date**	**Reasoning**
John[a]	29 June 1780	Listed as Thomas Owston of Potter Brompton; Thomas's and Mary's fathers' names.
Thomas[b]	29 October 1782	Listed as Thomas Owston of Windlebeck; Father's name. Vickerman name in progeny.
William[c]	16 March 1784	Listed as Thomas Owston of Windlebeck.
Francis[d]	12 August 1793	Listed as Thomas Owston of Ganton Walk. Elizabeth is aged 56 at the time.

a. St. Nicholas's Church (Ganton, Yorks), "Ganton Parish Register, 1553–1794," unnumbered p. 144, John Owston baptism, 29 June 1780; p. 129, John Owston and Mary Emin marriage, 5 February 1744. St. Hilda's Church (Sherburn, Yorks), "Sherburn Parish Register, 1717–1777," unnumbered p. 9, John Vickerman and Ann Clifford marriage, 17 April 1743.

b. St. Nicholas's Church (Ganton, Yorks), "Ganton Parish Register, 1553–1794," unnumbered p. 148, Thomas Owston baptism, 29 October 1782. Ohio Department of Health, "Ohio Death Records 1908–1932," p. 830, Charles Vickerman Owston, 25 August 1912; digital images (https://www.ancestry.com). "Delaware County Burials," Delaware County [Ohio] Genealogical and Historical Societies (http://www.rootsweb.ancestry.com/~ohdchs/cemetery/burials.htm).

c. St. Nicholas's Church (Ganton, Yorks), "Ganton Parish Register, 1553–1794," unnumbered p. 149, William Owston baptism, 16 March 1784.

d. St. Nicholas's Church (Ganton, Yorks), "Ganton Parish Records (Bishop's) Transcripts. 1601–1867," unnumbered p. 196, Francis Owston baptism, 12 August 1793.

Of the six tested Ganton Owston men, four completed the Big Y-500. The remaining two only tested with a Y-43 test and are deceased. Ganton03 replaced Ganton01, his deceased second cousin once removed, and Ganton06 replaced his deceased father, Ganton02.[65]

Our documentary research placed Sherburn08, the closest related tested participant to the Cobourg men, as either an eighth cousin or eighth cousin once removed.[66] At the time, we believed that the Ganton and Cobourg participants were probably thirteenth cousins once removed to fourteenth cousins depending on individual relationships (see Case Two for more on these project members). The close genetic distance at 111 markers questioned

64. Owston, *Owston/Ouston DNA Project.* Ouston, *2003 Directory of Ouston/Owston Families.*

65. James M. Owston, "Owston/Ouston DNA Study's 2015 in Review," *Owston/Ouston One-Name Study,* 30 December 2015 (https://owston.wordpress.com/2015/12/30/owstonouston-dna-studys-2015-in-review/). Owston, *Owston/Ouston DNA Project.*

66. Owston, *Owston/Ouston DNA Project.*

whether we placed William Owston (1778–1857) with the correct parents. Since he was baptized three months after the marriage of Thomas Owston and Mary Vickerman, I began doubting our original hypothesis—an unexpected pregnancy may have hastened a marriage to legitimize a child.[67]

Beginning in spring 2015, I began reviewing Ganton parish registers and bishop's transcripts, but found nothing conclusive supporting either parentage possibility. In late 2015, I decided to test Cobourg01, Ganton04, and Sherburn08 with the Big Y-500; Thornholme04 was also included as a control. All four men shared the I-A10206 SNP and eleven other SNP markers. As STR testing had previously shown, SNP testing confirmed that the three perceived Owston/Ouston families shared a common ancestor.[68]

I was concerned that not enough men were tested to provide conclusive evidence of William Owston's parentage. YFull analysis of Big Y-500 BAM[69] files needed at least two individuals to determine if unique SNPs existed for the families. In 2016 and 2017, I added Cobourg05, Cobourg08, Ganton03, and tested a new subject Ganton05. Sherburn12 and Sherburn02 were added as control subjects. Additionally, Sherburn01, Thornholme05, and two new participants (Ganton06 and Sherburn14) were also tested. Sherburn04, a former Y-43 participant, tested with the Big Y-500 in June 2018. Cobourg06's results were added in October 2018. We currently have sixteen Big Y-500 participants.[70]

All three of the Cobourg men were descended from different sons of William Owston (1778–1857): Cobourg08 from James Wilson Owston (1809–1858), Cobourg05 from Charles Paget Herbert Owston (1817–1858), and Cobourg01 and Coburg06 from John Gillon Owston (1826–1901).[71] Ganton03 and Ganton04 were each descended from two different sons of Thomas Owston (1782–1869) and Ann Hunter (1783–1856): William Owston (1803–1866) and Anthony Owston (1805–1884). We were confident that this Thomas was descended from Thomas Owston (1755–1823) and Mary Vickerman, as the Vickerman surname appeared as a middle name twice among Thomas and Ann's descendants.[72]

Ganton05 descended from Francis Owston (1793–1870) and Mary Smith (1793–1854). Our placement of Francis as the son of Mary (Vickerman) Owston followed conventional assumptions, as Elizabeth (Walker) Owston was fifty-six in 1793 and past the typical age of childbearing. In contrast, Mary (Vickerman) Owston was forty in 1793. Ganton06, another descendant of Francis and Mary (Smith) Owston, had Big Y-500 results posted in early 2018.[73]

67. Owston, *Owston/Ouston DNA Project.* Ouston, *2003 Directory of Ouston/Owston Families.*

68. Owston, *Owston/Ouston DNA Project.*

69. Binary Alignment Map (BAM) is a file format commonly used for compressed DNA data.

70. Owston, *Owston/Ouston DNA Project.*

71. James M. Owston, "Owston/Ouston Family Lineage Charts," (http://www.owston.com/family/owston/Owston_Family_Charts.pdf), 8.

72. Owston, *Owston/Ouston DNA Project.* Ouston, *2003 Directory of Ouston/Owston Families.* Ohio death certificates were accessed; details are omitted to protect the privacy of recent generations.

73. Owston, *Owston/Ouston DNA Project.* Ouston, *2003 Directory of Ouston/Owston Families.* St. Nicholas's

If all four Cobourg men shared a unique SNP, it would have been shared by their common ancestor William Owston (1778–1857). If the descendants of Thomas and Ann (Hunter) Owston and those of Francis and Mary (Smith) Owston shared the same unique SNP, it would have been shared by their common ancestor, Thomas Owston (1755–1823).

All four Cobourg men shared both A10921 and A10923. All four Ganton men shared A10208. Since the Cobourg men did not share A10208 that Thomas Owston of Ganton (1755–1823) would have most certainly carried, William Owston (1778–1857) could not be his son. Therefore, our original conclusion that William Owston was the son of Thomas Owston and Elizabeth Walker of the Sherburn family was confirmed. Nearly thirty years after formulating our original conclusion, DNA evidence suggested our original hypothesis was correct. But, there were other issues.[74]

Case 2: disconfirming a previous conclusion

During the early 1990s, the three Owston/Ouston researchers assumed that Thomas Owston (1755–1823) of Ganton was descended from the Gyles Owston (d. 1641) of Ganton—there was no indication otherwise in the records. Big Y-500 results, however, questioned our original assignment of this family.[75]

As noted in the previous case, I began testing Cobourg and Ganton men and a handful of other Owstons with the Big Y-500. Ganton04 and Cobourg01, the first two participants both shared A10206 and eleven other SNPs. This was congruent with the Owston/Ouston families having matching STR results and, therefore, having a common ancestor.[76]

When Sherburn08's results were returned a week later, a problem developed, as he and Ganton04 were reported as both sharing A10216 in addition to A10206. A10216 was placed downstream from A10206 on both YFull and FamilyTreeDNA's haplotrees. Additionally, Cobourg01 and Thornholme04 did not share A10216; although all four men shared A10206 and its companion SNPs.[77]

Since the Ganton family was believed to be distinct from the Sherburn family, of which both Sherburn08 and Cobourg01 were members, this seemed incongruent. This puzzle would plague me for eighteen months. Somehow, I forgot one of the cardinal rules of genetic genealogy: DNA never lies—when properly interpreted.

As a year and a half passed, I wrestled with this seeming impossibility. It appeared that I

Church (Ganton, Yorks), "Ganton Parish Register, 1553-1794," unnumbered p. 98, Elizabeth Walker baptism, 14 November 1737.

74. Owston, *Owston/Ouston DNA Project.* Ouston, *2003 Directory of Ouston/Owston Families.*

75. Owston, *Owston/Ouston DNA Project.* Ouston, *2003 Directory of Ouston/Owston Families.*

76. Owston, *Owston/Ouston DNA Project.* Ouston, *2003 Directory of Ouston/Owston Families.*

77. YFull YTree v6.05, *YFull* (https://yfull.com/tree/I1).

was following several of Elisabeth Kübler-Ross's stages of grief. Denial was first, and was followed by anger—"this could not possibly happen."[78] Knowing that STRs often change values, I read numerous articles, blogs, and posts about SNPs to see if back mutations were possible. I also questioned several experts about the possibility of SNPs changing values. The answers were consistent: "no," "I don't think so," and "I guess it's possible, but we've never observed it happening."

By spring 2017, I was able to test another Ganton family member descended from a collateral line. FamilyTreeDNA initially listed Ganton05 with the terminal SNP of A10206, while listing Ganton04 with A10216. YFull, however, assigned both Ganton04 and Ganton05 with both A10216 and A10208 SNPs.[79]

I contacted YFull about the discrepancy between its SNP assignment and that of FamilyTreeDNA. They responded that Ganton05 had four quality reads on A10216 and that was sufficient to be considered positive; they also explained that FamilyTreeDNA needed ten positive reads.[80] I was convinced that A10216 was an unreliable SNP.

In December 2017, FamilyTreeDNA began identifying Ganton05 as positive for both A10216 and A10208. This was found in his matching results and on his assignment of the Y-DNA Haplotree. A10216 was not listed as a derived marker on Ganton05's named results. I eventually found it when searching for "all" instead of "derived." It was identified by a question mark as seen in figure 4.7. This was apparently due to it having only four positive reads as seen in figure 4.8. I contacted FamilyTreeDNA about this. I thought it should be reported as either questionable or presumed, as it was not showing in Ganton05's named variants. FamilyTreeDNA held firm and allowed A10216 to appear positive in Ganton05's results despite having only four positive reads.[81]

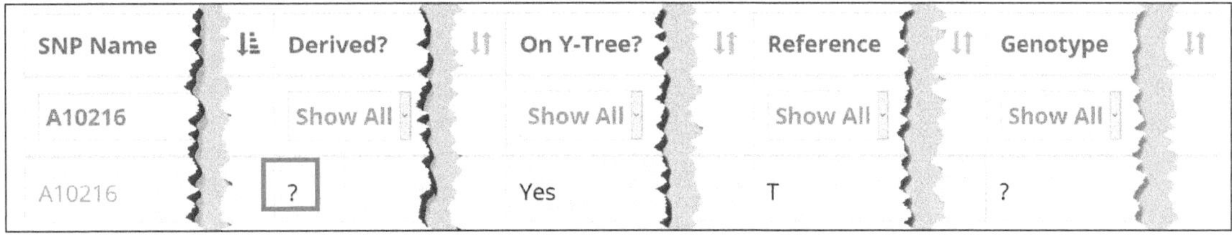

SNP Name	Derived?	On Y-Tree?	Reference	Genotype
A10216	Show All	Show All	Show All	Show All
A10216	?	Yes	T	?

Figure 4.7. Ganton05's questionable read on A10216

78. Elisabeth Kübler-Ross, *On Death and Dying: What the Dying Have to Teach Doctors, Nurses, Clergy & their own Families* (New York: Simon & Schuster, 1969).

79. Owston, *Owston/Ouston DNA Project.* YFull YTree v6.02, *YFull.*

80. Vadim Urasin, to James Owston, email, 14 September 2017; privately held by Owston.

81. Owston, *Owston/Ouston DNA Project.* Michael Sager, to James Owston, email, 4 and 5 January 2018; privately held by Owston. Kübler-Ross, *On Death and Dying.*

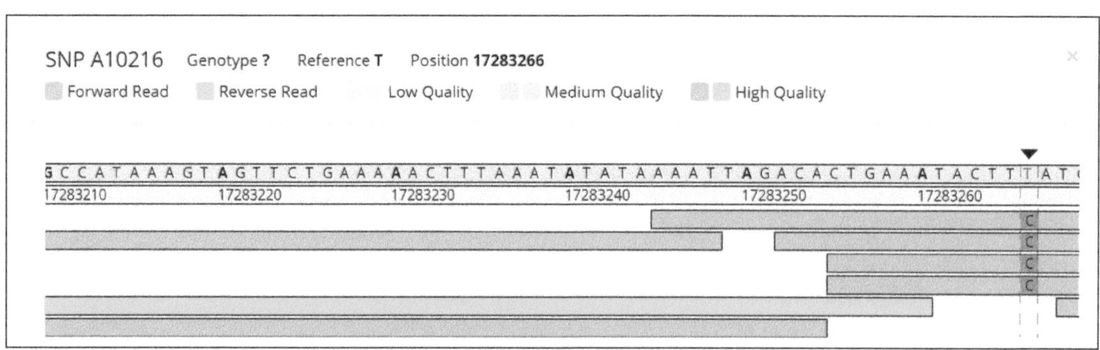

Figure 4.8. Ganton05's four positive reads on A10216

When the Big Y-500 results were reconfigured by FamilyTreeDNA for the Build 38 upgrade, I charted the matching SNPs and my eighteen-month dilemma became a revelation. Table 4.3 charts all participants' Big Y-500 results. An additional SNP, Y130060, discovered by YFull in 2018 was added. This was later added to the FamilyTreeDNA results.[82]

Table 4.3. Shared SNPs in the Owston/Ouston Project

Participant	A10206	BY31751	A15739	Y130060	A18911	A10216	A10921	A10208
Ganton03	Positive	Positive	Negative	Negative	Positive	Positive	Negative	Positive
Ganton04	Positive	Presumed	Negative	Negative	Positive	Positive	Negative	Positive
Ganton05	Positive	Positive	Negative	Negative	Positive	Questionable	Negative	Positive
Ganton06	Positive	Positive	Negative	Negative	Presumed	Presumed	Negative	Positive
Sherburn08	Positive	Positive	Negative	Negative	Positive	Positive	Negative	Negative
Cobourg01	Positive	Positive	Negative	Negative	Positive	Negative	Positive	Negative
Cobourg05	Positive	Positive	Negative	Negative	Positive	Negative	Positive	Negative
Cobourg06	Positive	Presumed	Negative	Negative	Positive	Negative	Positive	Negative
Cobourg08	Positive	Positive	Negative	Negative	Positive	Negative	Positive	Negative
Sherburn12	Positive	Positive	Negative	Negative	Negative	Negative	Negative	Negative
Sherburn04	Positive	Positive	Negative	Negative	Negative	Negative	Negative	Negative
Sherburn02	Positive	Positive	Negative	Negative	Negative	Negative	Negative	Negative
Sherburn01	Positive	Positive	Negative	Positive	Negative	Negative	Negative	Negative
Sherburn14	Positive	Positive	Negative	Positive	Negative	Negative	Negative	Negative
Thornholme04	Positive	Negative	Positive	Negative	Negative	Negative	Negative	Negative
Thornholme05	Positive	Negative	Positive	Negative	Negative	Negative	Negative	Negative

After playing with the arrangement of SNPs, the solution was obvious. We got it wrong. The family of Thomas Owston and Mary Vickerman of Ganton was not descended from the older Ganton family of Gyles Owston, but appeared to be from the Sherburn family. In addition, the Ganton branch appeared to be closely related to the Yeoman Farmer branch of which Sherburn08 was a member, as both families bore the A10216 SNP.[83]

82. Owston, *Owston/Ouston DNA Project.* YFull YTree v6.02, *YFull.*
83. Owston, *Owston/Ouston DNA Project.*

The SNP data also revealed that the Sherburn and Thornholme families had unique SNPs: BY31751 for the Sherburn participants and A15739 for Thornholme Owstons. In addition, the Holderness branch's participants shared Y130060, but others did not.[84]

The newer Ganton branch, as suggested by Y-SNP testing, also shared a close common ancestor with the Cobourg line; therefore, the lack of variations between some Ganton and Cobourg men with 111 Y-STRs, and even with the Big Y-500's 561 STR markers, became clearer.[85] Not only was the A10216 issue seemingly solved, but the question that began in February 2015 of why Cobourg and Ganton men seemed so closely related appeared to be answered. The key was the newly reported upstream A18911 SNP that showed positive or presumed positive for the Ganton branch, Cobourg line, and Sherburn08.[86]

The shared ancestor for both the Cobourg family and Sherburn08, George Owston (1636–1676), seemed to be the likely source. Since four (Sherburn01, 02, 04, and 14) Big Y-500 tested descendants of George's brother Thomas (1627–1681) did not have this SNP, it appears to have originated with George. Therefore, SNP evidence suggested that the Ganton branch was descended from George Owston as well.[87]

The Cobourg family, who descended from George's son John (1665–1739), did not have A10216; therefore, this SNP appears to have originated with George's son Francis (1667–1733). The key now was to find our nearly thirty-year-old error based on documentary research.[88]

Since the error appeared to be in the Ganton family, we worked backward from the common ancestor of the newer Ganton branch—Thomas Owston (1755–1823). In rechecking parish records, the identification of Thomas Owston, born 1755, being Mary Vickerman's husband appeared to be correct.

While two Thomas Owstons were born in Ganton in previous years, Thomas Owston, son of John Owston, blacksmith, baptized on 13 April 1755 was the only one who lived to a majority age and could have married Mary Vickerman in 1777. At this point, I found a possible key to our error.[89]

John Owston had married Mary Eman (Emin in the marriage record) on 5 February 1744. The bishop's transcripts confirm that John and Mary were "both of the parish of Galmpton," an older name for Ganton. "Mary, wife of John Owston, farmer" was buried on 18

84. Owston, *Owston/Ouston DNA Project.* Ouston, *2003 Directory of Ouston/Owston Families.*
85. Owston, *Owston/Ouston DNA Project.*
86. Owston, *Owston/Ouston DNA Project.*
87. Owston, *Owston/Ouston DNA Project.* Ouston, *2003 Directory of Ouston/Owston Families.*
88. Owston, *Owston/Ouston DNA Project.* Ouston, *2003 Directory of Ouston/Owston Families.*
89. St. Nicholas's Church (Ganton, Yorks), "Ganton Parish Register, 1553-1794," unnumbered p. 100, Thomas Owston baptism, 13 April 1755.

September 1777. "John Owston of Windlebeck" was buried 26 June 1782. From 1781 to 1786, Thomas and Mary Vickerman Owston are listed as living in Windlebeck. Thomas, born eleven years after his parents' marriage, was the only child of this union.[90]

We had ascribed John to the Gyles Owston of Ganton family who had lived in the parish for at least two centuries, probably because John was listed as being from Ganton. The parish records in Ganton during the seventeenth and eighteenth centuries sometimes spelled the Owston surname as Ouston. This was not uncommon, and Ouston had become the permanent spelling of two branches of the family—the Holderness branch and the extinct Kirby Misperton branch. Earlier Ganton records have the name also spelled as Owstaine (1556–1646), Oustaine (1574–1603), and Owstayne (1559).[91]

John Owston/Ouston (1684–1741) of the Gyles Owston of Ganton family seemed the likely father of the younger John who married Mary Eman. The elder John was the only Owston male living in the parish during the span of years consistent with younger John's birth. Since the younger John and his bride were both listed as being from Ganton parish in his 1744 marriage record, the connection seemed reasonable.[92]

Additionally, the elder John's uncle was Anthony Owston (d. 1728). The name Anthony was a distinct forename found thrice in the early Ganton family and it appeared five times in the later Ganton family. It was not, however, found in either the Sherburn or Thornholme Owston families. Nor was the name found in the family of Thomas Owston's (1782–1869) wife, Ann Hunter (1783–1856). It seemed logical to suggest that the name was another connection to the family of Gyles Owston.[93]

Since the elder John married Elizabeth Read in Ganton on 4 June 1718,[94] the younger John's birth would need to follow; otherwise, he would have carried his mother's surname. At the time, we relied heavily on the International Genealogical Index (IGI) in sorting out the various Owston/Ouston families. The IGI and bishop's transcripts listed a John Ouston, the son of John Ouston, who was baptized in Langton, Yorkshire, on 20 January 1719.[95]

90. St. Nicholas's Church (Ganton, Yorks), "Ganton Parish Records (Bishop's) Transcripts. 1601–1867," unnumbered p. 121, John Owston and Mary Emin marriage, 5 February 1744; digital images, *FindMyPast* (https://www.findmypast.com). St. Nicholas's Church (Ganton, Yorks), "Ganton Parish Register, 1553–1794," unnumbered p. 162, Mary Owston burial, 18 September 1777; p. 165, John Owston burial, 26 June 1782; pp. 139–154, Owston baptisms 1773–1793. Joshua Fawcett, *Church Rides in the Neighborhood of Scarborough, Yorkshire* (London: Simpkin, Marshall & Company, 1848), 122.

91. Ouston, *2003 Directory of Ouston/Owston Families.* St. Nicholas's Church (Ganton, Yorks), "Ganton Parish Register, 1553–1794," unnumbered pp. 86–96, Ouston baptisms 1693–1731. Search of Owstaine, Oustaine, and Owstayne in Ganton Parish Records, database search, *FindMyPast* (https://www.findmypast.com).

92. Ouston, *2003 Directory of Ouston/Owston Families.*

93. Ouston, *2003 Directory of Ouston/Owston Families.* Search of 'Anthony' in Wintringham Parish Records, database search, *FindMyPast* (https://www.findmypast.com).

94. St. Nicholas's Church (Ganton, Yorks), "Ganton Parish Register, 1553–1794," unnumbered p. 123, John Ouston and Elizabeth Read marriage, 4 June 1718; digital images, *FindMyPast* (https://www.findmypast.com).

95. St. Andrew's Church (Langton, Yorkshire, England), "Langton Parish Records (Bishop's) Transcripts. 1600–1871," unnumbered pp. 63, John Ouston baptism, 20 January 1719; digital images, *FindMyPast* (https://www.findmypast.com).

Additional information in the parish register indicated that the family was from the hamlet of Kennythorp that lies within the confines of Langton parish.[96] No other records of either father or son appear in this or nearby parishes. Therefore, we assumed that the elder John had left Ganton for Langton, eighteen miles to the west; this was the beginning of our misattribution.

While it was not unusual for members of the Owston clan to relocate at such distances during the period, typically these were permanent moves. However, John is found in Ganton in 1720 when his daughter Mary is baptized on 6 November. Three other daughters, Anne (1723), Elizabeth (1725), and a second Mary (1831), born after the first Mary's death in 1830, were all baptized in Ganton.[97] It is highly unlikely that John and Elizabeth (Read) Owston, both from Ganton, would travel to Langton, a distant parish with no apparent connection to either family and reside in this place for a period under two years.

The passing of John Owston/Ouston (1684–1741), if he was not the father of John (1719), would result in the surname lineage of the Gyles Owston of Ganton family becoming extinct.[98] The line would have simply daughtered out. In addition, it appeared that we made the most common error in genealogy—we confused two men with the same name. But, who was the correct John Owston?

A clue to the correct John's identity may be found in his occupation that is noted in Thomas Owston's baptismal record: John was identified as a blacksmith.[99] The Ganton Owston men in John Owston's (1684–1741) lineage were identified as husbandmen. We've established that Francis Owston (1667–1733) was probably the source of the A10216 SNP. An indenture dated 1707 lists Francis Owston also as a blacksmith.[100]

> This indenture made the thirteenth day of September in the sixth year of the reigne of Our Soveraign Lady Ann by the grace of God of Great Brittain France and Ireland Queen Defender of the Faith in Anno Dom 1707. Between John Owston the Younger of Sherburn in the County of York Yeom of the one part and Francis Owston of Sherburn aforesaid in the said County of York Blacksmith of the other part.[101]

---

96. St. Andrew's Church (Langton, Yorkshire, England), "Langton Parish Register. 1653–1725," unnumbered pp. 10, John Ouston baptism, 20 January 1719; digital images, *FindMyPast* (https://www.findmypast.com).

97. St. Nicholas's Church (Ganton, Yorks), "Ganton Parish Register, 1553–1794," unnumbered pp. 93–96, Baptisms of John Ouston's children 1720–1731; p. 113, Mary Ouston burial, 10 May 1730.

98. St. Nicholas's Church (Ganton, Yorks), "Ganton Parish Register, 1553–1794," unnumbered p. 116, John Owston burial, 25 December 1741; pp. 93–96, Baptisms of John Ouston's children 1720–1731.

99. St. Nicholas's Church (Ganton, Yorks), "Ganton Parish Register, 1553–1794," unnumbered p. 100, Thomas Owston baptism, 13 April 1755.

100. Ouston, *2003 Directory of Ouston/Owston Families*. Owston, *Owston/Ouston DNA Project*.

101. Timothy J. Owston, "The Descendants of Peter and Petronel Owston," *Owston Family: Sherburn Based Branch of the Family, East Yorkshire with Links to Other Branches* (March 2013), especially transcript of the 1707 indenture between John Owston and Francis Owston (http://freespace.virgin.net/owston.tj/owstonln.htm : accessed 9 March 2015).

Both men are listed in the record as being sons of George Owston (1636–1676). The indenture is the transfer of a cottage, outbuildings, and surrounding lands for £10. John (1665–1739) is the ancestor of the Cobourg line, while Francis (1767–1733) is the ancestor of Sherburn08 through son Peter (1709–1765). The appellation "John Owston the Younger" distinguishes him from an older man named John who lived in Sherburn; the older John's origins are unknown.[102]

A search of the Sherburn parish records provides the final clue: the baptism of "Johanes filius Francisci Owston" on 4 Feb 1704 (O.S.).[103] John is never listed again in the Sherburn parish register or the bishop's transcripts. He may have been missed during previous analysis of the records; the information for all of 1704 is missing from the bishop's transcripts for Sherburn. Additionally, Thomas Owston (1755–1823) named his youngest son Francis, a name that is never found in the early Ganton family but is found eight times in the Sherburn family from the sixteenth through the eighteenth centuries.[104]

With the Ganton men sharing A18911 that originated with George Owston (1636–1676), the A10216 SNP that originated with Francis Owston (1767–1733), the close STR genetic distance with the Cobourg family, and a walking distance between Sherburn and Ganton, a conclusion can be drawn. John Owston, the father of Thomas Owston of the newer Ganton branch, was probably the same person as John Owston (b. 1704 O.S.), the son of Francis Owston (1667–1733), and grandson of George Owston (1636–1676). Therefore, the two Thomas Owstons in Case One were second cousins. Only Next Generation Sequencing testing of numerous men could have corrected our error of assignment.[105]

Conclusion

Y-DNA STR and SNP testing have both aided our one-name study in providing answers where the documentary evidence was inadequate. By testing various family members, we were able to reaffirm a hypothesis that was constructed nearly thirty years ago. This testing also led to the identification of an incorrect attribution and our eventual discovery of the promising solution. Y-DNA testing made both possible.

---

102. Owston, "The Descendants of Peter and Petronel Owston."

103. St. Hilda's Church (Sherburn in Harford Lythe, Yorkshire, England), "Sherburn Parish Register, 1653–1719," unnumbered p. 18, Johanes Owston baptism, 4 February 1704; digital images, *FindMyPast* (https://www.findmypast.com). "O.S." indicates an Old Style date.

104. St. Nicholas's Church (Ganton, Yorks), "Ganton Parish Register, 1553–1794," unnumbered p. 154, Francis Owston baptism, 12 August 1792. Ouston, *2003 Directory of Ouston/Owston Families.*

105. Owston, *Owston/Ouston DNA Project.* Ouston, *2003 Directory of Ouston/Owston Families.*

Unknown and Misattributed Parentage Research

Melissa A. Johnson, CG

What is Unknown Parentage?

"Unknown parentage" describes the circumstance in which an individual does not know the identity of one or both biological parents. "Misattributed parentage" refers to the situation where parentage has been attributed to the wrong person. The biological parentage then becomes unknown until revealed through research or family discussions. There are numerous reasons for unknown parentage such as adoption, donor conception, abandonment by one or both parents, memory loss, medical errors and malpractice, criminal acts, lack of disclosure between birth parents, and more. Many individuals with unknown parentage, such as some adoptees or individuals raised by one parent, are knowledgeable about their family circumstances. Others discover unknown parentage later in life, contradicting what the person has always believed to be true. This sometimes happens as a result of DNA testing, typically when a person's test results either show genetic relationships that are not expected or do not show genetic relationships that are expected.

DNA testing has become an essential tool for individuals with unknown parentage who seek to identify their biological families. When DNA testing was in its infancy, unknown parentage cases were difficult to solve; testing was expensive and the testing company databases included fewer individuals. For test takers, this resulted in few DNA matches that were often distant in relationship. Today, DNA testing is extremely popular and much more affordable. The company databases have grown larger, and test takers now have closer DNA matches. It is not uncommon for an individual with unknown parentage to take a DNA test and have several second cousin matches. At the same time, technological advances and new tools have resulted in techniques that are useful to help identify and work with DNA matches. These improvements have made unknown parentage cases much easier to solve.

Overview of Unknown Parentage Methodology

The process of identifying unknown parentage requires time, dedication, and patience. When using DNA, the level of difficulty depends on several factors, mostly surrounding the quality of the DNA matches (whether there are close relatives or more distant relatives) and the ability to identify the ancestors of those matches. The methodology used to solve unknown parentage cases involves tying together (1) genetic evidence from DNA test results and (2) documentary evidence from traditional sources. The documentary aspect is two-fold—it involves the individual with unknown parentage, as well as his or her DNA matches and their ancestors.

Documentary Research Strategies

Prior to using DNA to resolve unknown parentage, research should begin with traditional documentary sources related to the individual whose parentage is in question. Information from these sources could potentially identify a biological parent or parents by name, but even if the given information falls short of an identification, it can be extremely useful in terms of ruling in or ruling out individuals found within the families of DNA matches. Some documentary sources that can be useful are non-identifying paperwork, original birth certificates, adoption judgments, oral histories, newspaper articles, court records, and other sources. Laws concerning the confidentiality of records vary by state, and not all of these records may be available in all locations.

Non-identifying information

Non-identifying information includes descriptive (but not personally identifying) details about an adoptee's birth family. Typically provided by the birth parent(s) at the time of the birth or adoption, this information can include the age, marital status, ethnicity, race, religion, education, employment, physical description, medical history, and social history of one or both birth parents. Sometimes, the reason for the adoption is included. Often, information about extended family members is provided, such as whether the birth mother had other children (siblings or half siblings of the adoptee), or the ages and medical histories of the birth mother's siblings and parents (the adoptee's aunts, uncles, and grandparents). The information that is collected varies on a case-by-case basis and is dependent on the informant's ability and willingness to answer the questions. While non-identifying information will not typically include names or locations, information such as the birth mother was seventeen years old, had two sisters and a brother, was Catholic, and was the daughter of a physician and a homemaker will establish a profile for the family being sought. When tracing relatives of key DNA matches, having a family profile developed from non-identifying information can be an extremely useful tool.

Non-identifying information is typically requested from the adoption registry or agency in the state where the adoption occurred. Some adoptees have had success with later attempts after requesting non-identifying paperwork on multiple occasions, especially if long amounts of time have lapsed between requests. The premise behind this is that the amount of information provided may differ based on who is extracting those details and determining what is considered "non-identifying."

Original birth certificates

An original birth certificate is the pre-adoption birth certificate that usually includes the adoptee's birth name and the name of at least the birth mother. In many instances, original birth certificates do not include the birth father's name, especially if the parents were not married. When a father's name is included, use caution—consider the possibility that the identification was made incorrectly by the mother or that the mother's legal husband was named as the child's father even if he was not.

The availability of original birth certificates varies by state. Some states have unrestricted access, while others require a court order, and some limit access completely. In recent years, legislation has changed in many states to allow adoptees access to original birth certificates, and efforts to change legislation continue. A comprehensive list of original birth certificate availability by state can be found on the American Adoption Congress website.[1]

Adoption judgments

Adoption judgments are the court records authorizing a legal adoption. The organization of these records varies by the state or county of jurisdiction. Availability varies by jurisdiction as well; in most cases, these records are sealed. However, some are open to the public or can be accessed through a petition from the adoptee. Adoption judgments are not original birth certificates—these are separate records, often with separate access restrictions.

Oral history

Oral history, including information from the adoptive parents, a known biological parent, family members within the adoptive family, and even friends and neighbors, can be helpful. Always contact those closest to the adoptee (his or her adoptive parents or siblings), but do not forget to reach out to more distant relatives such as the adoptee's grandparents, aunts, uncles, cousins, and other relatives. Small details that were not retained in official records may have been discussed between the adoptive parents and an attorney or social worker, and those bits of information may have been shared with family and friends.

Adoption registries

For adoptees, there are steps that can be taken prior to using DNA to determine if biological relatives can be identified. There are many adoption registries that allow biological relatives and adoptees to connect if there is mutual interest. Additionally, many states have confidential intermediary programs, where a liaison serves as a point of contact between the adoptee and members of the biological family, to determine if both sides wish to make contact.

All URLs were accessed 27 January 2019 unless otherwise indicated.

1. American Adoption Congress (https://www.americanadoptioncongress.org/state.php).

Other historical records

Depending on the circumstances surrounding the unknown parentage, historical records such as orphanage records, religious records, divorce records, court records, newspapers, and other sources may also be useful. Known information about the situation should be evaluated, and sources of potential interest should be sought out and mined for clues. For example, if a birth mother was married at the time of her child's birth, divorce records may name a man with whom she was suspected of having an affair. A child who was placed for adoption at age two may have been the subject of custody or child support proceedings prior to the adoption—seeking out family court records involving the birth mother may help identify the birth father. Attention to detail with regard to handwriting, strike-outs, write-ins, filing and recording dates, and other attributes of records can be crucial and can provide useful hints.

Information provided by non-genetic sources can be invaluable, but should also be used with caution as it has the potential to be inaccurate, particularly with regard to oral histories and non-identifying paperwork. Even if very little is discovered from documentary sources, small details can be used in collaboration with DNA testing to narrow down candidates and identify the correct biological family.

DNA Testing

Resolving unknown parentage using DNA evidence relies on the quantity and quality of the test taker's DNA matches. As a result, individuals with unknown parentage should ensure that they have tested with each of the main testing companies and taken every type of DNA test that could help identify the birth parent or parents. At the time of publication, the primary DNA testing companies are 23andMe, AncestryDNA, FamilyTreeDNA, and MyHeritage. In the future, more companies may exist, and successful outcomes for resolving unknown parentage may rely on testing with those companies as well. Regardless of how many companies enter the market, the overall goal will always be to test with as many relevant companies as possible so that key DNA matches are not overlooked. Many adoptees will have significant matches to close relatives; a recent study showed that nearly one-third of adoptees who took DNA tests in 2015 and 2016 matched to a relative who was a half first cousin or closer, and two-thirds of adoptees matched to a relative who was a half second cousin or closer.[2] The chances of matching to close relatives increases as more people take DNA tests.

2. Blaine T. Bettinger, "Adoptee Testing: A Study," *The Genetic Genealogist*, 8 January 2017 (https://thegeneticgenealogist.com/2017/01/08/adoptee-testing-a-study/).

The following DNA testing plan is recommended for adoptees:

1. Test with AncestryDNA (https://www.ancestry.com/dna/). Additionally, sign up for a subscription to Ancestry to view family trees for DNA matches.

2. Download the AncestryDNA raw data file and upload it to FamilyTreeDNA using the "Autosomal Transfer" feature (https://www.familytreedna.com/autosomal-transfer). Upgrade to allow access to tools. (Some recently processed AncestryDNA kits have raw data files that will not upload into FamilyTreeDNA's system. If this is the case, purchase a FamilyTreeDNA Family Finder test to achieve the same result. There is also a small fee to access some tools.)

3. Upload the AncestryDNA raw data file to MyHeritage (https://www.myheritage.com/dna).

4. Upload the AncestryDNA raw data file to *GEDmatch* (https://gedmatch.com/).

5. Test with 23andMe (https://www.23andme.com/). The test needed to work with DNA matches is the Ancestry-only option; however, the Health + Ancestry Service also provides the information needed.

6. For males trying to identify paternal roots, take a 37-marker Y chromosome STR test with FamilyTreeDNA (https://www.familytreedna.com/products/y-dna). If significant results appear at the 37-marker level (matches with a genetic distance of 0 or 1), upgrade to the 67-marker or 111-marker test.

This testing plan puts the individual with unknown parentage into all of the main autosomal DNA testing company databases for a few hundred dollars (pricing as of December 2018) and, for men, provides a test that can help identify the biological father's surname.3

Identifying Unknown Parentage

DNA match lists identify genetic relatives, and genealogical research helps connect those relatives to each other and to the test taker through shared ancestry. Resolving unknown parentage using DNA typically involves the following steps:

1. Separate DNA matches into "family clusters" and identify shared ancestry among them.

2. Predict relationships to DNA matches and develop hypotheses about the common ancestral couple.

3. For more information on transferring data between companies see Leah LaPerle Larkin, "What's New in Autosomal DNA Transfers," *The DNA Geek*, 12 Sep 2017 (http://thednageek.com/whats-new-in-autosomal-dna-transfers/). See also Leah LaPerle Larkin, "How to Transfer your AncestryDNA Test to other databases," *The DNA Geek*, 9 June 2018 (http://thednageek.com/how-to-transfer-your-ancestrydna-test-to-other-databases/).

3. Trace descendants of a hypothesized common ancestral couple in search of candidates for the birth parent.
4. Target test individuals that are more closely related to the hypothesized biological parent to confirm the identification.

This process will help separate DNA matches into distinct families, identify multiple ancestors of the individual with unknown parentage, and determine where the separate families converge, likely with the marriage of a descendant from one family to a descendant from another family. The individual with unknown parentage will fall within the descendants of that converging couple.

Family Clusters and Shared Ancestry

To resolve unknown parentage, the test taker's DNA matches need to be separated into groups of people who are related to each other—known as "family clusters." For example, each individual with unknown parentage has a maternal and paternal line—and that yet-to-be-identified mother and father each have their own paternal and maternal line, which comprises four distinct families that are probably not related to each other but are all related to the test taker. When considering the maternal and paternal lines for each of those four families, there are eight distinct family clusters related to the test taker, and so on.

DNA matches, especially those at the third and fourth cousin level, will fall into these different family clusters. One match might be related through a grandmother's paternal line and another through a great-great-grandfather's maternal line. In that instance, those DNA matches are not related to each other. It is possible to use one DNA match to hypothesize the generation where a common ancestral couple may lie, and trace downward on several of the match's ancestral lines to identify birth parent candidates. However, working with DNA to solve unknown parentage is easier if we can find several matches who are related to each other, then trace descendants of a common ancestral line they share.

Identifying family clusters

Separating the test taker's DNA matches into family clusters can be done in several different ways. Using any approach, the easy task is identifying family clusters; the difficult task is identifying noncommunicative DNA matches and their ancestors. Individuals who know the identity of one parent and are seeking to identify the other parent can eliminate a large number of their DNA matches from consideration by testing the known parent. If testing the known parent is not possible, testing a close relative on the known parent's side—preferably a half sibling (child of that parent), an aunt or uncle (sibling to that parent), or a cousin (niece or nephew to that parent)—is the next best option. Eliminating DNA matches on the known parent's side can also be accomplished by testing a more distant relative if no close relative can be tested—even a first cousin, aunt, or uncle of the known parent would help to eliminate some matches on known parent's lines. If no known relatives can be tested, having a well-documented family tree for the known parent is useful for identifying matches on that line and removing them from consideration. This process of elimination leaves what we will consider the "relevant" matches—those related to the parent whose identity is not known. For adoptees or those seeking to identify both parents, this process is not doable, but it is still possible to isolate DNA matches into family clusters.

Shared matches

Each of the main DNA testing company websites provides a tool that generates a list of relatives that the test taker and another DNA match both have on their match lists. This feature is known by different names depending on which database is being used:

- AncestryDNA: "shared matches"
- FamilyTreeDNA: "in common with"
- 23andMe: "relatives in common"
- MyHeritage: "shared DNA matches"
- GEDmatch: "people who match both kits, or 1 of 2 kits"

The "shared matches" or "in-common-with" feature is the easiest way to separate DNA matches into family clusters, and the closest relevant DNA match is an excellent place to start. Generating a list of relatives that are shared between the test taker and this match will help create the first family cluster. The primary focus should be on relatives that share the most DNA—usually up to and including predicted third and fourth cousins with the largest amounts of shared DNA. To create a different family cluster, use the next closest relevant DNA match that did not appear in the first family cluster created. Continue this process until all relevant DNA matches (through fourth cousins who share significant amounts of DNA) are placed into a family cluster. This typically needs to be done on a database-by-database basis; however, if any matches are in multiple databases, family clusters in which they appear can sometimes be combined.

Stop and consider the unlikely

In the majority of cases, shared DNA matches are people who are related to each other and to the test taker through a common ancestral couple. However, in some instances, test takers that appear on the match lists of two people could come from different ancestral lines. For example, Jim Brown and Joe Brown are first cousins—their fathers are brothers. If they have a shared match, Roberta, it may seem as though Roberta must be related to them through their shared Brown grandparents. However, that is not necessarily the case. Roberta's maternal second cousin is Jim's mother, Lena; and Roberta's paternal first cousin is Joe's mother, Ellie. In this case, Roberta would show as a shared match between Jim and Joe. However, the two men do not have a common ancestor with her—they both share ancestry with her on other lines.

Tina's parentage

Consider adoptee Tina's 163 centimorgan (cM) DNA match to Sophie. Since Tina is an adoptee just starting her search, we do not know whether her connection to Sophie is maternal or paternal. Sophie's family lines are well documented (the documentation is not shown here because of space restrictions), but without any additional context, the connection between the two women could be on any of Sophie's family lines shown in figure 5.1.

However, Tina's shared matches with Sophie include a 145 cM DNA match to Stanley. Stanley's lineage is added in figure 5.2. Sophie and Stanley's respective family trees show that their mothers were sisters, the daughters of Joseph and Tillie. This eliminates Sophie's paternal line from consideration, allowing the focus to remain on connections to either Joseph or Tillie. Tina's other shared DNA matches with Sophie or Stanley or both may be useful in determining whether her connection is through Joseph or Tillie's side, further narrowing down the possibilities.

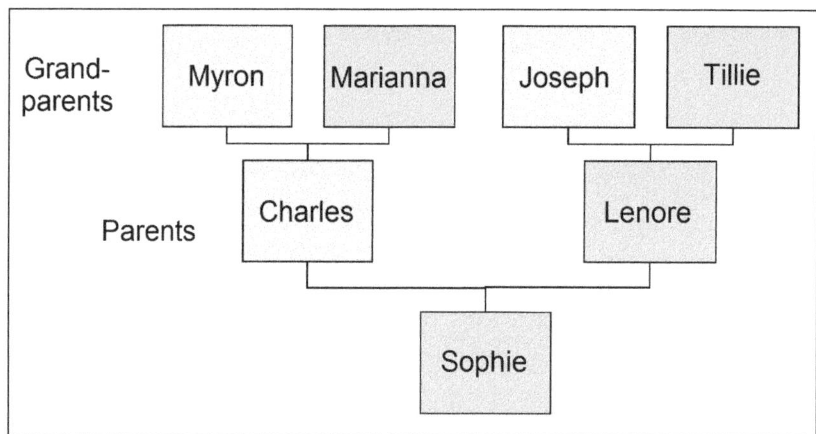

Figure 5.1. Sophie's Ancestry

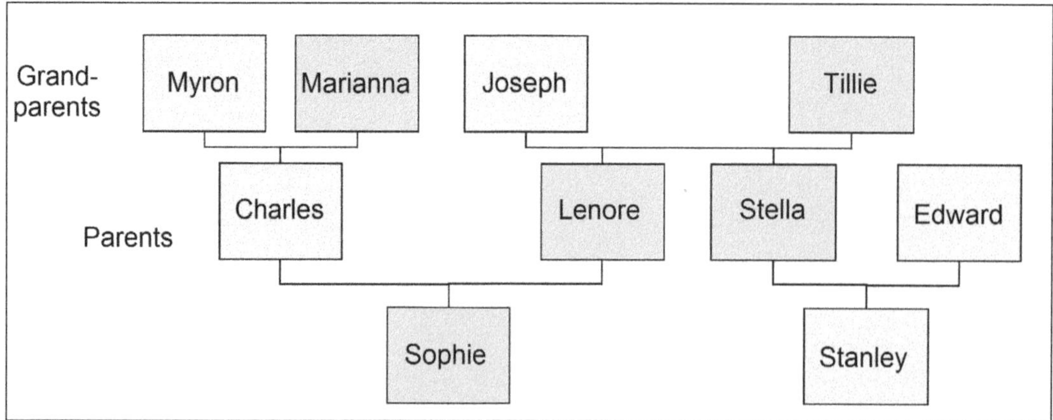

Figure 5.2. Sophie and Stanley's shared ancestry

Third-party tools GWorks and WATO

There are numerous third-party tools available that allow for different ways of viewing and working with DNA information, such as segment data, DNA matches, and trees of DNA matches. The most useful third-party tool for unknown parentage is GWorks,[4] which automates the process of identifying common ancestral couples among DNA matches. The "Compare All Trees" tool in GWorks generates a list of individuals found in the family trees of more than one match. Details about the matches are provided; it is important to confirm that these individuals have different lines of descent from the common ancestral couple, not simply three siblings that all tested and have the same family tree. If multiple DNA matches who have different lines of descent from an ancestor can be found, then these can be grouped into a family cluster, similar to the shared matches approach. The "What Are the Odds?" (WATO) tool on DNA Painter is also useful for DNA analysis in misattributed parentage cases.[5]

Other methodologies

Why use shared matches, rather than other methodologies, to create family clusters? The use of DNA for unknown parentage cases has evolved over the past several years, mostly as a result of the increasing sizes of the testing company databases. Early on, when the databases were small and the matches were more distant, techniques such as mirror trees and segment triangulation were more useful. However, at the time of this writing, the shared matches feature is the lowest hanging fruit and is the most useful for identifying unknown parentage.

Mirror trees are tools that can be used only on AncestryDNA, primarily to identify unknown parentage. The basic premise behind a mirror tree is that the tree "mirrors" a DNA match's tree, but identifies the individual with unknown parentage as the home person. The purpose of this is to determine who the individual with unknown parentage matches in the mirrored tree, and thus who a shared common ancestor may be with that DNA match—the same basic premise as looking at shared matches. This technique was more useful when the testing company databases were smaller and had fewer features; however, the utility of mirror trees has diminished over the past several years with the increased use of the shared matches features.

Another technique for creating family clusters is segment triangulation—when one or more DNA segments are shared among three or more individuals. The individual with unknown parentage can be one of these individuals, and the other two should be DNA matches who have a known common ancestral couple. Working under the same theory as the shared matches methodology, if two DNA matches share a common ancestor, then the individual

4. GWorks, *DNAGedcom* (http://dnagedcom.com/).

5. "What Are the Odds?" *DNA Painter* (https://dnapainter.com/tools/probability). For tips on using this tool, see also Ann Raymont, "What Are the Odds? New help in finding birth parents," *DNASleuth*, 1 August 2018 (https://dnasleuth.wordpress.com/2018/08/01/what-are-the-odds-new-help-in-finding-birth-parents/).

with unknown parentage is likely related to those DNA matches through that common ancestral line.

The use of segment triangulation for unknown parentage—although not likely to confuse DNA matches that appear as shared but are actually from different family lines—presents its own challenges and limitations. When considering three individuals who are DNA matches to each other, there is a smaller chance that two individuals would match a third in the same location than that those two individuals would each match the third in different locations.

In cases where endogamy, pedigree collapse, and intermarriage exist within families, identifying family clusters may take a multi-pronged approach using more than one of these different types of methodologies. However, for most unknown parentage research problems, shared matches will provide reliable data, which will ultimately be tested for accuracy in the targeted-testing phase, regardless of the initial methodology or approach used. The approach used to identify a candidate for a birth parent is less important than testing that hypothesis later on.

Identifying DNA matches and their ancestors

In some cases, identifying a DNA match is not difficult—many people post family trees or ancestral surnames and make it easy to discover several generations of their ancestors. Some others may not share a tree but are open to providing information, especially if shared DNA matches can help provide clues as to which family lines are of interest. However, many people have taken DNA tests as a result of advertising that touts ethnicity estimates and countries of origin; many of these individuals do not visit the testing company websites frequently and are not knowledgeable about DNA match lists and their features. Unfortunately, many of these individuals do not have family trees associated with their profiles and do not respond to messages.

When working with DNA matches, take notes or screenshots of their usernames, profiles, family trees, shared matches, and other identifying information. Since identifying unknown parentage depends so much on knowing the identities of DNA matches, it can be extremely detrimental when a DNA match or family tree disappears. Capture this information prior to contacting DNA matches, especially those that are closely related to the biological parent being sought.

Identifying information about noncommunicative DNA matches is challenging but possible. In most cases, the main goal is to identify the ancestry of the DNA match—if this can be done, communicating with the person may not be necessary. Be sure to check many different locations for the person's family tree instead of only browsing on Ancestry. Remember that trees can be found on the other testing company websites and on *GEDmatch*, on other family tree or family history websites such as Geni.com or MyHeritage. If a family tree cannot be found and contacting the DNA match is the goal, determine if the DNA match can be found in another testing company database. It may be possible to make contact using a different messaging platform, or the DNA match may have a FamilyTreeDNA or *GEDmatch* account with an email address. Typically, there are higher success rates when contacting DNA matches via email rather than through the testing company's messaging platforms. However, if contact cannot be made via the testing company websites and an email address cannot be found through other testing companies, some general research on the DNA match can generate results.

When contacting a match through Ancestry's message system there are two choices: the orange button or the green button. Rather than using the green "Send Message" button on the main page of a DNA match, click through to the match's profile by clicking on their name or username. On this page, use the orange "Contact Name" button (where Name is the person's Ancestry account name). Both buttons send a message to the user's Ancestry inbox; however, some users report that only the orange button will send an email notification to the person's email address on file with Ancestry.

Between genealogy websites, social media, digitized newspapers, and other sources, most people have left a trail of breadcrumbs on the internet that makes them easy to identify and contact. On AncestryDNA, a user's profile page can include a name or username and a photograph as well as the user's age group, education, field of employment, occupation, religion, research interests, and message board posts. Similar information can sometimes be found on other DNA testing company websites. If the DNA match's name is known, a simple Google search or a search on social media sites such as Facebook, LinkedIn, Instagram, Twitter, or other platforms may yield results sufficient to identify and contact the person. If the DNA match's name is more common, it may be useful to search for the name along with keywords such as genealogy, family tree, or ancestry. It can also be helpful to initiate a Google search for the person's name along with one or more of their ancestral surnames; this may result in finding message board posts, family trees, obituaries, and other material.

Sometimes a DNA match does not use his or her full name, but displays an Ancestry username. Many people use the same or similar usernames to log in and register for multiple websites. A general Google search for the DNA match's username can often result in finding profiles on general interest websites that may provide a person's age, residence, or other useful details. Likewise, many people use the first part of their email address as a username. It may be possible to reach the DNA match by adding @yahoo.com, @gmail.com, or @aol.com to their username. For example, the Ancestry user mjohnson1983 may have the email address mjohnson1983@gmail.com. For the initial inquiry, it is not necessary to include a large amount of identifying information, in case you have reached the wrong person—simply inquire about whether you have reached the person who took a DNA test and is listed as mjohnson1983.

Contacting the DNA matches that you share with a particular match of interest can also be helpful. These individuals may have had prior contact with the DNA match or may know his or her identity. It can also be helpful to know how much DNA other matches share with the DNA match of interest—this can help place the match more specifically within family clusters and narrow down his or her identity.

The purpose of these research techniques is to identify a means of contacting the DNA match and, as an alternative, learn enough about the DNA match, through publicly available records, to be able to document the person's ancestry and connect them to others in the family cluster. Identifying DNA matches and researching their ancestry can be challenging, but it is a necessary step toward hypothesizing the common ancestral couple and the first hurdle in the journey to identifying unknown parentage.

A word about online family trees

As we review family trees for our DNA matches, it can be easy to gather information on several generations of a DNA match's family simply by trusting their online family tree. In most cases, information in an online family tree will be accurate for at least the first two generations—the individual's parents and grandparents—based on the idea that most people living today can identify their parents and grandparents. For younger DNA matches, this may extend to grandparents and great-grandparents, as those ancestors may have lived in recent years. However, in order to draw accurate conclusions and reliably prove the unknown parentage, it may require some effort to verify that these trees are accurate and well documented. Keep in mind that family trees are subject to input errors, misattributed parentage, and other mistakes.

Predicting Relationships to DNA Matches

Centimorgan data is one of the most reliable pieces of evidence for problem solving using DNA. The single most important tools related to centimorgan data are several charts developed using information from The Shared cM Project, a crowd-sharing endeavor undertaken by Blaine T. Bettinger to collect centimorgan data for known relationships. The charts in figures 5.3 and 5.4 are extremely valuable for predicting relationships and visualizing the variety of possible relationships for different shared centimorgan values.[6]

In addition to these charts, DNA Painter[7] offers the "Shared cM tool," which provides relationship probabilities for centimorgan values based on the data collected as part of the Shared cM Project. Between these three tools, predicting relationships to DNA matches is simpler, although there are usually multiple relationship possibilities.

For individuals with unknown parentage, it is important to consider a variety of relationship possibilities, including half relationships, especially for close DNA matches who could be half siblings or half nieces and nephews. Considering context—about the adoptee, the DNA-match of interest's family and ancestry, and other aspects unique to each unknown parentage case—can help eliminate some relationships or at least render them less probable.

6. Blaine T. Bettinger, "The Shared cM Project — Version 3.0 (August 2017)," PDF report (https://thegeneticgenealogist.com/wp-content/uploads/2017/08/Shared_cM_Project_2017.pdf). Full size versions of the charts can be printed from the PDF file. The project charts are periodically updated.

7. *DNA Painter* (https://dnapainter.com/).

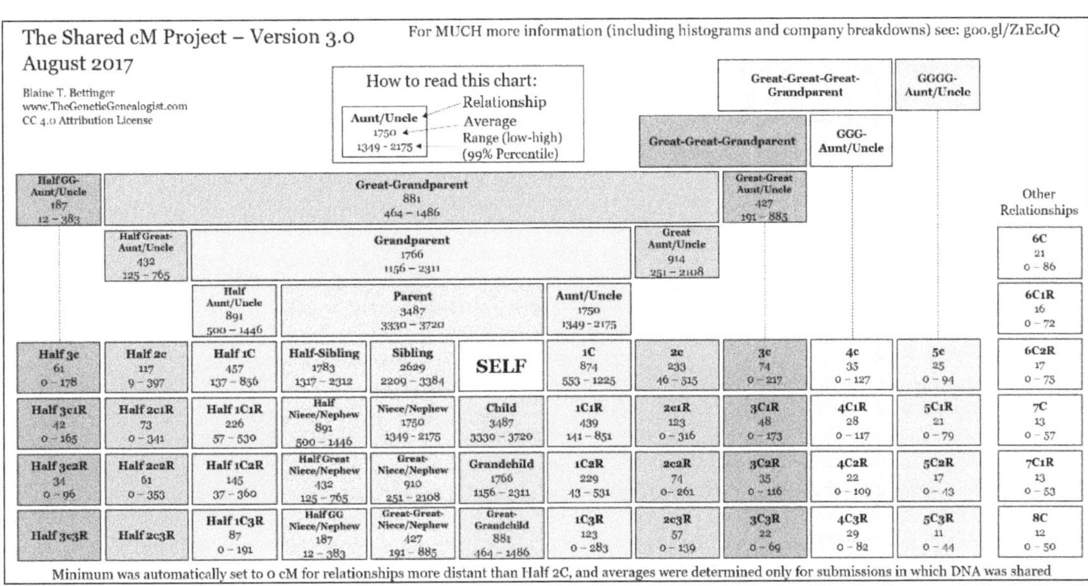

Figure 5.3. The Shared cM Project Chart (access PDF for full size image)

The Shared cM Project – Version 3.0
August 2017

Blaine T. Bettinger
www.TheGeneticGenealogist.com
CC 4.0 Attribution License

For MUCH more information (including histograms and company breakdowns) see: goo.gl/Z1EcJQ

Cluster	Relationships	Total #	Average	Range (95th Percentile)	Range (99th Percentile)	Expected
Cluster #1	Siblings	1345	2629	2342 - 2917	2209 – 3384	2550
Cluster #2	Half Sibling, Aunt/Uncle/Niece/Nephew, and Grandparent/Grandchild	2473	1760	1435 – 2083	1294 – 2230	1700
Cluster #3	1C, Half Aunt/Uncle/Niece/Nephew, Great-Grandparent/Great-Grandchild, and Great-Aunt/Uncle/Niece/Nephew	2261	884	619 – 1159	486 – 1761	850
Cluster #4	1C1R, Half 1C, Half Great-Aunt/Uncle/Niece/Nephew, and Great-Great Aunt/Uncle/Niece/Nephew	1842	440	235 – 665	131 – 851	425
Cluster #5	1C2R, Half 1C1R, 2C, and Half Great-Great-Aunt/Uncle/Niece/Nephew	2224	232	99 – 397	47 – 517	213
Cluster #6	1C3R, Half 1C2R, Half 2C, and 2C1R	2284	123	0 – 236	0 – 317	106
Cluster #7	Half 1C3R, Half 2C1R, 2C2R, and 3C	2492	75	0 – 158	0 – 229	53
Cluster #8	Half 2C2R, 2C3R, Half 3C, and 3C1R	1864	49	0 – 114	0 – 175	27
Cluster #9	Half 3C1R, 3C2R, and 4C	1528	36	0 – 84	0 – 122	13
Cluster #10	Half 3C2R, 3C3R, Half 4C, and 4C1R	1040	29	0 – 67	0 – 118	7

Figure 5.4. The Shared cM Project Cluster Chart

The idea behind predicting relationships is two-fold—first, it helps to identify the generation in which a common ancestral couple is likely to be found; and second, it helps to predict where the adoptee may fall within the family of the relevant DNA matches.

Revisiting adoptee Tina's DNA matches to first cousins Sophie (163 cM) and Stanley (145 cM), the charts in figures 5.3 and 5.4 and the DNA Painter "Shared cM tool" can be used to predict a variety of possible relationships between Tina and the Sophie-Stanley family. The centimorgan values fall closest to clusters 5 and 6 in figure 5.4, and encompass a variety of different types of relationships, including: first cousin three times removed, half first cousin twice removed, half second cousin, second cousin once removed, first cousin twice removed, half first cousin once removed, second cousin, and half great-great-aunt/uncle or niece/nephew.

In this unknown parentage case, the amount of shared DNA between Tina and Sophie and between Tina and Stanley is very similar, suggesting that Tina may have the same relationship to Sophie as she does to Stanley. Sophie was born in 1927 and Stanley was born in 1929; adoptee Tina was born in 1994. This context—the age differences—can factor into relationship predictions, sometimes making relationships impossible and sometimes making "removed" relationships (first cousin twice removed, second cousin three times removed, and so on) more likely.

Researching the DNA match's family as a whole (rather than just direct ancestral lines) can provide additional context that may drive the focus to some relationships that are more likely than others. Stanley and Sophie's grandparents, Joseph and Tillie, had no known children other than Lenore and Stella as seen in figure 5.5. Thus, there were no other grandchildren of Joseph and Tillie who were first cousins to both Stanley and Sophie—this makes it less likely that the first cousin twice removed and first cousin three times removed relationships apply to Tina and both Sophie and Stanley. However, Joseph was married previously and had two children, Anna and Anton, by his first wife. This context may result in focusing more on some of the half relationships—if Tina was one of Anton or Anna's descendants, she would be half first cousins, probably at least two or three times removed, to Sophie or Stanley.

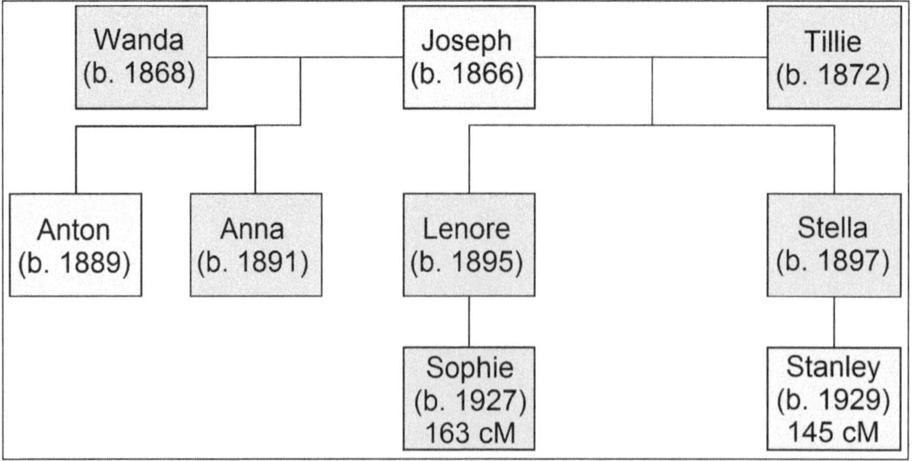

Figure 5.5. Joseph's first marriage

Predicting relationships also involves considering the absence of shared DNA. Sophie and Stanley's match lists include several descendants of Joseph's daughter Anna—these individuals do not match Tina. Given that they would be fairly close relatives to Tina if she was a descendant of Anton or Anna, the absence of shared DNA suggests that the half first cousin twice removed and half first cousin once removed relationships are less likely. Tina's lack of shared DNA to Anton and Anna's descendants also makes it less likely (but not impossible) that Tina is related to Sophie and Stanley through their grandfather Joseph's line. Given this, the second-cousin-level relationships on Tillie's side seem the most promising, even though there are other relationship possibilities.

All of the predicted relationships within clusters 5 and 6 are second cousin or closer, indicating that Tina's biological parent is likely a descendant of Stanley and Sophie's great-grandparents. Evidence suggests that the connection is not on Joseph's side of the family, so Tillie's parents, Walter and Maryanna, are the hypothesized common ancestral couple, based on the amount of shared DNA and context clues related to the family.

Figure 5.6 shows that Tillie had a sister and a brother who lived to adulthood—Veronica and Timothy. Grandchildren of Veronica and/or Timothy would be second cousins to Sophie and Stanley. Based on the birth years of Veronica and Timothy, one possibility is that Tina is one of their great-grandchildren. This seems probable given the amount of shared DNA with Stanley and Sophie (who in this scenario would be Tina's second cousins once removed).

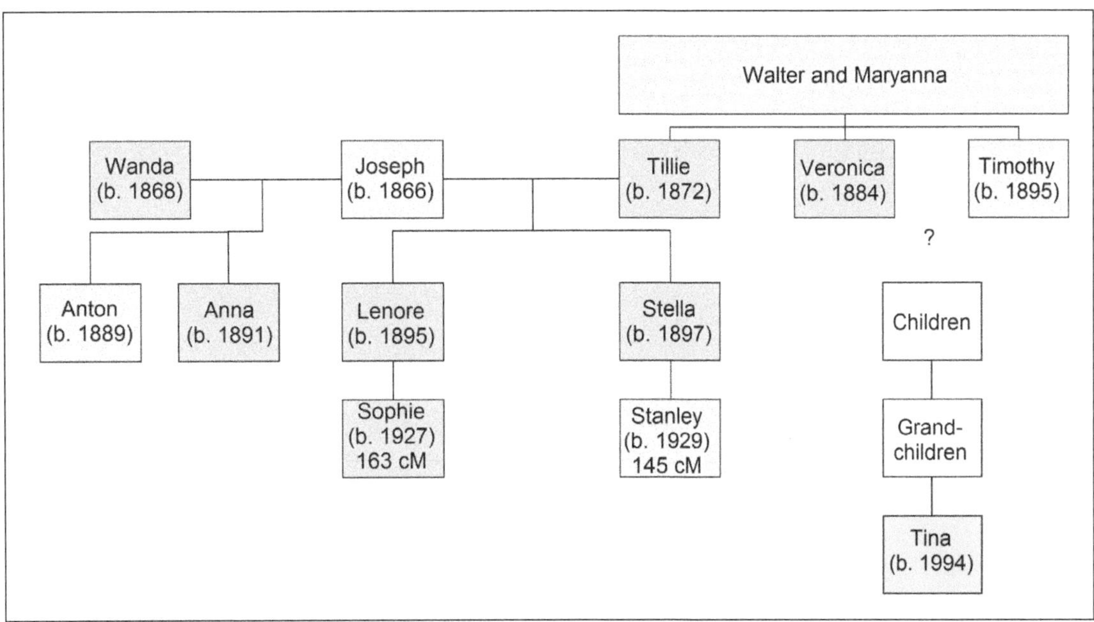

Figure 5.6. Tillie's parents and siblings

X-DNA and Y-DNA

The methodology used for identifying unknown parentage focuses heavily on ruling in and ruling out key DNA matches' ancestral lines. There are several additional approaches that can help accomplish this goal, including Y-DNA and mtDNA testing, and X-DNA analysis.

The use of X-DNA to rule in or rule out a DNA match's ancestral lines can be extremely valuable in unknown parentage research. The use of X-DNA is rooted in the ability to focus on ancestral lines through which X-DNA is inherited, while eliminating ancestral lines from which X-DNA could not be inherited. Since individuals with unknown parentage do not know their ancestry, the focus is on ruling in or ruling out a DNA match's ancestral lines. There are several important considerations to remember when using X-DNA. First, an X-DNA match must be a significant size to eliminate a DNA match's ancestral lines from consideration. A small 1.4 cM segment of shared DNA on the X chromosome likely does not indicate a true X-DNA match. The same segment size parameters should be used to evaluate X-DNA as are used to evaluate other autosomal DNA. Second, the absence of X-DNA, with the exception of some very close relationships, cannot be used as evidence to prove relationships on a specific ancestral line.

For males with unknown parentage, Y- DNA testing can sometimes help pinpoint a paternal line surname. When a test taker has meaningful Y-DNA matches, the results can suggest the birth father's surname. If autosomal DNA test results have honed in on a specific family, but it is difficult to determine whether that family belongs to the birth mother or the birth father, Y-DNA testing could help.

Estimated or predicted haplogroups

Y-DNA and mtDNA haplogroups, when available, are useful for identifying unknown parentage. 23andMe is an autosomal DNA testing service; however, the company provides test takers with their mtDNA haplogroups (all test takers) and Y-DNA haplogroups (males only). FamilyTreeDNA offers Y-DNA and mtDNA tests that provide haplogroups.

Many haplogroups are extremely common, so while a common haplogroup does not necessarily indicate a connection on the direct paternal or direct maternal line, a lack of a shared haplogroup can help rule out the possibility of a connection on a direct paternal or direct maternal line. However, the depth of testing on Y-DNA and mtDNA can result in different haplogroups assignments even between test takers who share a haplogroup when both have taken high-resolution tests. This requires understanding how sub-branches of the Y-DNA and mtDNA haplogroups may be assigned. For example, one test taker may be assigned to mtDNA haplogroup U5 after a low-resolution test and another assigned to haplogroup U5b1d1c after a high-resolution test. The U5b1d1c haplogroup is a sub-branch of U5. These two test takers may share a recent matrilineal ancestor that could be overlooked if this sub-branching is not understood or unless both test to the same level.

Tracing Descendants of a Hypothesized Ancestral Couple

Once a common ancestral couple is predicted, it is often necessary to trace descendants of the ancestral couple. While researching descendants, the main goals are to find the converging couple—where one family cluster connects to another—and to identify a candidate for one birth parent who was in the right place at the right time (and fits known information, if applicable).

Researching the living—or reverse genealogy—involves using many of the same sources as traditional genealogical research. Depending on when the hypothesized common ancestral couple lived, descendant research can be a large undertaking, involving the use of census records, wills, vital records, obituaries, and other sources to make connections between generations and document all of the couple's children, grandchildren, great-grandchildren, and so on. As the research trickles downward into the late twentieth century and into the present day, some different sources and tactics will be necessary.

When seeking out living individuals, background check websites—often known as "people search" websites—are extremely useful to ensure that the right person has been found (based on their connections to family members, and past and current residences) and to identify current contact information. Many of these websites offer some information for free, and more information can be obtained for a small fee. People search websites include

- AnyWho (https://www.anywho.com/)
- BeenVerified (https://www.beenverified.com/)
- Classmates (https://www.classmates.com/)
- InstantCheckmate (https://www.instantcheckmate.com/)
- Intelius (https://www.intelius.com/)
- MyLife (https://www.mylife.com/)
- PeekYou (https://www.peekyou.com/)
- PeopleFinders (https://www.peoplefinders.com/)
- Pipl (https://pipl.com/)
- Spokeo (https://www.spokeo.com/)
- Veromi (https://www.veromi.com/)
- Whitepages (https://www.whitepages.com/)
- ZabaSearch (https://www.zabasearch.com/)

Each of these websites receives information from different sources, such as tax records, magazine subscriptions, third-party mailing lists, and more. As a result, each provides different information. Search for individuals using several of these sites—do not limit the investigation to one or two or even three.

For living individuals, records such as deeds and mortgages can help identify current addresses. Do not skip searching state or county recordings if the descendant being sought is not believed to be a homeowner—there may be tax liens, power of attorney filings, court records, and other sources not related to home ownership. Additionally, search for local public records databases, such as those providing salaries of public employees or retirees, driver's license or vehicle registration information, voter registration details, and other types of information. Some states have databases of professional registrations for occupations such as nurses, doctors, engineers, insurance professionals, and other professions. These registries often include home addresses.

Search social media sites such as Facebook, Instagram, LinkedIn, and others, including "friend," "connection," and "follower" lists. These lists may identify other relatives or groups of individuals from a specific school, town, or employer that can help confirm or refute that the person is the correct individual. Read a person's social media profile completely, including pages, groups, education, employment, and other information. Be sure to read comments on posts and photographs; the comments can provide details about the person's affiliations and whereabouts both today and in the past, how they know the people they are communicating with, and other important details. When searching for a person, do not look only at their profile—for example, on Facebook, also search for Melissa Johnson comments; Melissa Johnson posts; and Melissa Johnson photos. Each of these searches will bring up different interactions that the person was involved in somehow, but not necessarily as the creator.

Digitized newspapers are an important part of descendant research. These sources can be used to identify individuals in obituaries and other articles. The most useful digitized newspaper collections are *GenealogyBank*[8] and *Newspapers.com,*[9] although there are many local libraries and historical societies that have digitized local newspaper collections. Newspaper obituaries are extremely useful for making links between generations and locating living descendants of an ancestral couple. For general Google searches, use name variants and add in search terms such as the individual's home town, current town, college, occupation, place of employment, hobbies, ancestral surnames, relatives' names, and other key terms.

As descendants are traced, it may become more clear how previously unidentified or unlinked DNA matches within a family cluster fit into the overall family. Additionally, if more than one family cluster was identified, descendant research will likely (and should) show how another family cluster ties in.

Targeted Testing

For unknown parentage research, the methodology used to identify a family of interest is not as important as target testing—seeking out specific individuals for DNA testing—within

8. *GenealogyBank* (http://www.genealogybank.com/).
9. *Newspapers.com* (http://www.newspapers.com/).

that family to confirm theories and prove parentage. The DNA of people who have already tested can provide enough information to hone in on a family of interest; however, DNA from specific descendants is often needed to identify the birth parent as a specific individual within that family.

Targeted testing should get as close as possible to the individual believed to be the birth parent—if not the birth mother or father themselves, then another child of the birth parent is the best option. Other close relatives, such as the birth parent's grandchildren, parents, or siblings can be used as alternatives. However, often when a family of interest is identified, the birth parent needs to be differentiated among several siblings who are candidates. If some of the siblings are deceased or not willing to test, their children can be tested to either rule in or rule out their parent.

Sometimes, the DNA of people who have already tested is not enough to hone in on a specific group of siblings, and a larger family unit must be considered. For example, revisit Tina's unknown parentage case illustrated in figure 5.7. The focus is on the family of Tillie, grandmother of key DNA matches Sophie and Stanley, and specifically on descendants of Tillie's sister Veronica and brother Timothy. Based on evidence related to the family, Tina is possibly a great-grandchild of either Veronica or Timothy. However, Veronica had seven children, and Timothy had nine children. Each of those children had between two and eight children, resulting in twenty-six grandchildren for Veronica and thirty-four grandchildren for Timothy—all candidates to be one of Tina's biological parents. One strategy for identifying which lines of descent may be of most interest is to look at the families of individuals who married into the focus family (such as Veronica's husband, Timothy's wife, or some of Veronica or Timothy's children's spouses). Tracing these spouses' lines back even two generations may result in identifying ancestors that appear in the family trees of Tina's DNA matches that are part of different family clusters.

Timothy's daughter Geraldine married Bob. Bob had a sister Laurie, whose great-grandson Garrison shares 105 cM of DNA with Tina. The marriage of Geraldine and Bob is where two separate families with whom Tina shares DNA come together, making it extremely likely that Tina is a descendant of Geraldine and Bob. In this case, Geraldine and Bob have three sons, suggesting that this is Tina's birth father's family. Targeted testing could be necessary in order to identify which son may be Tina's biological father.

Targeted testing is often the last and most difficult step toward confirming the identity of a birth parent because it involves reaching out to the birth family—either directly to the birth parent or to their close relatives—to establish contact and to ask for a DNA test. Asking strangers to take a DNA test can be challenging. Considerations when making contact with birth families and the ethical issues surrounding targeted DNA testing are covered in the "Uncovering Family Secrets: The Human Side of DNA Testing" and "Ethical Underpinnings of Genetic Genealogy" chapters.

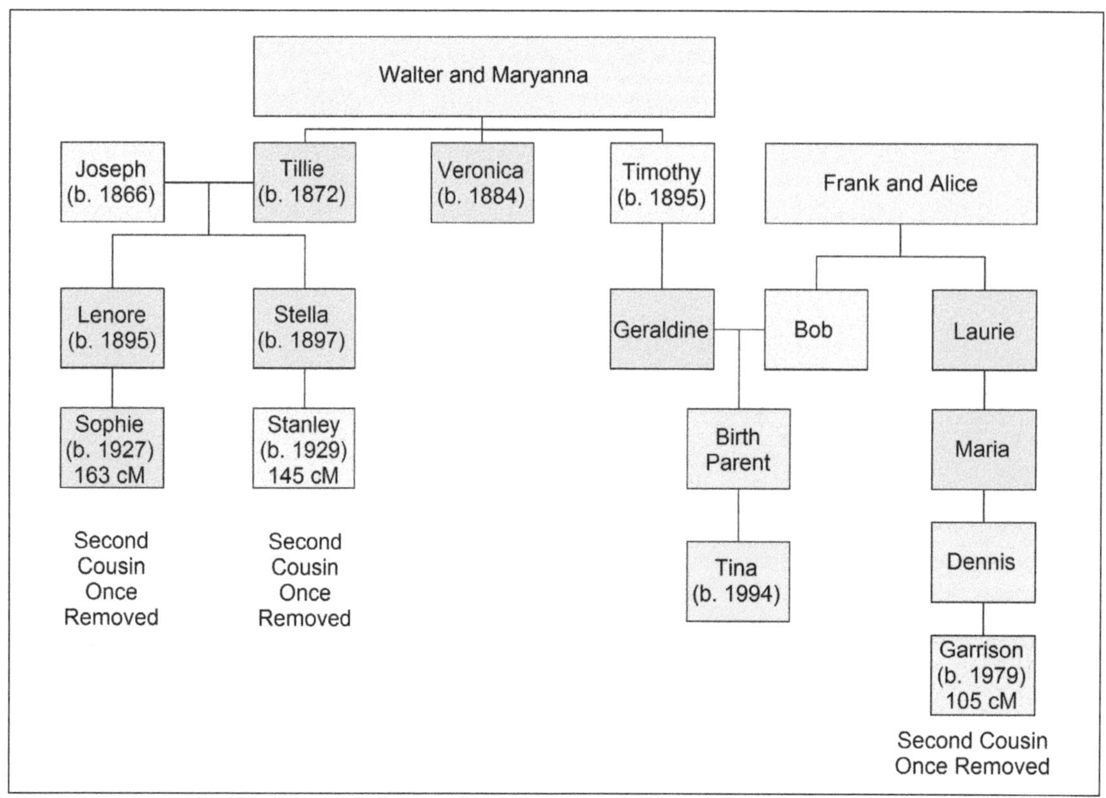

Figure 5.7. Tina's parentage

Conclusion

Unknown parentage research is a growing area and an important part of the field of genealogy. The research requires persistence and creativity, as each case has different information, quirks, and nuances. However, through careful analysis of both DNA and documentary evidence, individuals with unknown parentage can identify and connect with their birth families and learn about their histories.[10]

10. For more information on researching misattributed parentage and online training courses on methods and tools see *DNAAdoption* (http://dnaadoption.org/).

The Challenge of Endogamy and Pedigree Collapse

Kimberly T. Powell

Endogamy or Pedigree Collapse: What is the Difference?

From the Ancient Greek *endon* (in, within) and *gamos* (marriage, matrimony), *endogamy* is the practice of individuals marrying within the same community or group over a long period encompassing numerous generations. This genetic isolation eventually results in many, if not all, members of the community being distantly related to each other, often in multiple ways. Endogamous groups might be ethnic, cultural, or religious (e.g., Acadians, Ashkenazi Jews, and Low-German Mennonites). Endogamy may also develop for economic, social, or geographic reasons. Island populations, such as those in Hawaii or Newfoundland, and some rural pre-twentieth-century American communities fall into this latter category.

Pedigree collapse is a related but separate concept. Pedigree collapse occurs when an individual descends from related parents. The parents might share a close relationship such as first cousins, or the connection may be many generations back in their tree. The result is a smaller or "collapsed" pedigree; some individuals appear more than once, reducing the total number of unique ancestors. Over time all endogamous populations develop some degree of pedigree collapse, but collapse can appear in non-endogamous pedigrees as well. The colored blocks in figure 6.1 represent ancestors duplicated within a pedigree.

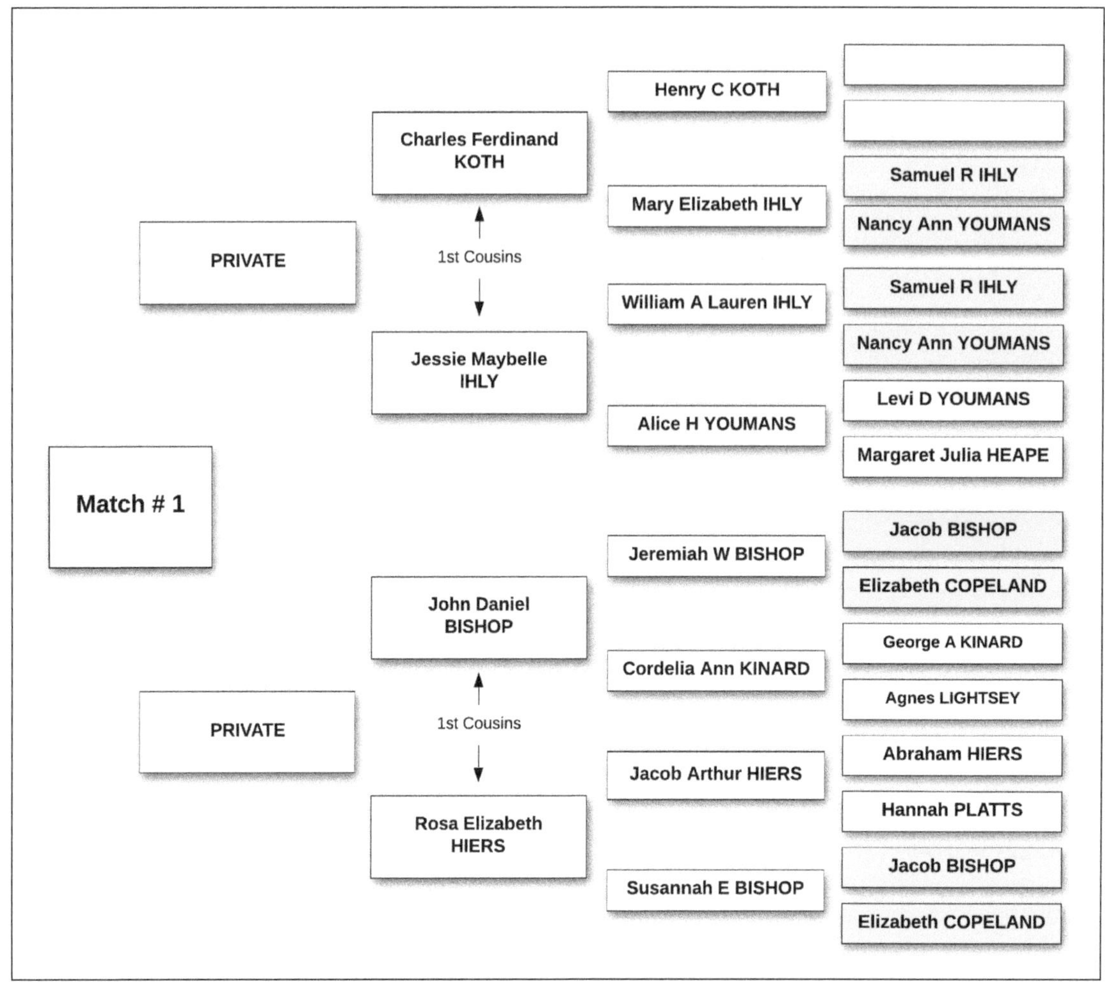

Figure 6.1. Pedigree collapse—this individual descends from twelve (12) great-great-grandparents, rather than the expected sixteen (16), due to descent from two sets of married first cousins.

What it Means for DNA Analysis

Endogamy and pedigree collapse affect our DNA pool for many reasons. Individuals descended from the same isolated genetic pool will frequently share multiple ancestors in common. Pedigree collapse, in both endogamous and non-endogamous situations, implies that the individual descends from the same ancestral couple more than once. Both of these scenarios complicate DNA analysis.

DNA matches may predict as closer than they are

While the amount of autosomal DNA inherited from any single ancestor is approximate due to the random nature of recombination, an individual generally inherits about half of their DNA from each parent, 25% from each grandparent, 12.5% from each great-grandparent, and so on, with each generation inheriting approximately half the amount of the previous generation.

Following this inheritance pattern, two second cousins will typically inherit 12.5% of their common great-grandparent's DNA. However, due to recombination, they will each inherit a *different* 12.5%, so only about 3.125% of their DNA will be in common.[1] These percentages, however, assume that each ancestor appears only once in the given pedigree.

Pedigree collapse, whether due to endogamy or not, can skew these expected percentages. An ancestor who appears more than once in an individual's pedigree contributes DNA through multiple lines, generally increasing the amount of DNA inherited by the descendant above the typical expected amount. While an individual descending through independent ancestral lines typically inherits about 1.56% of their third-great-grandparent's DNA, if the individual's parents are third cousins through that same distant grandparent, the individual instead inherits approximately 3.13% of that third-great-grandparent's DNA—1.56% through each parent. This roughly doubles the amount of expected shared DNA with a fourth cousin descendant of that same common ancestor (from 0.2% to 0.4%), making it appear as if the relationship is closer than it is—nearer third cousin once removed than fourth cousin.

Individuals descended from endogamous populations face a more significant challenge. The complications of pedigree collapse are magnified as genetic matches may not only descend from the same ancestor multiple times but also frequently share multiple common ancestors. Thus, the estimated genetic relationship with an individual may be inflated (appear closer than it is) due to numerous sources contributing to the DNA shared by the two matches. While complicating relationship predictions, the increased level of shared DNA does have the benefit of increasing the chance that autosomal DNA from more distant ancestors has survived recombination in descendants to the present day.

Proving shared connections is often more complicated

A collapsed pedigree also means there is more than one path back to the shared ancestor, complicating the genealogy research necessary to prove the shared connection. This is especially true when common ancestors appear on the test taker's maternal *and* paternal lines as shown in figure 6.2. The individual inherits a higher percentage of DNA from ancestors in common with both of their parents, increasing the difficulty of determining exact relationships with genetic matches.

All URLs were accessed on 25 January 2019 unless otherwise noted.

1. Leah LaPerle Larkin, "The Limits of Predicting Relationships Using DNA," *The DNA Geek Blog* (https://thednageek.com/the-limits-of-predicting-relationships-using-dna/).

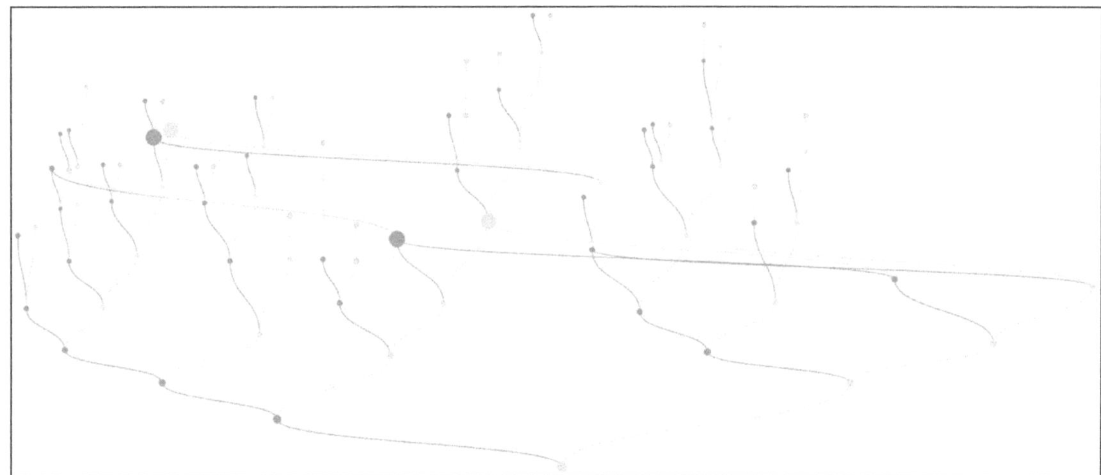

Figure 6.2. This individual's parents are related to each other through at least two sets of shared ancestral couples. Tree created with the Exploring Family Trees visualization tool (https://learnforeverlearn.com/ancestors/) by B.F. Lyon.

Shared or "In Common With" (ICW) matches can be more deceiving

Shared or ICW matches can be a powerful tool for identifying patterns of shared ancestry, including narrowing down which branch may be the source of the connection. When dealing with endogamous population, these types of connections, while still useful, must be approached with greater skepticism, especially if both individuals have a significant number of matches from that same population. This is especially true when close relatives such as parents and grandparents are from the same endogamous community.

It is also possible that two individuals both descend from the same endogamous population on one or more lines but share DNA from another line entirely. For example, Marshall, a match on 23andme, is identified as a predicted fifth cousin "on the paternal side" to an individual whose paternal grandmother descends from the same endogamous community as Marshall. The "paternal" prediction made by 23andme is likely because twenty-seven of the test taker's top thirty "shared" matches are from the paternal side, which is known because the test taker's father also tested at 23andme. However, on closer examination, none of these "shared matches," including the test taker's father, share DNA on the 15.5 centimorgan (cM) segment that the test taker shares with Marshall. The significant size of the shared segment combined with a lack of paternal matches strongly suggests a connection through some other shared ancestry on the *maternal* side.

Table 6.1 shows matches that Marshall shares with the test taker and the test taker's close relatives. While Marshall shares numerous DNA segments with the test taker's paternal relatives, including 90 cM across two segments with the paternal grandmother, his match with the test taker is a 15.5 cM segment on chromosome 10, which is not shared with any of the paternal relatives but instead matches the test taker's mother and maternal aunt. The distant connection is on the mother's paternal side, a different, somewhat endogamous community with many members who migrated to and settled in the area where the paternal ancestors lived around the turn of the nineteenth century. This is one reason why testing as many close relatives as possible is important!

Table 6.1. DNA segments shared by match Marshall with multiple relatives of the test taker

Match	Chr	Start Point	End Point	cM
Paternal Cousin 1	3	48.1	63.3	15.8
Paternal Great Aunt	3	57.2	64.4	11.6
Paternal Grandmother	6	23.5	106.3	65.6
Paternal Cousin 2	7	13.4	31.1	25.7
Paternal Cousin 3	10	8.2	12.9	8.8
Mother	10	57.9	72.3	15.5
Maternal Aunt	10	57.9	72.6	16.5
Paternal Cousin 4	12	6.1	20.4	24.4
Paternal Grandmother	12	6.3	20.4	24.4
Paternal Great-Aunt	19	13.3	38.7	22.4
Paternal Cousin 3	20	35.5	41.2	7.2

Triangulation and clustering are complicated

Triangulation can be complicated when two people who married are closely related, primarily when you reach the stage of determining the ancestor or ancestral lines shared by members of the group. The lack of a match on a specific paternal segment between two full siblings, for example, indicates that one sibling inherited that segment from the paternal grandfather, while the other sibling received it from the paternal grandmother. When those grandparents are themselves related, however, these *paternal* segments could still ostensibly trace back to a single ancestral couple—although not the same ancestor due to lack of shared DNA—with one sibling receiving the segment from the male half of the pair and the other receiving it from the female. Significant endogamy, especially if it is present on all lines, can even make it challenging to establish overlapping triangulated groups as maternal or paternal.

Test takers with related parents or grandparents may find it difficult or impossible to cluster matches into groups using techniques such as the Leeds Method[2] shown in figure 6.3, or tools like DNADNA,[3] the Collins' Leeds Method 3D tool in the DNAGedcom Client,[4] and Genetic Affairs.[5]

2. Dana Leeds, "DNA Color Clustering: The Leeds Method for Easily Visualizing Matches," *DNA with DANA LEEDS* (http://www.danaleeds.com/).
3. *DNADNA* (https://www.dnadna.uk/).
4. *DNAGedcom* (http://dnagedcom.com/).
5. Evert-Jan Blom, *Genetic Affairs* (http://geneticaffairs.com/).

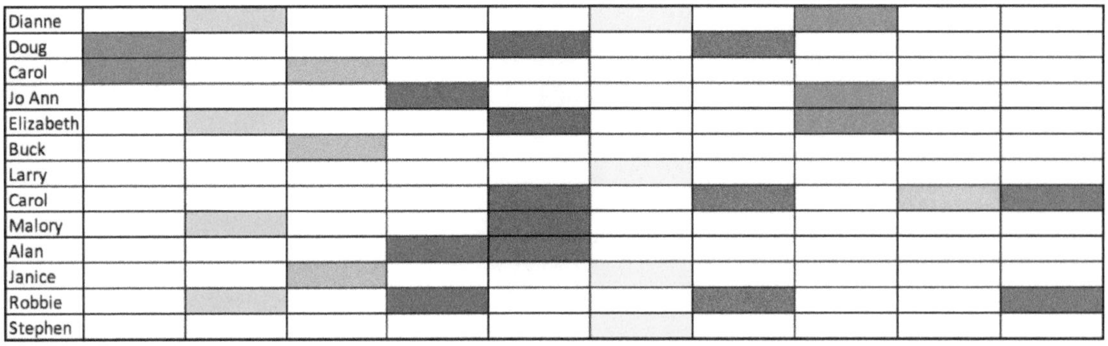

Dianne										
Doug										
Carol										
Jo Ann										
Elizabeth										
Buck										
Larry										
Carol										
Malory										
Alan										
Janice										
Robbie										
Stephen										

Figure 6.3. Recent cousin marriages within an endogamous community complicate cluster coloring of AncestryDNA matches using the Leeds Method.

Surnames can be a distraction and a detriment

Matches from endogamous communities often share multiple surnames in common. Some of these surnames may belong to common ancestors, who may or may not have provided the shared DNA. Others may be prevalent in the community—the surnames of collateral relatives and neighbors who could be potential, yet-to-be-identified ancestors, or just a case of distracting coincidence. These familiar surnames often increase the work needed to find common ancestry and may also lead research in the wrong direction. One positive, however, is that individuals endogamous on only one side or one line can often sort matches likely belonging to this line based on these similar surnames, even if they are not surnames in common.

On the other hand, a lack of shared surnames between matches is also a common occurrence in individuals dealing with endogamy, especially those with endogamy on only one line. Autosomal DNA from distant ancestors is more likely to persist in descendants from endogamous populations, increasing the likelihood that the connection is further back than either individual has researched. Individuals who see no surnames in common with a match may be less likely to engage or share information.

Some endogamous populations present a more significant challenge for genealogists than others, based on factors including degree of intermarriage within the population, number of generations impacted, and genetic isolation of the population. However, the same basic approaches generally apply.

Evaluate Shared DNA

Overcoming the challenge of inaccurate relationship predictions due to endogamy takes work and time. For this reason, the most straightforward approach is often to assume such relationships are at least a half or full step further than predicted unless a more precise calculation is necessary (for example, to support proof of descent through a specific common ancestor by ruling out the possibility of other genetic connections). Because calculating the expected percentage of shared DNA between two individuals who share descent through multiple lines requires knowledge of all (or as many as possible) relationships between the two individuals, this is also often the only practical approach.

In cases where an accurate projected relationship is needed, in-depth analysis of the exact amount of shared DNA between two matches may be useful.

Check for runs of homozygosity

In cases where an individual (either the test taker or a match) descends from parents from the same endogamous population, it can be useful to begin by checking to see if they have a "run of homozygosity" or a region of DNA where the maternal copy is identical to the paternal copy. If a genetic cousin overlaps in this same region of DNA, then the percentage of DNA the two share in common is effectively doubled for that particular segment, as they match on both the maternal and paternal copy.

Runs of homozygosity (ROH) are not uncommon in individuals from endogamous populations where spouses often share some type of cousin relationship. It can, therefore, be worthwhile to investigate the possibility of ROH not only for the test taker, but also for matches. In cases where a match shares more than the expected amount of DNA this is especially important.

If the individual has DNA uploaded to *GEDmatch*,[6] the "Are your parents related?" tool is an easy way to check for runs of homozygosity. The blue bar in figure 6.4 represents a homozygous region of DNA. David Pike also offers a number of utilities which compute regions of homozygosity.[7] Access to unzipped autosomal DNA raw data files from 23andme or FamilyTreeDNA is necessary to use Pike's tools.

Chr	Start Location	End Location	Centimorgans (cM)	SNPs
13	94009539	107031597	24.9	4622

Figure 6.4. A run of homozygosity in the "Are your parents related" tool at *GEDmatch*.[8]

6. *GEDmatch*® (http://www.gedmatch.com/). As of January 2019, *GEDmatch* Classic calls this the 'One-to-one' compare tool, while *GEDmatch* Genesis calls it the One-to-One Autosomal DNA Comparison tool. This refers to a tool that compares one selected test taker's genome to that of another selected test taker.

7. David Pike, "Search for Runs of Homozygosity (ROHs)," Autosomal Utilities (https://www.math.mun.ca/~dapike/FF23utils).

8. "Are your parents related," *GEDmatch: Tools for DNA and Genealogy Research* (https://www.gedmatch.com/).

Calculate expected shared DNA percentage for *each unique relationship*

One method of determining the expected shared DNA between individuals related through multiple lines is to total the expected shared DNA percentage, or expected total shared cM, for each distinct relationship between the two individuals. The "distinct relationship" is critical in endogamous situations as descendants of married cousins are expected to inherit double the DNA from that common ancestral couple, half through each of the married cousins. Thus, when an individual descending from two married cousins has a match to someone descended from that same ancestral couple, there are two possible relationships to account for—one through the paternal cousin (B) and one through the maternal cousin (A). This is shown in figure 6.5.

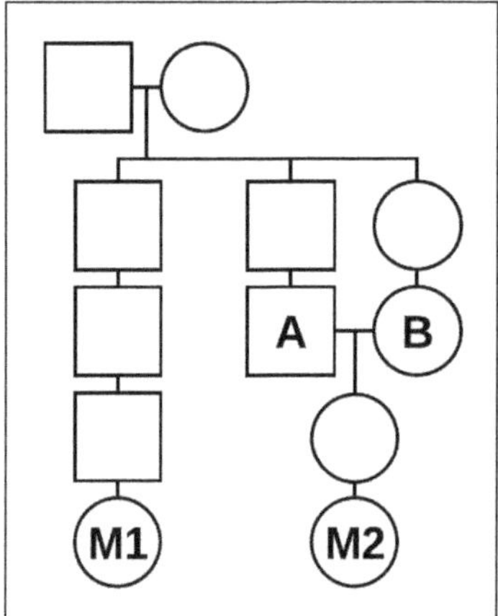

Figure 6.5. Match One (M1) and Match Two (M2) share two possible genetic relationships from the common ancestral couple—one through each of the married first cousins A and B on M2's line.

Figure 6.6 represents a fairly typical situation in a family tree affected by both pedigree collapse and endogamy. The test taker and her predicted second to third cousin match share DNA on multiple lines. The closest traced relationship between the two matches, however, is third cousin once removed, which means they are expected to share about 48 cM (range 0–173 cM).[9] However, pedigree collapse combined with the endogamous nature of the community results in a higher actual shared amount of 205 cM, which most closely predicts a relationship of second cousin or second cousin once removed.

Because Jacob Bishop and Elizabeth Copeland appear twice in both the test taker's tree and the match's tree, the expected percentage of shared DNA and expected amount of total shared cM is calculated for each of the four possible relationship combinations involving that ancestral couple as demonstrated in table 6.2. The relationship due to common descent from George Kinard and Agnes Lightsey is also included.

9. Blaine T. Bettinger, "Shared cM Project," *The Genetic Genealogist* (https://thegeneticgenealogist.com/). Search the blog posts for the most recent update to the project such as the August 2017 update at https://thegeneticgenealogist.com/2017/08/26/august-2017-update-to-the-shared-cm-project/. The project charts are periodically updated.

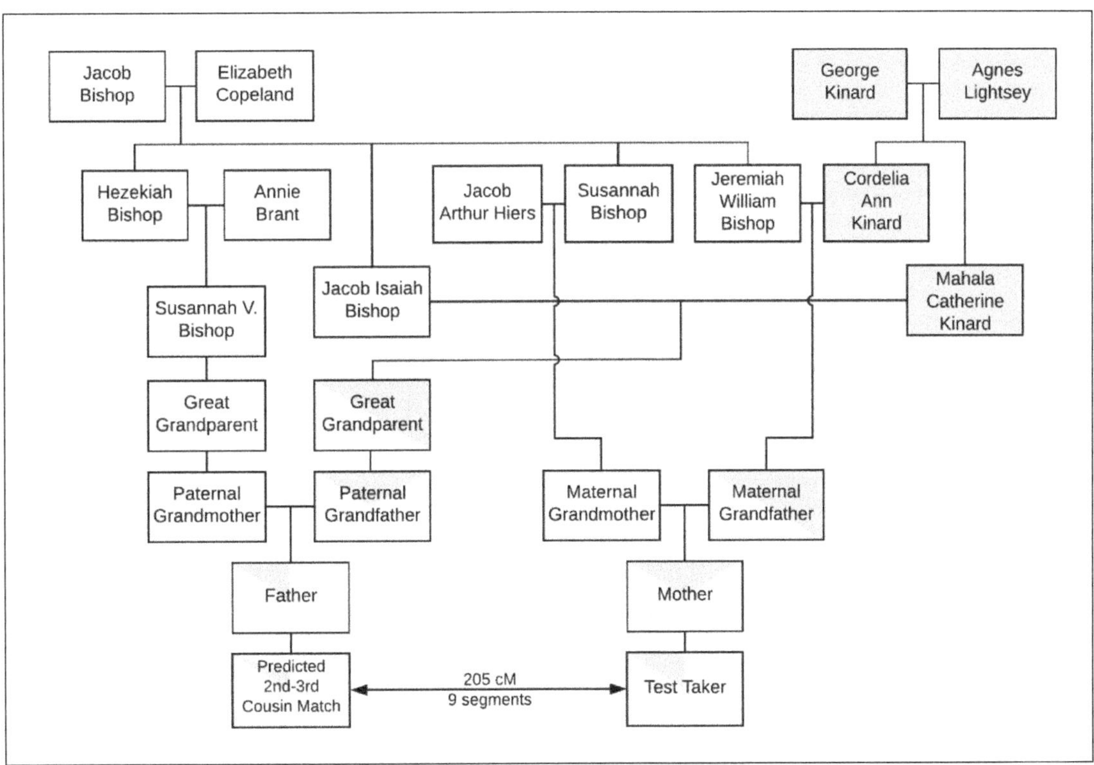

Figure 6.6. The presence of common ancestors on multiple lines increases expected DNA shared between two matches.

Expected DNA percentages come from *The ISOGG Wiki*.[10] Average cM are taken from the Shared cM Project by Blaine T. Bettinger.[11] Statistical average cM could be used with slightly different results. It is important to remember that these calculations only predict the possible *expected* shared DNA between two individuals, rather than the actual amount shared.

Combining the expected shared DNAs for all relationships seen in figure 6.6 results in a total of 1.56% of shared DNA (about 116 cM). This expected shared DNA calculation is not as high as the 205 cM they actually share but is very close at second cousin once removed. Using the average cM from the Shared cM Project results in a higher total of 214 cM. This is in the same range as the actual shared DNA. The answer may lie somewhere between the expected shared DNA calculation and the total cM determined from the Shared cM Project as there is at least one more shared common ancestral couple one generation further back in the family tree who may or may not contribute to their shared DNA.

10. "Autosomal DNA Statistics," *ISOGG Wiki* (https://isogg.org/wiki/Autosomal_DNA_statistics#Table).
11. Blaine T. Bettinger, "August 2017 Update to the Shared cM Project," *The Genetic Genealogist.*

Table 6.2. Calculating shared DNA for two matches with multiple common ancestors				
		Test taker		
		Jacob Bishop and Elizabeth Copeland (child: Susannah)	Jacob Bishop and Elizabeth Copeland (child: Jeremiah)	George Kinard and Agnes Lightsey (child: Cordelia)
Match	Jacob Bishop and Elizabeth Copeland (child: Hezekiah)	3C2R 0.195% shared DNA (35 cM)	3C2R 0.195% shared DNA (35 cM)	
	Jacob Bishop and Elizabeth Copeland (child: Jacob)	3C1R 0.391% shared DNA (48 cM)	3C1R 0.391% shared DNA (48 cM)	
	George Kinard and Agnes Lightsey (child: Mahala)			3C1R 0.391% shared DNA (48 cM)

Calculate the coefficient of relationship

Another method for calculating the expected amount of DNA shared between two individuals is a measure commonly used by geneticists called the "coefficient of relationship."[12] The final coefficient ranges from 0 (unrelated) to 1 (identical twins); the higher the number, the greater the amount of shared common ancestry between the two individuals.

$$(R_{XY}) = \Sigma(1/2)^n$$

The coefficient of relationship R between two relatives X and Y is calculated by raising ½ to the "n" power, where n equals the total number of degrees of relationship, or generational distance, separating two individuals. If the individuals share multiple relationships, the coefficient is summed (Σ) for each independent relationship path $(1/2)^n$. This includes separate paths through each of the two ancestors who make up a common ancestral couple.

To determine the coefficient of relationship, first identify all of the most recent common ancestors of X and Y. Then count the generational distance or degree of relationship by counting the steps up the family tree from X to the common ancestor and then back down the tree to Y. The total number of steps up the tree and then down again is the generational distance. Assume, for example, that X and Y are half first cousins, sharing a single grandparent in common. Starting at X, climb up two generations to reach the common ancestor and then descend two generations back down the tree to get to Y. Therefore, the degree of relationship, or genetic distance, is 2 + 2 = 4.

12. F. M. Lancaster, "Calculation of the Coefficient of Relationship," *Genetic and Quantitative Aspects of Genealogy* (http://www.genetic-genealogy.co.uk/Toc115570135.html).

Having found the generational distance between X and Y through their common ancestor, next calculate $(1/2)^n$ where n is the generational distance between X and Y. This determines the part of their relatedness (coefficient of relationship) for which that ancestor is responsible.

If X and Y share more than one common ancestor, such as full first cousins, count the steps (n) needed to determine the coefficient of relationship for each possible relationship path, then add the coefficients together. Because full first cousins share two grandparents in common, for example, the relatedness of the cousins from each grandparent is calculated separately, and the two paths are then added together $(1/2)^4 + (1/2)^4 = 1/16 + 1/16 = 1/8$.

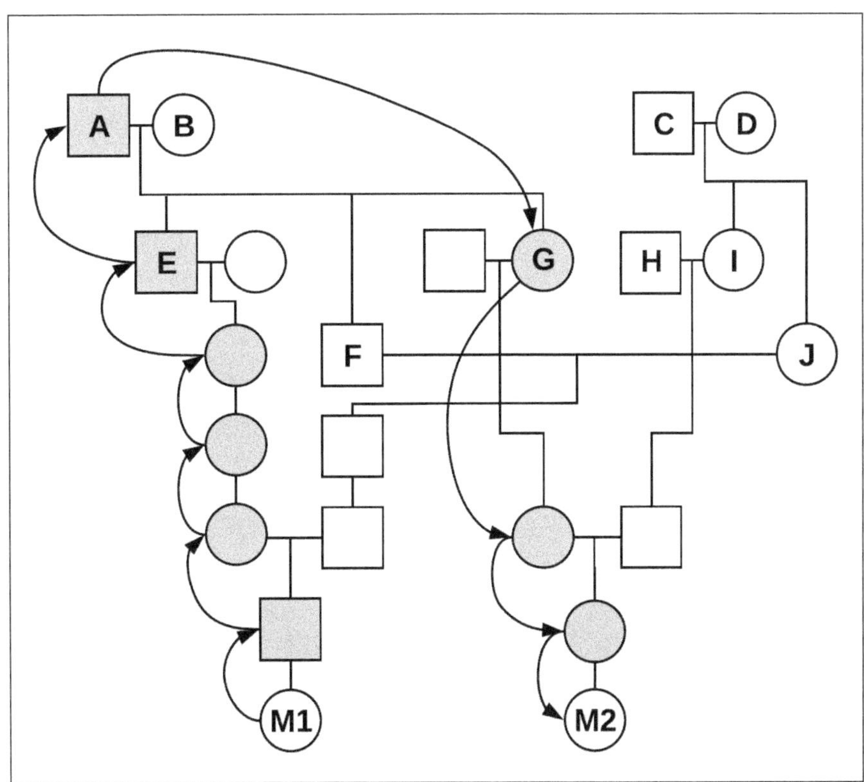

Figure 6.7. Each arrow in this pedigree represents a step, or degree of relationship between Match One (M1) and Match Two (M2) through common ancestor A along one of ten possible relationship paths (detailed in the text) connecting these two individuals.

To calculate the degrees of relationship (n), trace the path from one cousin back to the common ancestor and then forward to the other cousin, counting the steps or connections as you go. Each path must be independent; no individual can appear in the same path more than once. Figure 6.7, for example, depicts a ten-step relationship, resulting in a coefficient of relationship of 1/1024, or 0.097%.

$$(1/2)^{10} = 1/1024$$

However, that is only one of ten distinct relationship paths between these two individuals:

1. through E, A, and G (10) $(1/2)^{10} = 1/1024$
2. through E, B, and G (10) $(1/2)^{10} = 1/1024$
3. through F, A, and G (9) $(1/2)^{9} = 1/512$
4. through F, B, and G (9) $(1/2)^{9} = 1/512$
5. through E, A, and H (10) $(1/2)^{9} = 1/512$
6. through E, B, and H (10) $(1/2)^{9} = 1/512$
7. through F, A, and H (9) $(1/2)^{9} = 1/512$
8. through F, B, and H (9) $(1/2)^{9} = 1/512$
9. through J, C, and I (9) $(1/2)^{9} = 1/512$
10. through J, D, and I (9) $(1/2)^{9} = 1/512$

Σ (Sum) = 9/512 = 0.017578 = 1.76%

A coefficient of relationship of 1.76% converts to about 131 cM; this is in the same second-cousin-once-removed range as the predicted shared DNA using the first method, and much higher than the shared DNA expected with the closest traced relationship (48 cM).[13]

Maximize Coverage

Test as many people as you can. Other relatives, including siblings, will have inherited different segments of DNA from shared ancestors and have different matches as a result. Many DNA matches, especially in cases of pedigree collapse and endogamy, will be distant matches on a single segment. In such cases it is not at all uncommon for one sibling to inherit the full segment and other siblings to inherit no part of it. Even cousins who are related through multiple lines should be tested; these results can be used to further filter matches.

Kits of close relatives, including aunts and uncles, nieces and nephews of non-tested siblings, and first or second cousins, are useful for sorting out maternal/paternal matches when parents and grandparents are not available to test. Sibling results can be invaluable, as siblings inherit different pieces of a mutual ancestor's DNA—which often proves crucial when trying to evaluate against other matches, as well as for mapping grandparent DNA and visual phasing.

Testing close relatives also amplifies the coverage of shared ancestors' DNA in genetic databases, increasing the chance of locating additional matches and improving the odds of mapping segments to specific ancestors. In the case of a close match, each segment that you

13. Larkin, "The Limits of Predicting Relationships Using DNA," *The DNA Geek Blog.* Jonny Perl, "The Shared cM Project 3.0 tool v4," *DNA Painter* (https://dnapainter.com/tools/sharedcmv4).

can assign to a distant ancestor or ancestral couple will leave fewer possibilities that you need to consider for other close relatives who do not match on that same segment.

It is essential, however, to access or create family trees for each tested relative who is not in your direct family line—aunts, uncles, nieces, nephews, cousins. This will help ensure that there are no additional independent common ancestors on the non-shared side of the family that could impact DNA segment analysis. Creating these family trees is especially important in endogamous communities where shared ancestors a few generations back on the in-law side are not uncommon.

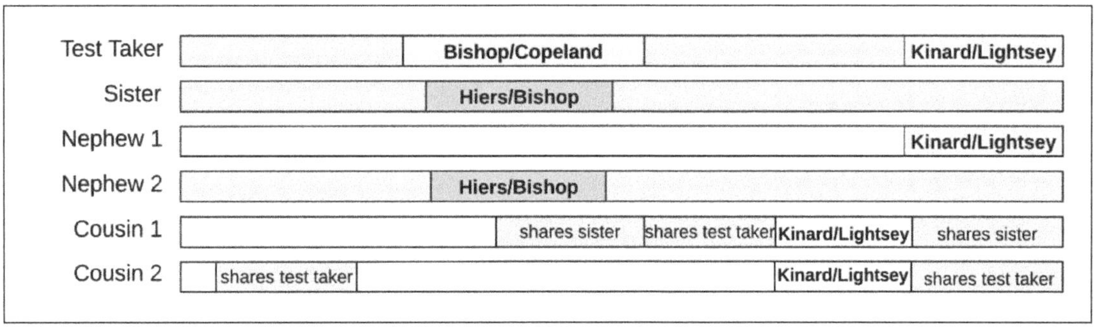

Test Taker		Bishop/Copeland			Kinard/Lightsey
Sister		Hiers/Bishop			
Nephew 1					Kinard/Lightsey
Nephew 2		Hiers/Bishop			
Cousin 1		shares sister	shares test taker	Kinard/Lightsey	shares sister
Cousin 2	shares test taker			Kinard/Lightsey	shares test taker

Figure 6.8. Identified segments on chromosome 13 of close relatives on the test taker's maternal line help to eliminate several MRCA possibilities. Grey shading indicates additional shared DNA on test taker's maternal line.

Figure 6.8 demonstrates two of several possible ways in which identification of segments for close relatives can help to rule out possible scenarios for the Test Taker.

1. In scenarios where an ancestral couple occupies more than one box on the pedigree chart, identification of overlapping segments for multiple relatives may help to eliminate possibilities. Due to pedigree collapse, the segment of chromosome 13 that the Test Taker received from Jacob Bishop and Elizabeth Copeland could have been inherited through either of her maternal grandparents. However, a previously identified segment in this same location passed down from Jacob Hiers and Susannah Bishop on the maternal grandmother's line to the Test Taker's Sister and Nephew 1 (son of one of the Test Taker's brothers), along with the fact that neither the Sister nor Nephew 1 shares DNA on this segment with the Test Taker, tells us that this particular Bishop/Copeland segment had to come from the Test Taker's maternal grandfather.

2. Two first cousins once removed, Cousin 1 and Cousin 2, have a confirmed match from a shared ancestral couple—George Kinard and Agnes Lightsey—between positions 98 Mbp and 106 Mbp on chromosome 13. The cousins (full siblings) likely share only paternal DNA with the Test Taker and her close relatives; the mother of the cousins is of French-Canadian descent and her ancestors were not part of the Test Taker's endogamous community. Because the cousins are unlikely to share

> maternal DNA with the Test Taker, her Sister, Nephew 1, or Nephew 2 in this location, this eliminates George Kinard and Agnes Lightsey as the source of the DNA that the Test Taker did inherit on this portion of chromosome 13. Also, because the ends of the cousins' segments align neatly with the start of the next segment of chromosome 13 inherited by the Test Taker and Nephew 1 from the same ancestral couple, this indicates a possible crossover point for the Test Taker at this point, from maternal grandmother to maternal grandfather.

When possible and with permission, get DNA test results for each relative in as many genetic databases as possible. This increases the chance of connections with crucial matches and maximizes access to valuable analysis tools. Contact close matches, especially those who tested with AncestryDNA or FamilyTreeDNA, and encourage them to transfer their results to either *GEDmatch* or MyHeritage, which both offer segment and triangulation data. While FamilyTreeDNA does not provide triangulation data, it is still a good option for AncestryDNA matches who have concerns about uploading into a public database such as *GEDmatch*; FamilyTreeDNA offers free transfers and a chromosome browser.

Evaluate Segment by Segment

Each segment of DNA comes from a single ancestor. In individuals with multiple identical ancestors due to pedigree collapse or endogamy, each shared segment still only comes from a single ancestral position on the pedigree chart. For this reason, a focus on analysis and triangulation of individual segments is essential when dealing with endogamy.

While the total amount of shared DNA and number of shared DNA segments are both typically increased by endogamy, the length of shared segments is usually not similarly affected.[14] There may, however, be an increased chance of both large and small segments surviving recombination from a distant common ancestor who occupies multiple positions in the pedigree; each line of descent from that common ancestor offers an additional opportunity for a specific segment to be passed down, either intact or in part, to the next generation.

Incorporate results from multiple services

The ability to combine, organize, and sort DNA segment data across numerous profiles and services is essential for dealing with endogamy. The tool of choice varies by user. Many use a tool created for this purpose called Genome Mate Pro, seen in figure 6.9.[15] Some create combined spreadsheets using files downloaded through the *DNAGedcom* Client. Other *DNAGedcom* tools—KWorks, JWorks, and the Autosomal DNA Segment Analyzer (ADSA)—can also be used with DNA from multiple services to identify overlapping segments among a group of individuals and to assign ICW status.[16]

14. Jim Bartlett, "Endogamy PART I," *segment-ology* (https://segmentology.org/2015/12/02/endogamy-i/).
15. Rebecca Walker, *Genome Mate Pro* (https://www.getgmp.com).
16. *DNAGedcom* (http://dnagedcom.com/).

Begin with the largest segments

Because the vast majority of genetic matches will match on only a single segment, one of the best ways to approach sorting matches into groups is to begin with the largest segments. These are statistically more likely to be closer relatives than other individuals who match on a single smaller segment—and can often be key to anchoring triangulated groups. Sorting by segment size is available for *GEDmatch* and FamilyTreeDNA ("longest block"), as well as in Genome Mate Pro (by profile and chromosome).

Chr	Start	End	cMs ⌄	SNPs	Graphic of Base Pair Start and End Position
4	86.3	146.7	49.1	6,447	
4	86.3	146.7	49.1	6,447	
4	23.4	70.0	42.8	4,908	
4	86.3	139.3	41.2	9,435	
4	150.9	183.3	39.0	4,437	
4	90.4	139.8	37.4	3,785	
4	86.3	134.5	36.6	5,318	
4	96.6	140.7	36.2	8,600	
4	86.3	127.6	31.9	7,580	
4	109.4	147.0	31.2	6,844	
4	110.1	146.4	31.1	7,000	
4	65.9	100.5	29.9	7,057	

Figure 6.9. A group of matches (excluding targeted test takers) in Genome Mate Pro sorted by largest segment.

Create Triangulated Groups

Triangulation, covered in detail in the "Lessons Learned from Triangulating a Genome" chapter, is the process of assembling and confirming a group of three or more people who share DNA in common with all others in the group along the same or overlapping segment on a particular chromosome. As long as this segment is of significant size, typically 15 cM or higher to be safe, the shared DNA indicates that all members of the triangulated group may have inherited that segment from a shared common ancestor. If the subject descends from the same common ancestor more than once, this particular segment originates from only one of those boxes in the pedigree chart. Thus, segment triangulation is often necessary for sorting out the tangle of shared common ancestors in an endogamous population.

A fully triangulated group requires three things:

1. At least two other people who are preferably not close relatives (not siblings, aunts/uncles, first cousins) who all match the test taker on the same chromosomal segment
2. That all individuals also match each other on the same segment
3. That the individuals share a common ancestor or ancestral line

Because there are two copies of each chromosome—one maternal and one paternal—two triangulated groups can exist for each segment, or partially overlapping segment, of each chromosome.

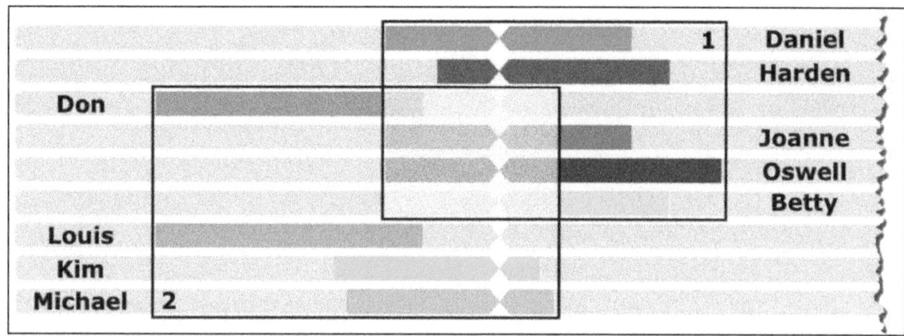

Figure 6.10. Two possible triangulated groups of FamilyTreeDNA matches on chromosome 7.

The two groups of matches in figure 6.10 do appear to triangulate. Using the FamilyTreeDNA matrix tool, five individuals in group one all share DNA in common with each other and not with the individuals in group two. The reverse is also true. However, FamilyTreeDNA does not provide segment-level detail of how DNA matches match each other, so it is unknown whether these individuals are "in common with" each other on this particular segment; their "shared" DNA could be on a completely different segment. As these segments are all about 20 cM or greater, so unlikely to be IBS, *and* there are two overlapping, distinct groups—one paternal and the other maternal—it is very likely that these are indeed triangulated groups, even if the common ancestor is not yet known.

Determine paternal and maternal sides for overlapping groups

Once you collect enough matches on a particular chromosomal segment, they will begin to resolve into two overlapping groups, one from the father's side and the other from the mother's side, as in figure 6.10. Because the individuals in Group 2, especially Kim and Michael, all overlap the individuals in Group 1, but do not share any DNA in common, this must mean that one group is maternal and the other paternal.

Close relatives—parents, aunts, uncles, and first and second cousins—can often make quick work of determining on which side a match might connect, but this is not always an option, especially for older test takers. This approach is also not always conclusive in cases where the test takers' parents are related.

Significant endogamy, especially if it is present on all lines, can complicate the process of determining maternal and paternal groupings. However, in cases where the parents are not significantly related to each other, shared or ICW matches can usually help to resolve which group belongs to which side.

The majority of the matches in figure 6.10 have not tested or uploaded anywhere other than FamilyTreeDNA, making it difficult to confirm parental groupings or triangulation. However, a first cousin once removed match on AncestryDNA agreed to upload her DNA to other services and provide access to her results. Mary's segment (not shown in figure 6.10) overlaps the entire Group 1 on FamilyTreeDNA and is in common with all of them. Because Mary is a paternal cousin with no known ancestry in common from her maternal side within five generations, this makes it likely that Group 1 is paternal and Group 2 is maternal.

Confirm triangulation

Contact each match individually when confirming triangulation using FamilyTreeDNA to ask if they also match the other identified group members on that particular segment. Alternatively, additional matches can be explored through companies which do provide triangulated segment data—including *GEDmatch*, 23andme, and MyHeritage.

Joanne is the only member of Group 1 also at *GEDmatch* and two legs do not make a triangle. However, a one-to-one comparison with cousin Mary, who agreed to upload to *GEDmatch,* shows that Joanne and Mary match each other on the 21.5 cM segment that both Joanne and Mary share with the test taker. The triangulation would be stronger with at least one additional member less closely related to the test taker, however.

In Group 2, both Don and Louis are also in *GEDmatch*, where one-to-one comparisons of the test taker and both matches indicate that all three are in common with each other on this approximately 45 cM segment of chromosome 7. Kim and Michael, who overlap on a portion of this same segment, are also at *GEDmatch* and are in common with each of the others, as well as each other. Don and Louis are the only ones in this group who share more than this single segment of DNA in common, so this forms a strong triangulated group.

Expand the group

With both paternal and maternal groups verified, the next step is to expand the number of group members. Focus first on matches with larger or overlapping segments, who share multiple segments in common with the test taker, or whose connection is known.

A potential match (Ken) on MyHeritage ticks off several of these boxes for Group 2—a large segment (56.7 cM) and a closer match (121.8 cM total shared DNA). Research identifies him as a fourth cousin once removed to the test taker. Both descend from George Bishop—one generation back from Jacob Bishop in the pedigree shown in figure 6.1. The expected average amount of shared DNA for this relationship is only 28 cM, however.[17] Even doubled to account for George Bishop occupying two positions in the test taker's pedigree, this is only half of the shared DNA amount, so it is possible there are other unknown relationships to account for.

Paternal cousin Mary is also at MyHeritage and shares no DNA in common with Ken; thus, Ken is accurately attached to the maternal triangulated group. Don from the maternal group at FamilyTreeDNA also agreed to upload DNA to MyHeritage and shares DNA with both Ken and the test taker on this segment.

17. Blaine T. Bettinger, "August 2017 Update to the Shared cM Project," *The Genetic Genealogist.*

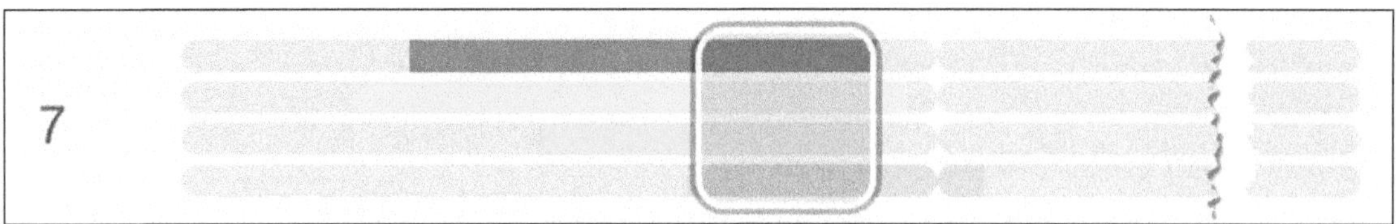

Figure 6.11. Triangulated matches in the chromosome browser at MyHeritage.

The quickest way to identify additional matches for this new triangulated group at MyHeritage is to use the DNAGedcom client to download the MyHeritage chromosome browser file, then sort by chromosome and start value to find individuals who share DNA on this same segment. Then cross-reference those results with the shared matches between previously identified members of the triangulation group (Ken, Don, and the test taker) to locate additional individuals who also triangulate on this segment using the MyHeritage chromosome browser as illustrated in figure 6.11.

Add additional members to the triangulated group following this same general method, including matches from as many testing companies as possible. Use your method of choice to identify additional individuals who overlap on this segment; options include Genome Mate Pro, a spreadsheet, "Matching Segments" in *GEDmatch* Tier One, or a chromosome browser file downloaded through the DNAGedcom client. Then confirm those who match the others in the identified triangulated group using tools such as *GEDmatch* One-to-One Comparison, "Shared DNA" at 23andme, or triangulated segments in the Chromosome Browser—One to many at MyHeritage.

The test taker in this example does not have a kit at 23andme, but her son does. Any of his matches that share DNA with this maternal triangulated group should represent at least a portion of the segment inherited by his mother from the common ancestor; thus, they are also included in the triangulated group shown in figure 6.12. This approach could be problematic if the son's father was from the same endogamous community, but in this case his paternal line is Eastern European and easily distinguished based on shared matches and ethnicity.

Chr 7	Kit Locations	Start	End	cMs	0 20 40 60 80
Test Taker	A/F/G/M	0.1	80.2	94.6	
George	23 (son)	2.9	30.9	46.7	
Christopher	23 (son)	11.1	18.9	20.1	
Marvin	23 (son)	11.3	18.9	12.6	
Ken	M	13.5	57.4	56.7	
Louis	F/G	18.0	54.7	45.3	
Don	A/F/G/M	18.0	54.8	21.5	
Laurie	F/G	18.7	35.6	22.9	
Janene	M	20.3	48.5	37.4	
Jerry	A/G	21.9	43.3	27.5	
Sherry	M	22.2	44.1	28.6	
Major	M	25.8	45.4	25.3	
Mary	M	28.8	55.1	31.2	
Kim	F/G	40.7	66.4	18.5	
Michael	F/G	42.3	67.0	16.5	
Jennifer	M	40.3	63.5	17.9	

Figure 6.12. Combined set of staggered, overlapping triangulated matches compared with the segment inherited by the test taker from one of her maternal grandparents.

Narrow Down the Common Ancestor

Once you have a strong triangulated group and a hypothesis of which line might lead to the common ancestor, it is time to dig deep into your triangulated matches, exploring not only their relationship with the test taker but also their connections with each other. Some of them may also share DNA from this same common ancestor on other segments—segments which may not have passed down to the test taker—and other triangulated matches on those segments may hold the key to the common ancestor.

1. *How much DNA do each of the individuals who triangulate on this segment share with each other?* Follow up on any close connections to determine how they are related. In the Bishop example, Don and Louis share 201 cM in common. Although neither knew how they were related to the other, research identifies them as second cousins once removed from a common Bishop ancestor, providing additional evidence that the common ancestor may be on the test taker's Bishop line.

2. *Identify individuals in the triangulated group who share more than one segment with the test taker (if any).* Their connections may be more recent, making it easier to identify the possible common ancestral couple or at least narrow down the line. Be careful, however, of those with multiple ancestors from the same endogamous community, as this increases the odds that they share DNA with the test taker through more than one ancestor.

3. *Explore individuals who share DNA with both the test taker and each match being investigated.* Include those who do not share DNA on the same segment, using tools such as "People who match both, or 1 of 2 kits" at GEDmatch, "Shared DNA Matches" at MyHeritage, "In Common With" at FamilyTreeDNA, and "Relatives in Common" at 23andMe.

4. Delve into shared matches or "cluster groups" associated with any members of the triangulated group who also tested at AncestryDNA. This assumes that your test taker has a kit at AncestryDNA. Genome Mate Pro simplifies the process of searching for AncestryDNA matches by name, partial name, nickname, common surnames, etc. For hard-to-locate individuals, narrow the search using shared matches of known individuals who triangulate on that same segment. If you are not positive that the AncestryDNA match is the same individual, contact them and ask.

5. Keep an eye out for individuals in the triangulated group who share DNA on the X chromosome, even when that is not the chromosome being investigated. The limited inheritance pattern of the X chromosome can be particularly useful for identifying, or at least narrowing down, the possible candidates for the shared DNA.

6. *Bring in Y-DNA or mtDNA, when appropriate.* The possible Bishop triangulated group includes two direct-line male Bishops who have no known connection with the test taker's Bishop line, or with each other. When compared with Y-DNA results from relatives on the test taker's Bishop line, the matches' Y-DNA results could confirm or disprove a Bishop connection. It would not, however, rule out other possible sources of shared DNA, including a female ancestor somewhere further back on that Bishop line.

7. *Proceed with caution when using shared matches.* Some of these shared matches may actually share by way of a different line than the one they each share with the test taker, especially when working with endogamous communities. This approach generally works best for individuals who match on a single segment and are known to have either never been a part of the endogamous community or who left the community very early. The group of shared matches should also be fairly small.

8. *Focus on matches whose traced ancestry does not appear to intersect with the endogamous community.* This approach is not always possible, especially for highly endogamous communities, but can be useful for early American colonial communities, for example. In the triangulated Bishop group, several of the matches with confirmed Bishop ancestors have no traced ancestry in the southern United States where the test taker's earliest known ancestor lived prior to the Revolution. The Bishop connection, if that is indeed the source of the common ancestry, likely traces back to at least early New England or, more likely, Europe.

Look at Trees as Well as Triangulations

Since endogamy and pedigree collapse increase the likelihood of matches being related through multiple family lines, it is important to explore all potential sources of shared DNA. Identification of a common ancestor is not enough to prove that the shared DNA comes from that ancestor. Even when a common ancestor is confirmed as the source of a particular DNA segment, additional work may be needed to verify the line of descent if that ancestor appears more than once in the pedigree. This generally necessitates either obtaining or building fairly complete family trees for focal matches.

The reality is that many matches either do not provide trees online or have only a few names in their online tree. It is also common that a match's profile may not provide enough information to even identify who they are or who the kit belongs to. Reach out to matches directly, when possible. If not a full family tree, they may be able to provide at least enough information to jumpstart the process of building their family tree or confirm that the connection is on their maternal or paternal line.

In situations where a match either does not know much about their ancestry or does not respond to your message, there may be no other option than to do the research yourself. Building family trees from scratch, especially when the genetic connection may be five or more generations back in the tree, can be incredibly time-consuming. The following strategies may speed up the process of identifying matches and narrowing down the potential common ancestral line.

- Search the internet by name, username, and email address (if available) as you may find online activity that provides additional clues to identity. If the match has a username that is not a first name or last name, or the first part of their email is not a name (for example, "grandmakate" instead of grandmakate@xyz.com), search for this username in quotes.
- Seek out on family-focused news such as obituaries, weddings, and reunions for any possible mention of parents and other close relatives.
- Look for matches on social media; public posts or pictures may lead you to additional relatives or help to confirm that you are following the correct family.
- Search out published family histories and family reunion lists for your ancestral lines, especially the ones focused on tracing all descendants of a couple (for example, Charles Lawrence Bishop, compiler, *The Bishop Family, 1785–1985: Jacob and Elizabeth Copeland Bishop and Their Descendants* (West Columbia, South Carolina: Professional Publishing Company, 1985).
- If the username or email is associated with a Find A Grave user, pay special attention to memorials they manage but did not contribute themselves; these may be relatives.

- If the match provides surnames but not trees, look for those surnames in combination with each other and identified locations using search engines, online newspapers, and genealogy databases.
- If the match is on *GEDmatch*, run a "One-to-many" analysis. If the individual administers multiple kits under the same email address, identify which kits do and do not match the test taker. This may help narrow down which of the match's parent or grandparent lines to focus on first. If you do not know who the match is, the closest matches may help or may have trees online.
- Use AncestryDNA Member Search to search for individuals who match the test taker at one of the other DNA testing services, especially those who did not also test at AncestryDNA. Do they have a public Ancestry Member Tree?
- Use available DNA analysis tools, such as "People who match both, or 1 of 2 kits" or "Matching Segment Search" at *GEDmatch* Genesis, to help narrow down the lines on which to focus when building family trees. Does the match have other close matches in the database? If so, do they also match you? Taking time to identify how close matches of your matches (both those who match the test taker and those who do not) are related to them means much less time spent chasing down ancestors who do not connect with the focus individual.

Map Your Chromosome

Chromosome mapping is the identification of which ancestor contributed each segment of your DNA. While the goal is to eventually assign segments of DNA to distant ancestors or ancestral couples, knowing which recent ancestors passed down the segments you inherited can also be very valuable, even when working in endogamous communities. For example, identifying which grandparent contributed each of the segments you inherited from your parents can be useful in narrowing down which family line you have in common with the shared matches on those segments.

If the test taker has both a paternal grandparent and a maternal grandparent also tested, it is possible, through the process of elimination, to identify the segments shared with each of their four grandparents. This is because any significant segment of DNA that the test taker shares with their grandparent was inherited from that grandparent. By process of elimination, any DNA not shared with that grandparent on the same side (paternal or maternal), has to come from the other grandparent on that side. It is impossible to have inherited a specific segment of DNA from both.

Most test takers are not fortunate enough to have living grandparents to test, however. This is where techniques such as the one discussed in the "Visual Phasing Methodology and Techniques" chapter come into play. Visual phasing is the process of assigning the DNA of three tested siblings—or sometimes two siblings plus a combination of other relatives—to each of their four grandparents using identified recombination points between the tested individuals.

The strategy of chromosome mapping to at least the grandparent level becomes even more useful when you also map the chromosomes of as many close relatives as possible. Close relatives will inherit different pieces of a mutual ancestor's DNA. Knowing which segments each relative has inherited from their shared ancestors, and which they have not, can be invaluable in helping to narrow down potential common ancestors with shared matches.

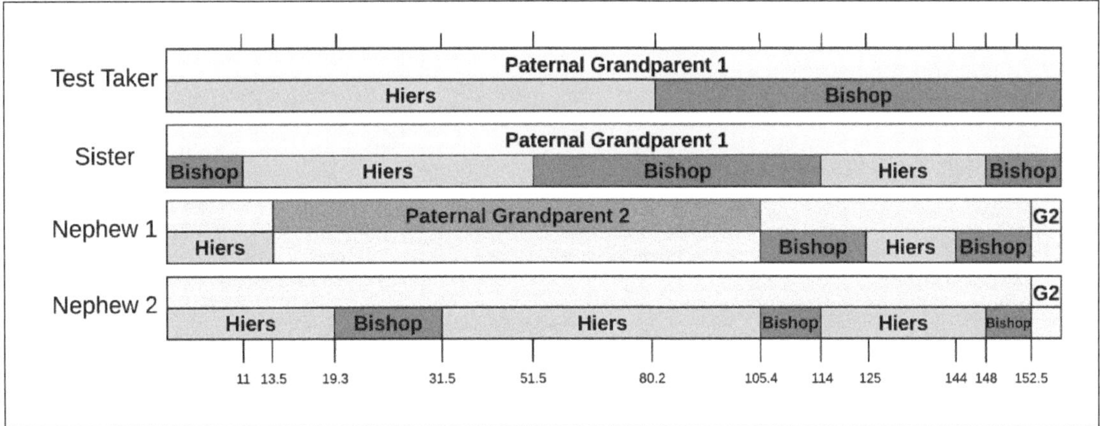

Figure 6.13. The use of visual phasing to map chromosomal segments to each of the four grandparents can help to confirm or eliminate hypothesized ancestral connections. Nieces/nephews of non-tested siblings can be used in conjunction with cousin matches when three siblings are not available.

Figure 6.13 illustrates chromosome 7 mapped using visual phasing for four close relatives. Don is a match of interest with the test taker on this chromosome; he is part of the triangulated group referenced in figure 6.12 hypothesized to share a distant, yet-to-be-identified Bishop ancestor. Don and the test taker share only this single 21.5 cM segment (18–54.8 Mbp). When compared to other close relatives of the test taker, Don is found to also share a portion of that same segment on chromosome 7 with the test taker's sister (18–51.5 Mbp) and a nephew (31.5–54.8 Mbp). Don shares no DNA, however, with other close relatives of the test taker, including another nephew (the son of a different sibling) and a set of three sibling cousins.

Because the test taker's maternal grandparents were first cousins through the Bishop line, a Bishop match could come from either the maternal grandfather or the maternal grandmother. In this scenario, visual phasing of the test taker, her sister, and two nephews—as demonstrated in figure 6.13—shows that the shared DNA segments with Don on chromosome 7 likely passed down through the maternal grandmother's line, through either Jacob Hiers or Susannah Bishop (see figure 6.1 for the pedigree). If there was no Bishop on this branch of the test taker's pedigree, chromosome mapping could also potentially put a wrinkle in the theory that this segment passed down from a distant Bishop ancestor or ancestral couple—a good thing if there is a possibility that the hypothesis is wrong!

Also investigate the DNA you did not inherit

The members of a group of triangulated matches who share only the single segment with the test taker may also match close relatives of the test taker on other segments. Determining this can be crucial in either supporting or disproving a theory. While Don shares only that single segment on chromosome 7 with the test taker and her sister, he also shares an additional 9.3 cM segment at 14.8–20.6 Mbp with the test taker's nephew on chromosome 19. While this is not a long segment, it is of note that the nephew is part of a group of triangulated matches overlapping this segment at 23andme who are known descendants of the test taker's maternal grandparents, John Daniel Bishop and Rosa Elizabeth Hiers. While neither the test taker nor Don tested with 23andme, visual phasing supports the theory that the nephew inherited DNA on this particular segment from the maternal grandmother (the same line through which the segment on chromosome 7 was inherited), while the test taker, her sister, and the remaining nephew received that portion of their DNA from the maternal grandfather.

This approach could be taken a step further as well when working with distant, single segment match groups. Other than known related matches (such as Don and Louis from our example seen in figure 6.12), do any other members of the expanded triangulated group share DNA with each other on any additional segments? Do any known cousins descended from the hypothesized common ancestral couple share DNA with any members of this triangulated group beyond the triangulated segment already explored? Examine additional shared DNA between group members using tools such as "People who match both, or 1 of 2 kits" or the One-to-One comparison at *GEDmatch*, shared match lists, FamilyTreeDNA's Matching Segment Matrix, and chromosome browsers (especially 23andMe which allows you to explore shared DNA between matches, not just with profile kits).

Explore DNA Network Visualization and Clustering

Clustering or network visualization is a method of organizing and grouping genetic matches into groups of related individuals. Depending upon the tool and strategy used, this can be as simple as grouping them into four grandparent clusters, or as involved as attempting to create groups related through a single ancestor or ancestral couple. AncestryDNA uses similar analysis of genetic networks in correlation with member trees to create DNA circles and New Ancestor Discoveries.

Network visualization or clustering is understandably complicated by pedigree collapse and endogamy. The intricate network of shared connections with multiple in-common-with ancestors can be difficult to computationally resolve into useful clusters. The best approach for endogamous lines will vary on a case-by-case basis but typically requires removing a majority of close matches from the analysis to eliminate some of the "clutter" caused by multiple highly interconnected matches. If there is recent pedigree collapse, useful analysis may also include removal of all matches who have DNA in common with more than two or possibly three other groups, or all matches who are estimated third cousins and above. What is left may include less than half of overall matches but offers a much better opportunity for clustering distant matches into groups of likely individuals related through a common ancestor.

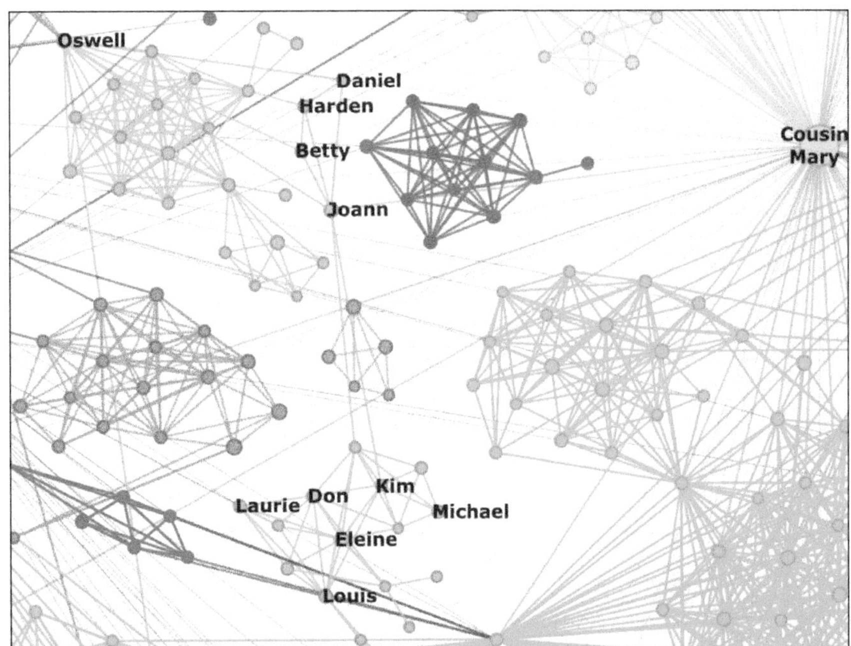

Figure 6.14. A small portion of an extended family graph created by ConnectedDNA demonstrates that the clustering of visual networks can accurately identify groups of related matches from a single common ancestor or ancestral couple, even within endogamous populations.

In figure 6.14 the teal green cluster in the FamilyTreeDNA extended family graph created by ConnectedDNA[18] for the test taker, her sister, two nephews, and two cousins clearly highlights the segment of interest on chromosome 7 as a likely cluster of related individuals, despite FamilyTreeDNA's lack of triangulation data. The graph is too complex to completely pull the maternal and paternal overlapping segments apart into two distinctly colored groups, but the division is still apparent by proximity, with the paternal group at the top and the maternal group at the bottom. Shared DNA between several members of the maternal group with members of the paternal group on different chromosomes due to endogamy explains the lack of complete separation. Identified names of FamilyTreeDNA matches are those from the triangulated segment groups identified in figures 6.10 and 6.12. It is significant that none of the previously identified individuals who tested with FamilyTreeDNA are missing from this cluster.

While AncestryDNA shared matches were reviewed while working with the triangulated group on Chromosome 7, they were not included in the final analysis due to lack of segment data and the time it would take to analyze the various connections between combinations of individuals. A network graph of the test taker's shared AncestryDNA matches was produced

18. Shelley Crawford, *Connected DNA* (https://www.connecteddna.com/).

by a disinterested third-party with no knowledge of the test taker's family tree beyond the endogamy factor. Over 57% of highly interrelated matches were left out of the final graph of estimated fourth cousin matches to produce a graph with enough distinct clusters to be useful. Among the remaining clusters is one which perfectly combines the members of the triangulated group known to also be in AncestryDNA, with a group of new connected individuals to contact and explore.

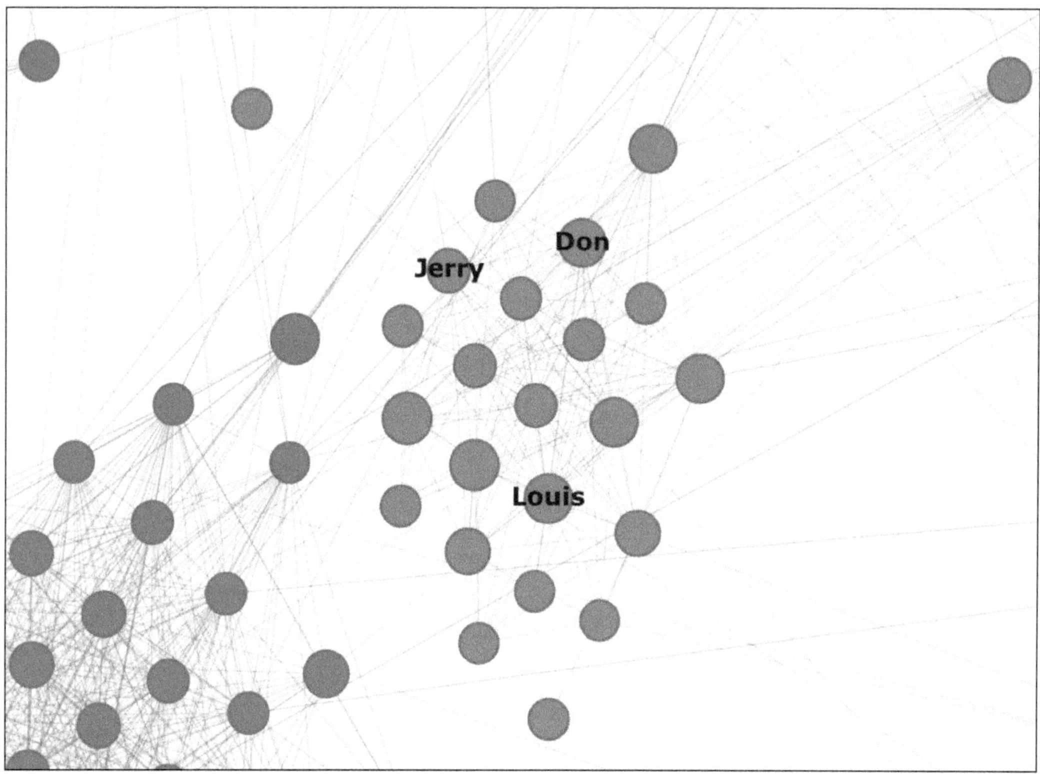

Figure 6.15. This small portion of an AncestryDNA single-profile graph, created by Connected DNA from the test taker's matches, includes a likely group of individuals (in grey-blue) who descend from the same common ancestor as the rest of our triangulated group on chromosome 7, including all three matches from the group known to have also tested with AncestryDNA. Additional names removed for privacy.

Resources and tools for DNA network visualization and clustering

The following tools and articles offer a starting point for researchers interested in further exploring the value of clustering and DNA network visualization in researching families impacted by endogamy or pedigree collapse:

ConnectedDNA
A commercial service providing network graphs from company match data to visualize clusters of matches. https://www.connecteddna.com/

DNADNA Visualization Software
DNADNA processes DNA files downloaded from various testing companies into Excel and uploads the resulting spreadsheet directly into Gephi to create visualizations of your DNA data. Runs on Excel 2016 or higher for Windows only. https://www.dnadna.uk/

DNAGedcom Client
Available to subscribers of DNAGedcom, the DNAGedcom Client not only enables downloading of DNA match, in common with (ICW), and ancestor files from AncestryDNA, FamilyTreeDNA, 23andme, and MyHeritage, but also offers genetic visualization tools. http://dnagedcm.com/Account/Subscriber.aspx

Genetic Affairs
This credit-based service automates the retrieval of new genetic matches for 23andme, FamilyTreeDNA, and AncestryDNA, and includes an automatic clustering feature. https://geneticaffairs.com/

Twigs of Yore: Analyzing AncestryDNA matches with NodeXL
This series of blog posts by Shelley Crawford walks through the steps of preparing and analyzing spreadsheets of DNA matches using the NodeXL plugin for Microsoft Excel (Windows only). http://twigsofyore.blogspot.com/2017/07/visualising-ancestry-dna-matchesindex.html

Conclusion

While DNA analysis in families or lines with significant pedigree collapse or endogamy is difficult, it is not impossible. Analysis at the segment level is frequently tedious and time-consuming; however, this systematic approach often yields evidence useful for answering specific genealogical questions. The increased likelihood that DNA from a distant ancestor (sixth great-grandparent or beyond) occupying more than one position on the pedigree chart has survived recombination to the present day is a potential unexpected benefit of endogamy, offering the opportunity to use autosomal DNA evidence in support of more distant ancestral connections than might otherwise be expected.

Approach endogamy as a challenge, not a curse. Consider how you might use current tools in innovative ways. Experiment with new ideas, tools, and techniques as they become available. Be patient, inquisitive, and tenacious. Be willing to guess. And be willing to be wrong. Even disproving a theory is one step closer to finding the correct answer!

Parker Study: Combining atDNA & Y-DNA

Debbie Parker Wayne, CG, CGL

A family legend linking one Henry Parker family in Texas[1] to Quanah Parker sparked the interest of a Parker genealogist. Quanah Parker and his mother, Cynthia Ann Parker, are Texas legends. Quanah was a famous warrior, the son of a Comanche chief and a woman captured as a child in the well-known Fort Parker Massacre in May 1836 in central Texas.[2] Any history buff would be happy to be related to this fascinating early Texas family. However, documentary research was not revealing any links to Quanah or to Cynthia Ann Parker's family.

DNA evidence collected over thirteen years, added to the documentary evidence collected over even more years, answers the question and debunks the relationship to Quanah and Cynthia Ann. The evidence also debunks stories of Native American ancestry on the patrilineal lines of Henry Parker.

Biological links between Henry Parker, his hypothesized father, and descendants of men named Henry Parker in multiple counties are explored in this study. DNA analysis began in 2005 with Y-DNA tests and continues today with autosomal DNA (atDNA) matches and new tools that periodically pop up.

This family study includes little targeted recruitment of DNA test takers. One brother and one cousin were recruited and their Y-DNA tests paid for; other Y-DNA matches were already in the database or tested themselves later. Several close family members and the

All URLs were accessed 21 September 2018 unless otherwise indicated.

1. Where names of test takers are used, permission has been granted by the test takers. Aliases are used where the test taker preferred to remain anonymous but gave permission for use of their test results. Screen shots and CSV files of match lists, shared DNA, genetic networks, and chromosome comparisons are archived in the author's files.

For a more detailed account of Henry and his families see Debbie Parker Wayne, "Bailing, Bigamy, Brother Love: The Family of Henry Parker and Nancy Black," Dallas Genealogical Society, *Pegasus 1* (2013):11–20; PDF (http://debbiewayne.com/pubs/pub_HenryParker2012_DPW.pdf).

2. Brian S. Hosmer, "Quanah Parker," *Handbook of Texas Online* (https://tshaonline.org/handbook/online/articles/fpa28). See also Margaret Schmidt Hacker, "Cynthia Ann Parker," *Handbook of Texas Online* (https://tshaonline.org/handbook/online/articles/fpa18).

same cousin asked for Y-DNA testing were then targeted for autosomal testing. Additional autosomal matches appear at regular intervals at multiple testing companies. Perhaps many of these Parker descendants also test in search of proof of the legend of Native American ancestry and links to Quanah.

The Henry Parker Family

Henry Parker Jr. was born in South Carolina probably in June 1823, or possibly June 1825; he died 9 April 1902 in Hood County, Texas.[3] He married (1), Nancy Black, in 1849 or 1850, probably in Milam County, Texas.[4] He "married" (2), Elizabeth (O'Neal) Kline Quarles, 5 May 1877 in Hood County.[5] Henry was still living with Elizabeth at his death.[6] "Jr." is used only on a few records when he is listed alongside a Henry Parker Sr.

Henry may have been born in 1823 or as late as 1825. The 1900 census is the only document that lists a specific month and year of birth for Henry, June 1825.[7] That year is supported by the earliest records. Henry Jr. was likely the one male under age five in 1830 and between age ten and fifteen in 1840 who was enumerated in the household of Henry Parker.[8] June 1825 may be the most likely date of birth if those age brackets were correctly assigned. Henry Parker Jr. appears on several tax rolls in Pope County, Arkansas, between 1843 and 1846, usually next to Henry Parker Sr. Men between ages twenty-one and fifty-five were assessed a poll tax during these years. This makes 1823 a more likely birth year as no one would likely pay taxes that were not required by law.[9]

---

3. 1900 U.S. census, Hood County, Texas, population schedule (pop. sch.), ED 91, p. 30A (stamped), dwelling (dw.)/family (fam.) 548, Henry Parker; National Archives (NARA) microfilm publication M623, roll 1645; for birth. For death, see Affidavit of A. F. McCoy and T. J. Abercrombie, 4 September 1902, Elizabeth Parker, widow's pension application no. 16665 (Rejected) for service of Henry Parker (Pvt., Co. A, Yell's Regiment, Arkansas Volunteers, Mexican War); Case Files of Rejected Pension Applications, Mexican War; Records of the Bureau of Pensions and Its Predecessors, 1805–1935; Department of Veterans Affairs, Record Group (RG) 15; NARA, Washington, D.C.

4. Milam County marriage records were destroyed in a fire in 1874. Nancy is named as Henry's wife in Milam County, Texas, Commissioners Court Minutes A:7; County Clerk's Office, Cameron. 1850 U.S. census, Milam County, Texas, pop. sch., Milam and Williamson District, p. 1A (stamped), dw. 231, fam. 243, Henry and Nancy Parker in Alexander Black household; NARA microfilm publication M432, roll 913; Henry and Nancy married within the year.

5. Hood County, Texas, Marriages A:167, Henry Parker and Mrs. Elizabeth Quarles, 1877; County Clerk's Office, Granbury.

6. Affidavit of McCoy and Abercrombie, Elizabeth Parker, widow's pension application no. 16665 (Rejected), Mexican War, RG 15, NARA, Washington, D.C.

7. 1900 U.S. census, Hood County, Texas, pop. sch., ED 91, p. 30A (stamped), dw./fam. 548, Henry Parker; NARA microfilm publication M623, roll 1645.

8. 1830 U.S. census, Clinton County, Illinois, pop. sch., p. 106 (stamped), line 27, Henry Parker; NARA microfilm publication M19, roll 25. 1840 U.S. census, Pope County, Arkansas, p. 133 (stamped), line 17, Henry Parker; NARA microfilm publication M704, roll 17.

9. Pope County, Arkansas, Tax Rolls, 1835-1859, unpaginated entries arranged alphabetically by first letter of surname; entries for Henry Parker Jr., Arkansas History Commission (AHC) microfilm roll 004039, Pope County roll 65; AHC, Little Rock. See also David Yancy Thomas, "A History of Taxation in Arkansas," *Publications of the Arkansas Historical Association* Vol. 2, 1908 (Fayetteville, Ark.: Arkansas Historical Association, 1908).

Henry Parker Sr. lived in Pope County, Arkansas, from 1835 through 1851. Henry Parker Jr. appears on the tax rolls there only in 1843 through 1846. The family owned no land and very little livestock.[10] In 1850, Henry Sr. lived near Edward and Redrick Parker.[11] A neighbor names the sons of Henry [Sr.] as George W., Edward, Ret, and Henry; daughters as Winny, Sally, and Hester; and mentions the existence of other sons and daughters not named.[12] Henry Jr. is not in Pope County after 1846. Henry's disappearance from Pope County records supports the hypothesis that he is the same Henry who mustered in for service in the Mexican War and left for Texas.

On 1 July 1846, Henry was mustered into Company A of the Arkansas Volunteers for the Mexican War.[13] The company arrived in hot, dusty San Antonio, Texas, in late August.[14] Henry was discharged on 23 September 1846 on a surgeon's certificate of disability.[15] Henry arrived in Milam County about 1849 or 1850.[16]

No marriage record exists for Henry Parker and Nancy Black in Milam County. Few records survived a courthouse fire in 1874.[17] However, records of the commissioners' court indicate Henry and Nancy were married.[18] By 1860, Henry owned 200 acres of land on Brushy waterway, near San Gabriel.[19] Children Alex, age nine; Waid, age seven; George, age six; Anderson, age four; and Elizabeth, age two; had joined the household.[20]

---

10. Pope County, Arkansas, Tax Rolls, 1835-1859; all years read for all Parker entries; Arkansas History Commission (AHC) microfilm roll 004039, Pope County roll 65; AHC, Little Rock

11. 1850 U.S. Census, Pope County, Arkansas, pop. sch., p. 287A (stamped), dw./fam. 643, Henry Parker, dw./fam. 641, Edward Parker, dw./fam. 642, Redick Parker; NARA microfilm publication M432, roll 29. No entries for a Parker or a Henry of the right age for Henry Jr. is found

12. N. D. Shinn affidavit, August 1896, George W. Parker [older, as distinguished from a nephew of the same name in the trees in this essay] claim no. 2405; Applications from the Bureau of Indian Affairs, Muskogee Area Office, Relating to Enrollment in the Five Civilized Tribes under the Act of 1896; Records of the Bureau of Indian Affairs, RG75; NARA microfilm publication M1650-40.

13. Compiled Military Service Record (CMSR), Henry Parker, Pvt., Co. A, Yell's Regt., Ark. Vols., RG 94, NARA, Washington, D.C.

14. Silas M. Shinn, "Recollection of the War" dated 1913; published in Desmond Walls Allen, comp. *Arkansas' Mexican War Soldiers* (Conway: Arkansas Research, 1988), 32.

15. CMSR, Henry Parker, Pvt., Co. A, Yell's Regt., Ark. Vols., RG 94, NARA, Washington, D.C.

16. Milam County, Texas, 1867 Voter Registrations, no. 502, Henry Parker entry, 13 July 1867; Texas State Library and Archives Commission (TSLAC) microfilm reel VR-9, Austin.

17. Texas Historical Records Survey, compiler, *Inventory of the County Archives of Texas, No. 166, Milam County (Cameron)* (Milam County, Texas: Works Progress Administration, 1941), 9.

18. Milam County, Texas, Commissioners Court Minutes A:7.

19. Milam County, Texas, Tax Rolls, 1846–1883; all years read for all Parker entries and variant spellings; TSLAC microfilm reel 1166-01, Austin. DeLorme, *Texas Atlas & Gazetteer* (Yarmouth, Maine: DeLorme, 1995), 70.

20. 1860 U.S. census, Milam County, Texas, pop. sch., p. 42A/B (stamped), dw./fam. 250, Henry Parker. Death certificates for some of these children name Henry as the father. See Texas Department of Health, death certificate no. 10068 (1927), G. W. Parker, Bureau of Vital Statistics, Austin. See also Texas Department of Health, death certificate no. 38890 (1925), Anderson Perry Parker, Bureau of Vital Statistics, Austin.

No record of Civil War service has been found for Henry.[21] He was home for at least part of that time as he had children born during those years: John Wesley born in 1862 and Laura or "Lena" born in 1863.[22] Henry was not assessed taxes in Milam County in 1861 to 1864 or 1866 to 1869. He was assessed Milam County property taxes in 1865 but owned no land.[23] Henry likely squatted on vacant land after selling the San Gabriel property. In 1866, Texas passed an act allowing settlers to claim preemption lands.[24] Henry applied under this act for land he had settled by 15 October 1866 on the headwaters of Yegua Creek. The Texas General Land Office approved the patent in November 1871.[25] Henry sold the land only a few months later.[26]

The Parker family is not found in 1870 census records.[27] This U.S. census is generally understood to be incomplete.[28] Henry was assessed taxes again in 1870 to 1873; he owned no land, but had small numbers of horses, cattle, and miscellaneous property.[29] Even though Henry was not found on the 1870 census it seems obvious he was in Milam County as he was assessed taxes for most of the years between 1850 and 1876 except the turbulent war years in the 1860s when the tax rolls appear to be incomplete.

By age thirty-five, Nancy displayed signs of insanity. In May 1874, the commissioners' court placed Nancy on the indigent list and authorized payment of $10.00 per month to her husband for 1 July 1873 to 1 June 1874.[30] This money was paid to Henry.[31] In January 1876, G.

21. James E. Williams, *Milam County, Texas in the Civil War* (Cameron, Texas: self-published, 1993); Henry Parker is not named on the rosters although several of his known neighbors are named in infantry and cavalry units. National Park Service, "Soldiers," database, *Civil War Soldiers & Sailors System* (http://www.itd.nps.gov/cwss/ : accessed 23 August 2009) and *Index to Service Records of Confederate Soldiers from Texas, 1861–1865*, 15 rolls (Austin: TSLAC, n.d.), roll 11; all candidates found were eliminated as not the Henry Parker of interest.

22. Earl Jones, "Earl's Genealogy Site," (http://www.earljones.net/aqwg6210.htm#14421 : accessed 26 September 2018); Earl is a great-grandson of Henry Parker Jr.'s son Alexander William. Documentary support for these children is sparse, but links will be shown with AncestryDNA DNA Circles.

23. Milam County, Texas, Tax Rolls, 1846–1883.

24. H. P. N. Gammel, compiler, The *Laws of Texas, 1822-1897,* 10 vols. (Austin: Gammel Book Co., 1898), 5:1191, "An Act Donating One Hundred and Sixty Acres of Land to Actual Settlers Upon the Public Domain," 12 November 1866; digital images, University of North Texas, *The Portal to Texas History* (http://texinfo.library.unt.edu/lawsoftexas/ : accessed 14 September 2005).

25. Henry Parker, Milam Land District, 3rd Class file no. 1954 /preemption, Texas General Land Office, Austin.

26. Milam County, Texas, Deeds 122:88-89, State of Texas to Henry Parker, 2 November 1871, and 122:89-90, Henry Parker to Claiborne Garner, 4 January 1872; County Clerk's Office, Cameron. Deed recorded 15 January 1914 otherwise it would have been lost in the fire of 1874.

27. 1870 U.S. census, Milam County, Texas, pop. sch.; NARA microfilm publication M653, roll 1598. All of the census pages are very legible. A page by page scan of the microfilm did not find the Parker family.

28. William Thorndale and William Dollarhide, *Map Guide to the U.S. Federal Censuses, 1790–1920* (Baltimore: Genealogical Publishing Co., 1987), xviii. See also United Nations, Department of Economic Affairs, *Population Studies, No. 23, Manuals on Methods of Estimating Population, MANUAL II, Methods of Appraisal of Quality of Basic Data for Population Estimates*, digital image (http://www.un.org/esa/population/pubsarchive/migration_publications/UN_1955_Manual2.pdf : 5 May 2008), 11.

29. Milam County, Texas, Tax Rolls, 1846–1883.

30. Milam County, Texas, Commissioners Court Minutes A:7.

31. Milam County, Texas, Commissioners Court Minutes A:29.

E. Boles was paid for conveying Nancy to the Texas State Lunatic Asylum.[32] In 1880, Nancy was in the asylum as a public pay patient with chronic dementia.[33] Sadly, she was still in the asylum twenty years later in 1900.[34] She died 29 April 1902.[35]

Henry disappears from Milam County records after 1876. He has not been found in 1880 census records. After disappearing from Milam County, a Henry Parker entered into a bigamous marriage to Elizabeth (O'Neal) Kline Quarles on 5 May 1877 in Hood County.[36] Nancy was still alive in the asylum.

Henry applied for and received a Mexican War pension.[37] He died 9 April 1902 in Hood County.[38] No probate was found.[39] Elizabeth possibly had no idea that her marriage to Henry was not legal until her widow's pension was denied.[40] Elizabeth spent the next months trying to prove she was Henry's legal wife, but never received a pension. She died by 3 October 1903.[41] All three principals in this bigamous triangle died within eighteen months of each other. Henry and Elizabeth had one child. Frank Parker was born 4 July 1875 or 1879; he died 1 April 1942 in Hood County.[42]

The chronology of the known facts of Henry in all these locations is consistent with representing the history of one man. Henry as a child is only represented by tick marks on the 1830 and 1840 census enumerations. He is named on a few Arkansas tax rolls, usually next to Henry Parker Sr., the head of household on those 1830 and 1840 census records.

32. Milam County, Texas, Commissioners Court Minutes A:73.

33. 1880 U.S. census, Travis County, Texas, pop. sch., ED 124, State Lunatic Asylum, p. 2, line 45, Nancy Parker. 1880 U.S. census, Travis County, Texas, "Defective, Dependent, and Delinquent Classes," Insane Inhabitants, p. 41, l. 46, Nancy Parker; NARA microfilm publication T1134, roll 50.

34. 1900 U.S. census, Travis County, Texas, pop. sch., ED 141, p. 107B (stamped), State Lunatic Asylum, line 75, Nancy Parker; NARA microfilm publication T623, roll 1673.

35. Parker researcher (name withheld), to Debbie Parker Wayne, e-mail, 26 May 2009, "Henry Parker - Nancy Black," Wayne Research Files. This researcher received a typed list of information from another Parker researcher. That second researcher viewed Nancy's records at the Austin State Hospital in the 1980s–1990s. The researcher was not allowed to photocopy, but made notes on the file contents. Since Nancy appears on the 1900 census, but is no longer found on the 1910 census, she probably died between those dates, allowing for the probability she died on the 29 April 1902 date stated in notes made from her asylum records.

36. Hood County, Texas, Marriages A:167.

37. "Declaration for Pension of Officer, Soldier or Sailor of Mexican War," 9 February 1887, Henry Parker Mexican War pension no. S.C. 5022, RG 15, NARA, Washington, D.C.

38. "Affidavit" of J. M. McCuan, M. D., undated, Henry Parker Mexican War pension no. S.C. 5022, RG 15, NARA, Washington, D.C.

39. Hood County, Texas, Index to Probate, Vol. 1, 1875-1936; TSLAC microfilm reel 1034520. Hood County, Texas, Index to Deeds, Vol. 1 & 2, 1856-1901; TSLAC microfilm reel 1034515. Milam County, Texas, Probate Minute Index, 1874–1934; TSLAC microfilm reel 981412.

40. Elizabeth Parker, widow's pension application no. 16665 (Rejected), Mexican War, RG 15, NARA, Washington, D.C.

41. Jacket summary, Elizabeth Parker, widow's pension application no. 16665 (Rejected), Mexican War, RG 15, NARA, Washington, D.C.

42. Texas Department of Health, death certificate no. 17910 (1942), Frank Parker, Bureau of Vital Statistics, Austin. 1900 U.S. census, Hood County, Texas, pop. sch., precinct 1, ED 91, p. 30A (stamped), dw./fam. 548, Henry Parker; NARA microfilm publication M623, roll 1645.

Henry "Jr." disappears from Arkansas just before he appears in Texas. Henry appears in Milam County just after he leaves military service. Henry appears in Hood County after he disappears from Milam County. However, Henry Parker is not an uncommon name. DNA evidence was sought to reduce the possibility of incorrectly identifying these many Henrys in different locations as the same man.[43] DNA matches offer much stronger evidence than a timeline showing a man with a common name disappearing from one location and appearing in another.

Y-DNA Evidence

The "Parker Y-DNA Project," administered by Dennis West and Gregory Parker, collects Y-DNA test results for all tested Parker males who join the project.[44] Based on shared Y-DNA STRs, the male test takers are organized into family groups. The descendants of Henry Parker "Jr." tested into Parker Family Group 1 (FG01). FG01 listed twenty members as of September 2018. A few more men have tested into FG01, but their tests done at AncestryDNA and DNAHeritage do not appear on the FamilyTreeDNA Project page.

Family groups consist of test takers who have a common ancestor based on closely matching Y-DNA STR test results, but there is no organization by common ancestor within the online listing.

Analyzing the Y-DNA data is not difficult for those who understand the basics of genetics and DNA inheritance factors. The methodology to collect and organize the data is, however, time-consuming. Few have taken the time to analyze and organize the data beyond the groups defined in the "Parker Y-DNA Project." The fact that many test takers have not tested at the same company or tested the same markers makes analysis more difficult.

Y-DNA haplogroup results

Haplogroup assignments are one of the most basic types of information provided by a Y-DNA or mtDNA test, but many do not understand how to analyze and use this information. Haplogroups represent the deep roots of the patrilineal (Y-DNA) and matrilineal (mtDNA) ancestry—the location of ancestors thousands of years ago. Two people in the same haplogroup share a common ancestor, but it might be thousands of years ago.

Haplogroups are assigned based on mtDNA and Y-DNA SNP analysis. Branching of the mtDNA and Y-DNA trees occurs when a SNP mutates and is then passed down to descendants. Because mtDNA and Y-DNA are not recombined before being passed down, the only changes are mutations. Testing of living populations with a known migration path and

43. For a more detailed account of Henry and his families see Debbie Parker Wayne, "Bailing, Bigamy, Brother Love: The Family of Henry Parker and Nancy Black," Dallas Genealogical Society, *Pegasus 1* (2013):11–20; PDF (http://debbiewayne.com/pubs/pub_HenryParker2012_DPW.pdf).

44. Dennis West and Gregory Parker, "Parker Y-DNA Project," *FamilyTreeDNA* (https://www.familytreedna.com/groups/parker/dna-results).

of ancient DNA samples allows human migratory history back tens of thousands of years to be revealed by haplogroups.

A few years ago, the haplogroup names used were of the form R1a, R1b, or I2a. These names were easy to remember when only a few digits were involved. As more Y-DNA SNPs were discovered, the Y-DNA tree split into more branches and the names expanded. They became more difficult to use. For example, R1a1a1g2 is harder to remember than R1a. Today, Y-DNA haplogroups are referenced by the first digit or two of the older name followed by the defining SNP for that branch of the Y-DNA tree. For example, a man determined to be in haplogroup R1a1a years ago is now in haplogroup R-M198 (occasionally seen as R1-M198). The test taker's haplogroup has not changed, his DNA has not changed, only the name of his assigned group has changed. For this reason, researchers must be sure to use naming conventions for the same version of the tree when comparing the haplogroups of two men.[45]

On the Y-DNA tree, only a few haplogroups are found in Native American men with no European ancestry in the Y-DNA line. These include several subclades or sub-branches of haplogroups Q and C.[46] Quanah Parker is presumed to have no European ancestry on his paternal line; therefore, he should be in one of these haplogroups.

Links to Cynthia Ann's son, Quanah Parker, on his patrilineal Y-DNA line were immediately refuted by the haplogroup assigned to Parker FG01 men. Parker FG01 men are assigned to the R1a branch of the Y-DNA tree. This branch is not a Native American branch. Only Native Americans who have a European ancestor on the Y-DNA line would share this haplogroup.

Most of the Parker FG01 men are in haplogroup R-M512 if the haplogroup is predicted based on STR marker values. Some of them, if a more extensive test has been performed, are assigned to confirmed haplogroup R-M198. Some who have taken tests such as FamilyTreeDNA Big Y-500 and National Genographic Geno 2.0 are in haplogroup R-S6241 and R-CTS8277, depending on the depth of testing. All of these SNPs represent sub-branches of the same main branch, previously known as R1a as shown in this list:

45. For more on haplogroup basics, see Blaine T. Bettinger and Debbie Parker Wayne, *Genetic Genealogy in Practice* (Falls Church, Va.: National Genealogical Society, 2016), 8, 33–37, 40–42, and associated exercises.

46. "Haplogroups," *Wikipedia* (https://en.wikipedia.org/wiki/Haplogroup). Roberta Estes, "Proving Native American Ancestry Using DNA," *DNAeXplained – Genetic Genealogy* (https://dna-explained.com/2012/12/18/proving-native-american-ancestry-using-dna/).

- R1a1a (R-M198 and R-M512 both represent this level)
    - R1a1a1b1a3a2b (R-CTS8277, formed 4100 years ago)
        - No long name assigned (R-S6244, formed 3600 years ago)
            - R1a1a1b1a3a2b1~ (R-S6242, formed 2900 years ago)
                - R1a1a1b1a3a2b1b~ (R-S6241, formed 2900 years ago)[47]

Based on the time each of these SNPs originated, men sharing the SNP could share a common ancestor 2,000 or more years ago. Studies are underway to further define sub-branches of the tree and discover more SNPs to assist in linking families in more recent time frames.

The Parker Y-DNA project includes many test takers who claim descent through the Parker Y-DNA line associated with Cynthia Ann's paternal line. Men descended from Cynthia Ann's grandfather, Elder John Parker, are members of the Parker Y-DNA Project, but they are assigned to Family Group 5. FG05 men are mostly assigned to haplogroup R-M269 and R-L48, both of which are R1b sub-branches.[48] The Parker FG01 men are assigned to haplogroup R1a. The R1a and R1b branches of the Y-DNA tree split over 20,000 years ago.[49] This eliminates the possibility that the Parker men in FG01 and FG05 share a common patrilineal ancestor after that time.

The haplogroup designation alone refutes any relationship to Quanah Parker through his patrilineal Comanche lineage or the patrilineal line of Cynthia Ann Parker's father's line. If there is a relationship between Henry Parker's family and Cynthia Ann's family it is not due to the Parker surname and the Y-DNA inheritance path. There might be a link through the women in Cynthia Ann's family, but nothing has been found yet in documentary research. However, links are found through both DNA and documents to the other men in Parker FG01.

Y-DNA STR marker results

The publicly available information on the Y-DNA STR results for those in Parker FG01 was collected from the project website. Added to this is the lineage and STR results provided to the author via email, interviews, and social media outlets for those who tested at other companies. Many of the participants in this project tested at a time when 25 Y-DNA STR markers were considered a sufficient number of markers for genealogical purposes. For many years now, most genealogists have recommended a minimum of 37 markers be tested, with 67 markers needed in many situations. Today, 12 to 25 markers are primarily used only to determine whether two test takers *might* have a common ancestor, not that they do have

47. Yfull YTree v6.05.09, *Yfull* (https://www.yfull.com/tree/R-CTS8277/). See also "Y-DNA Haplogroup Tree 2018," *ISOGG* (https://isogg.org/tree/index.html). For SNP assignments and haplogroups of the men in Parker FG01 see Dennis West and Gregory Parker, "Parker Y-DNA Project," *FamilyTreeDNA* (https://www.familytreedna.com/groups/parker/dna-results).

48. Dennis West and Gregory Parker, "Parker Y-DNA Project," *FamilyTreeDNA* (https://www.familytreedna.com/groups/parker/dna-results).

49. "Haplogroup R1," *Wikipedia* (https://en.wikipedia.org/wiki/Haplogroup_R1).

a common ancestor in a genealogical timeframe. Lineages were provided by the test takers and confirmed with documentary research. The documentary research is not described here because of space limitations.

Table 7.1 displays the lineage and differing marker values of descendants of Henry Parker Sr. All of the descendants of Henry Sr. who have tested marker DYS607 have a value of 14. This differs from other Parker descendants in the Parker FG01 who have a value of 15 for DYS607. This makes it likely that Henry Parker Sr. had a value of 14 for DYS607; it is a family-defining marker value for Henry Parker Sr.'s descendants. Henry Sr. passed this marker value to all of his sons. Those sons then passed it to their descendants.

All of the descendants of Henry Parker Jr. who have tested marker DYS464c have a value of 16. Descendants of other sons of Henry Parker Sr. have a value of 15 for DYS464c. This makes it likely that this marker mutation occurred first in Henry Parker Jr.; it is a family-defining marker value for Henry Parker Jr.'s descendants. Along with the DYS607 mutation inherited from his father, Henry Jr. passed the DYS464c mutation to his sons. Those sons then passed both mutations to their descendants.

Also, Bobby Parker has a value of 30 for marker DYS449. This mutation could have occurred anytime between George W. Parker (1853) and Bobby Parker.

Typically, so many marker differences in only four to six generations might seem excessive. Many families have lineages much longer than six generations where no mutations at all have been detected. However, DYS449 and DYS464c are fast-changing markers. More mutations are typical in fast-changing markers.[50] The marker values do not contradict the proposed lineage even though there are a larger number of mutations than is typical. Additional documentary evidence confirming the lineages adds credibility to the conclusion.

While many researchers hope to find exact Y-DNA matches, exact matches only prove test takers share a common ancestor. The specific ancestor cannot be identified. There are times, such as with this Parker family, where mutations can be mapped to the family tree, allowing more certainty that the parent-child links are correctly assigned.

Y-DNA future research

Few men in this group of FG01 have tested a sufficient number of STR markers and few lineages in FG01 have tested enough descendants in a lineage to strongly support a conclusion. Henry Parker Sr. could be the father of Henry Jr. and Henry Jr. could be the father of the five test takers claiming descent from Henry Jr. The DNA is consistent with the claimed and hypothesized relationships, but other relationships are possible. Only one man (Robert Parker) has tested 111 markers. Only two (Archie Gordon-Parker and Gerald Parker) have tested 67 markers. One has tested 37 markers (Parker-02). One has only tested 25 markers

50. "What do colors for the Y-DNA results chart headings mean?" The Family Tree DNA Learning Center, *FamilyTreeDNA* (https://www.familytreedna.com/learn/project-administration/gap-reference/colors-y-dna-results-chart-heading/).

(Bobby Parker) and one tested at Ancestry.com where fewer than 37 overlapping markers were tested (Parker-A01).

<table>
<tr><th colspan="6">Table 7.1. Y-DNA tested descendants of Henry Parker Sr.[a]</th></tr>
<tr><td colspan="6">Henry Parker Sr.</td></tr>
<tr><td colspan="5">Henry Parker Jr.</td><td>George W. Parker [older] b. 1836</td></tr>
<tr><td>George W. Parker [younger] b. 1853</td><td>Perry Anderson Parker</td><td colspan="2">John Wesley Parker</td><td>Frank Parker</td><td>Thomas Eugene Parker</td></tr>
<tr><td>Perry C. Parker</td><td>Alfred Parker</td><td>Thomas Allen Parker</td><td>Ezra Daniel Parker</td><td>Frank M. Parker</td><td>anonymous</td></tr>
<tr><td>William H. Parker</td><td>Robert C. Parker</td><td>B. Parker</td><td>Parker-A01[b]
Ancestry
43 markers</td><td>Gerald Parker
67 markers</td><td>Parker-02
37 markers</td></tr>
<tr><td>William R. Parker</td><td>Robert Parker[c]
111 markers</td><td>Archie Gordon-Parker
67 markers</td><td></td><td></td><td></td></tr>
<tr><td>Bobby Parker
25 markers</td><td></td><td></td><td></td><td></td><td></td></tr>
<tr><td>464c=16</td><td>464c=16</td><td>464c=16</td><td>464c=16</td><td>464c=16</td><td>464c=15</td></tr>
<tr><td>*</td><td>607=14</td><td>607=14</td><td>*</td><td>607=14</td><td>607=14</td></tr>
<tr><td>449=30</td><td>449=31</td><td>449=31</td><td>449=31</td><td>449=31</td><td>449=31</td></tr>
</table>

Bold names in shaded boxes represent Y-DNA test takers. Each row represents a generation. Each column represents the lineage to the test taker. Documentary support for parent-child links are not shown here because of space limitations. See figure 7.1.

* indicates a marker not tested; presumably, if tested, the value would match the other family members.

a. Additional members of Parker FG01 are included on the public project page that are not included here as no permission was provided to use the results in this publication. See Dennis West and Gregory Parker, "Parker Y-DNA Project" *FamilyTreeDNA*.

b. Tested at Ancestry.com when Y-DNA tests were available there. A copy of the test results is in the author's files.

c. Robert Parker is the sibling of the author.

Only two of the Parker FG01 Y-DNA test takers have done extensive SNP tests: Robert Parker and Bobby Parker. Neither has given permission for results to be uploaded to Yfull (which is in Russia)[51] or to other websites for analysis. Therefore, all analysis must be done using tools on the FamilyTreeDNA site or using tools that run on a personal device. As time

51. *Yfull* (https://www.yfull.com/).

allows, analysis similar to that shown in the "Y-DNA Analysis for a Family Study" chapter may lead to more discoveries about this Parker family.

Identification of the parents of Henry Parker Sr. is a primary goal. Additional men in Parker Y-DNA FG01 likely descend from the same lines as Henry Parker Sr. and can provide a clue to which Parker family Henry descends from. Additional SNP and STR analysis should help identify the parentage of Henry Parker Sr. and his patrilineal ancestors. The DNA information helps to focus the documentary research.

Autosomal DNA Evidence

Three ways genealogists may use atDNA are illustrated in this section. First, the hypothetical relationship between two test takers is determined using a family tree and comparing the total amount of shared DNA between the test takers to a chart (such as the "Shared cM Project"[52]). Second, the shared DNA segments are determined by identifying chromosome number with start and stop locations of the shared segments. If any segments can be triangulated (triangulated segments are shared by three or more descendants in different lines from a common ancestor), these segments may be mapped to the common ancestral couple. See the "Lessons Learned from Triangulating a Genome" chapter for more information. Third, chromosome maps and genetic networks can confirm links and may help link test takers who share DNA with those who did not receive the same DNA from their common ancestors. Mapped segments can help focus research on specific lines to search for common ancestors with others who share that same DNA segment.

Correlating the DNA data with family trees may answer genealogical questions. Test results and family trees help (1) confirm families are correctly assembled and (2) group closely related lines for collaboration in searching for the common ancestor and the earlier ancestors in the line.

With Y-DNA testing, genealogists were limited to testing only men. Once atDNA prices came down and the number of SNPs tested became more useful for cousin-matching, the DNA of women began answering just as many genealogical questions. Autosomal DNA removes the limitation of using DNA only for patrilineal and matrilineal lines and can be used for all ancestral lines. Although some of the Y-DNA Parker FG01 test takers never added atDNA tests, some did, and there are dozens of additional family members, both men and women, who have taken atDNA tests at every company serving genealogists. A new "Parker Family Group 1" atDNA project was created at FamilyTreeDNA.[53] The company displays little information publicly for atDNA tests, even in projects, because of privacy concerns.

52. Blaine T. Bettinger, "Shared cM Project," *The Genetic Genealogist* (https://thegeneticgenealogist.com/). Search the blog posts for the most recent update to the project such as the August 2017 update at https://thegeneticgenealogist.com/2017/08/26/august-2017-update-to-the-shared-cm-project/. The project charts are periodically updated.

53. Debbie Parker Wayne, Project administrator, "Parker FamGroup1 Project" (https://www.familytreedna.com/groups/parker-fam-group-1/).

The project administrators can analyze the DNA and share the findings, with permission of the project members.

Several factors affect atDNA analysis. The atDNA tests most used today check only 500,000 to 700,000 locations out of the 3 billion total base pairs or locations comprising our chromosomes. Each testing company tests different locations with some overlap. Algorithms then compute the start and stop locations for shared segments, with some estimates made for the locations that are not tested. This causes "fuzzy" or imprecise start and stop locations for the matching segments.

Random recombination naturally results in some cousins sharing DNA and some cousins not sharing DNA. The more closely related two test takers are, the more atDNA and the longer segments they normally share.

With the level of DNA testing available today, all second cousins and closer are expected to share some detectable amount of atDNA. Beyond the second cousin level, some genealogical cousins will not share enough DNA to be listed as matches. The lack of shared atDNA beyond the second cousin level should not be used as evidence that no relationship exists. It is possible two test takers are cousins even though no shared atDNA is indicated.

Because of random recombination, some family groups will have many cousins who share atDNA and have triangulated segments, while other family groups will not. In this study the Henry Parker group has many cousins who have atDNA test results and the descendants have many triangulated segments. Some test takers not shown as descendants of Henry Parker share DNA with group members. In this case, Henry Parker Sr. could be a shared ancestor not yet listed in the test takers' family trees. Otherwise, these test takers could share a non-Parker ancestor.

The publicly available information on the lineages for those in Parker FG01 was collected. Lineage information privately provided via email, interviews, and social media outlets was added. Some family trees required additional research to adequately document and to fill gaps where additional common ancestors may have been lurking unfound.

Each test taker was annotated with the total amount of atDNA shared with other project members, calculated (or traced) relationship based on the documented lineage, and total amount of shared atDNA.

Additional analysis steps included comparing the amount of DNA shared by two project members to the "Shared cM Project" chart to ensure consistency with the calculated relationship, deleting DNA segments smaller than 7 cM from the analysis as these may not be IBD,[54] and analyzing shared atDNA and X-DNA segments to determine if any of the seg-

54. "Identical by descent," *ISOGG Wiki* (https://isogg.org/wiki/Identical_by_descent).

ments triangulate. During the analysis and correlation process, as new tools were announced, they were used when applicable to the situations under study.

Methods of confirming a common ancestor and eliminating competing ancestor hypotheses

In the discussion of atDNA analysis, all relationships are abbreviated. For example, 3C1R is third cousin once removed, half-2C2R is half second cousin twice removed, and so on. Shared cM values include only segments larger than 7 cM (except those who tested only at Ancestry where the shared amount could not be verified).

"How much evidence is enough evidence? If a body of compiled evidence is strong enough such that any additional piece of potential evidence is unlikely to move the balance in favor of a competing hypothesis, then the currently compiled research is likely to be [reasonably] exhaustive."[55] Researchers like to check off a box that indicates, for example, two or three sources confirm a hypothesis. In truth, it is much more complicated, requiring determination as to whether those sources were created independently, whether the information came from an eyewitness or someone who heard it from another, whether the record is original or derivative, whether it was created by a scribe with no biases or someone with a reason to shade the truth, and more.[56]

When using DNA evidence, there are many methods to confirm analysis and correlation. The trees for some matches will almost certainly be created, extended, or documented by the thorough researcher. However, other methods also lend credibility to conclusions. Not all methods work for every situation, but possibilities are described here and in other chapters of this book. Tomorrow, additional possibilities are likely to appear as genealogists continue to expand our toolbox and our knowledge.

Tree depth, completeness or gaps, and accuracy

The depth, completeness or gaps, and accuracy of family trees must be considered during analysis.[57] This includes our own tree as well as those of our DNA matches. *Tree depth* relates to the predicted or hypothesized relationship of the test takers. If an atDNA test predicts test takers to be fourth cousins, then each person's tree should minimally go back to the shared third-great-grandparents and ideally a further generation or two. If the tree is not deep enough, it is impossible to identify a common ancestor. *Tree completeness* relates to whether there are any gaps in lines back to the depth needed. If there are gaps in the tree, there is the potential for two test takers to share more than one common ancestor, which complicates the DNA analysis. *Tree accuracy* relates to how thorough the documentary research is that underpins the parent-child links. If a tree is deep and complete but inaccurate,

55. Paul Woodbury, "Genetic Genealogy and Genealogical Proof—A Balancing Act," Association of Professional Genealogists, *Professional Management Conference* (Kansas City, Missouri: APG, 2018).

56. Thomas W. Jones, *Mastering Genealogical Proof* (Falls Church, Va.: National Genealogical Society, 2013), 23–26.

57. Board for Certification of Genealogists (BCG), *Genealogy Standards*, 2nd ed., 30–31, standard 52, "Analyzing DNA test results," bullets 1 and 2.

it leads to incorrect conclusions. Depth, completeness, and accuracy are all needed to form credible conclusions.

The author's own family tree is 100% complete back to the fourth generation—second-great-grandparents—which is the generation into which Henry Parker Jr. and Nancy Black fit. At the third-great-grandparent level, the tree is 81% complete—there are six unidentified individuals: two couples and two wives. For any relationship at the fourth cousin level or beyond, there is a possibility of multiple shared ancestors that cannot be identified because of gaps in the tree.[58] This tree is not complete. Neither are most of the trees of those sharing DNA.[59] In the author's opinion, her ancestors back to the second-great-grandparent level are well-documented and meet the research guidelines for proof as defined in *Mastering Genealogical Proof*.[60]

Confirming that the correct common ancestor has been identified and eliminating competing hypotheses for the common ancestor requires analysis of the trees of DNA matches. This can be difficult to do when a study includes many test takers, when some matches do not have a publicly-available online tree, and when those trees are incomplete or not well-sourced. Ideally, every online tree would be accurate, complete beyond the generation of a likely shared ancestor (based on the amount of shared DNA), and impeccably sourced to credible, original records. Such trees are rarely found.

In some situations it will be necessary for a researcher to create or expand the pedigree of DNA matches. This will be time consuming. Methods other than completing trees and looking for common ancestral names can also be used. Alternative techniques that add credibility can be found in these articles in the *National Genealogical Society Quarterly* (*NGSQ*):

> Thomas W. Jones, "Too Few Sources to solve a Family Mystery? Some Greenfields in Central and Western New York," *NGSQ* 103 (June 2015): 85–103. This article uses generational distance estimates correlated with total amount of shared DNA to eliminate other potential relationships between the test takers.
>
> Karen Stanbary, "Rafael Arriaga, a Mexican Father in Michigan: Autosomal DNA Helps Identify Paternity," *NGSQ* 104 (June 2016): 85–98. This article uses X-DNA patterns of inheritance and shared X-DNA segments, multiple shared ancestral lines, and shared amounts of atDNA to address pedigree gaps.

---

58. The author's genealogical research is contained in a genealogical database on a personal computer. With the advent of DNA testing, public trees were created on several company websites, but include only direct ancestral lines and no collateral relations. The online trees do not include all of the evidence contained in the private genealogy database. This problem is being remedied as time allows, but currently there is no well-sourced and complete public tree available online for access by the reader. This chapter is written as a teaching tool reducing the importance of tree verifiability. If this chapter were a genealogical proof argument a means of verifying the parent-child links in the tree might be available to the reader.

59. Catherine A. Ball et al., "DNA Circles™ White Paper," *AncestryDNA* (https://www.ancestry.com/corporate/sites/default/files/AncestryDNA-DNA-Circles-White-Paper.pdf), 15; fewer than 40% of trees on *Ancestry* are complete to the third-great-grandparent level.

60. Thomas W. Jones, *Mastering Genealogical Proof* (Falls Church, Va.: National Genealogical Society, 2013).

Patricia Lee Hobbs, "DNA Identifies a Father for Rachel, Wife of James Lee of Huntington County, Pennsylvania," *NGSQ* 105 (March 2017): 43–56. This article uses triangulation to address pedigree gaps.

Other methods of dealing with pedigree gaps and reducing the potential for unidentified shared ancestors are described in the following sections. In some cases, multiple methods may be needed to provide the most credible evidence to support a conclusion.

Compare shared amounts of DNA with hypothesized relationships

During analysis, shared amounts of DNA reported by the testing companies should be compared with statistical probabilities and empirical data reported by actual test takers.[61] Statistical probabilities can be found in the Wiki of the International Society of Genetic Genealogists.[62] Reports of shared amounts from actual test takers can be found in the Shared cM Project. Jonny Perl has converted Blaine Bettinger's printed chart to an interactive tool and added Leah LaPerle Larkin's probability predictions to the tool.[63] The output of these tools can be included in charts and tables in an essay. The meaning of the values should be discussed in the narrative to ensure the reader understands the reasoning that leads to a conclusion.

In Parker Family Group 1, several of the Y-DNA test takers shown in table 7.1 have also taken an atDNA test. The amount of atDNA shared by those test takers is displayed in table 7.2. The amounts of shared DNA are within the ranges in the Shared cM Project. However, random recombination results in a wide range of actual shared DNA amounts. These values are consistent with the hypothesized relationships shown in the tables and charts in this chapter, but other relationships are also possible. The same amount of shared DNA can be found in many different relationships. Also, credible conclusions for test takers sharing a common ancestor who was born in the late 1700s or early 1800s, more than 200 years ago, need DNA evidence from more than three or four test takers.

Descendants of Henry Parker (Sr.) who tested both Y-DNA and atDNA

Table 7.2 indicates the lineage and the amount of atDNA shared by four of the men who have closely matching Y-DNA STR values. The total amounts of atDNA shared by the hypothesized Parker cousins is consistent with the sharing amounts reported in the Shared cM Project.

61. BCG, *Genealogy Standards*, 2nd ed., 30–31, standard 52, "Analyzing DNA test results," bullet 3.

62. "Autosomal DNA statistics," International Society of Genetic Genealogists (ISOGG) *Wiki* (https://isogg.org/wiki/Autosomal_DNA_statistics). Also "Cousin statistics," ISOGG *Wiki* (https://isogg.org/wiki/Cousin_statistics).

63. Jonny Perl, Blaine T. Bettinger and Leah LaPerle Larkin, "The Shared cM Project 3.0 tool v4," *DNA Painter* (https://dnapainter.com/tools/sharedcmv4). Check for updated and alternate versions at https://dnapainter.com/tools.

Bobby Parker shares

- 0 cM with his third cousin once removed Robert Parker (average 48 cM, but may share from zero to 173 cM)
- 16 cM with his half second cousin twice removed Gerald Parker (average 61 cM, but may share zero to 353 cM)
- 8 cM with his third cousin twice removed Parker-02 (average 35 cM, but may share zero to 116 cM)

Robert Parker shares

- 0 cM with his third cousin once removed Bobby Parker (average 48 cM, but may share from zero to 173 cM)
- 33 cM with his half second cousin once removed Gerald Parker (average 61 cM, but may share zero to 353 cM)
- 62 cM with his third cousin once removed Parker-02 (average 48 cM, but may share zero to 173 cM)

Gerald Parker shares

- 16 cM with his half second cousin twice removed Bobby Parker (average 61 cM, but may share zero to 353 cM)
- 33 cM with his half second cousin once removed Robert Parker (average 61 cM, but may share zero to 353 cM)
- 0 cM with his third cousin Parker-02 (average 74 cM, but may share zero to 217 cM)[64]

Sizes and locations of shared segments

This Parker group is a family full of genealogists who have become DNA test takers. As of September 2018, over thirty-two of the author's DNA matches have been identified as descendants of Henry Parker Sr. Many more show up as shared matches and in common with these identified matches, but research on those trees is still in process. Figure 7.1 illustrates the portion of the Parker family tree that includes test takers discussed in this study and lists at the bottom the relationships and amount of atDNA—including only segments 7 cM or larger in size— shared by these test takers.

Numerous tools, even old-fashioned paper and pencil, can create charts and tables for analysis. The chart in figure 7.1 was created with Microsoft Word Smart Art; a Word table formats the data shown at the bottom of the figure. Table 7.3 contains the same shared cM data as figure 7.1 but in a different format. Each researcher should use the format that seems most logical to him or her during analysis and follow editor recommendations when publishing.

64. Blaine T. Bettinger, "Shared cM Project," *The Genetic Genealogist* (https://thegeneticgenealogist.com/).

Table 7.2. Y-DNA and atDNA tested descendants of Henry Parker Sr.				
Henry Parker Sr.				
	Henry Parker Jr.			George W. Parker [older] b. 1836
	George W. Parker [younger] b. 1853	Perry Anderson Parker	Frank Parker	Thomas Eugene Parker
	Perry C. Parker	Alfred Parker	Frank M. Parker	anonymous
	William H. Parker	Robert C. Parker	**Gerald Parker**	**Parker-02**
	William R. Parker	**Robert Parker**[a]		
	Bobby Parker			
Bobby Parker	self	3C1R 0 cM	half-2C2R 16 cM	3C2R 8 cM
Robert Parker	3C1R 0 cM	self	half-2C1R 33 cM	3C1R 62 cM
Gerald Parker	half-2C2R 16 cM	half-2C1R 33 cM	self	3C 0 cM
Parker-02	3C2R 8 cM	3C1R 62 cM	3C 0 cM	self

Bold names in shaded boxes represent Y-DNA and atDNA test takers. Each row represents a generation. Each column represents the lineage to the test taker. Documentary support for parent-child links are not shown here because of space limitations. Most of these lines are documented in figure 7.1.

The gray-shaded boxes with relationships and cM amounts duplicate the information in the unshaded boxes. In some charts created by programs, the gray-shaded area will be blank. Here we duplicated the information.

a. Robert Parker is the sibling of the author.

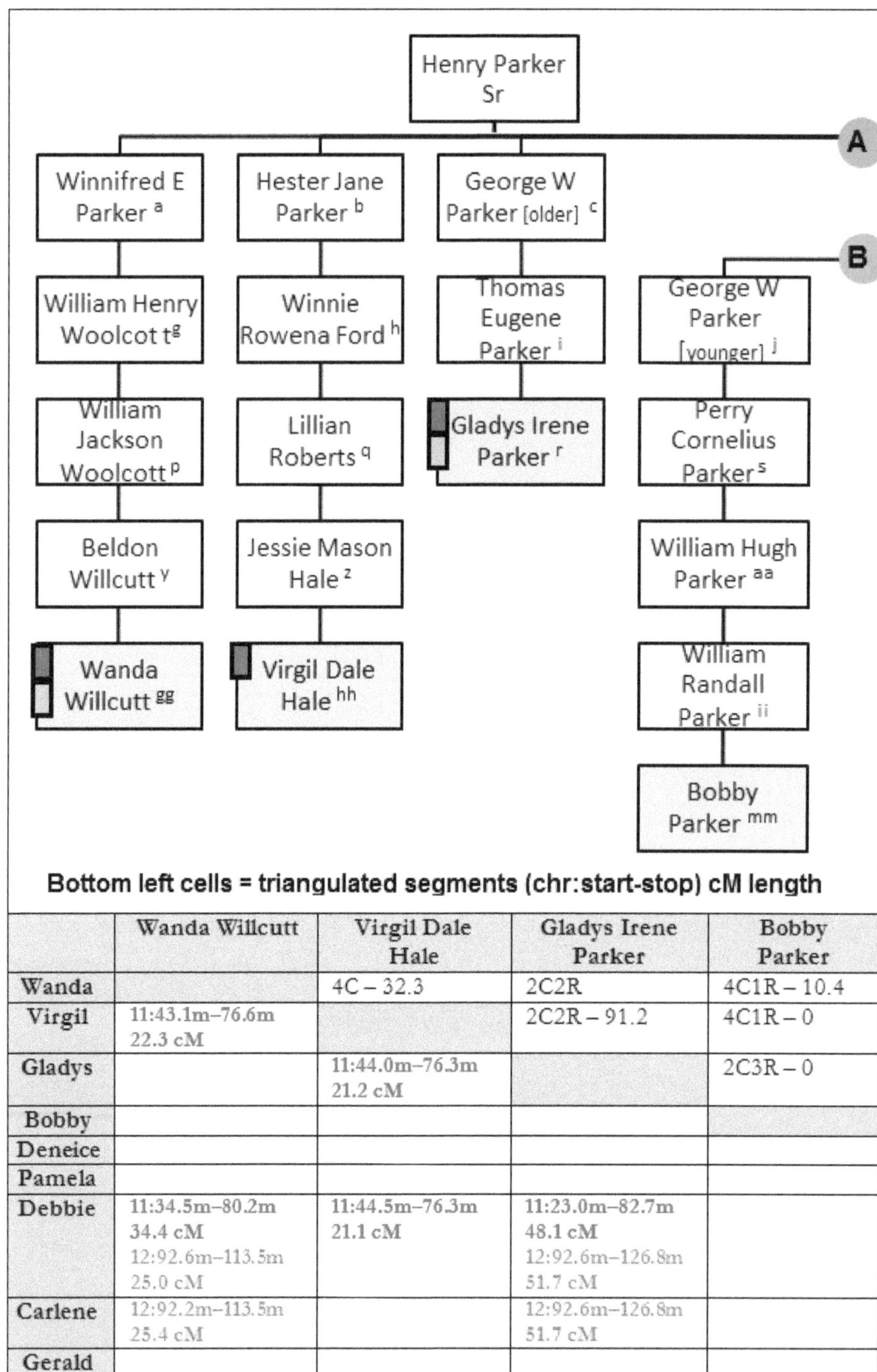

Bottom left cells = triangulated segments (chr:start-stop) cM length

	Wanda Willcutt	Virgil Dale Hale	Gladys Irene Parker	Bobby Parker
Wanda		4C – 32.3	2C2R	4C1R – 10.4
Virgil	11:43.1m–76.6m 22.3 cM		2C2R – 91.2	4C1R – 0
Gladys		11:44.0m–76.3m 21.2 cM		2C3R – 0
Bobby				
Deneice				
Pamela				
Debbie	11:34.5m–80.2m 34.4 cM 12:92.6m–113.5m 25.0 cM	11:44.5m–76.3m 21.1 cM	11:23.0m–82.7m 48.1 cM 12:92.6m–126.8m 51.7 cM	
Carlene	12:92.2m–113.5m 25.4 cM		12:92.6m–126.8m 51.7 cM	
Gerald				

Figure 7.1.a. Parker descendants, atDNA test takers (extended on the facing page)

The author thanks Archie Gordon-Parker (from table 7.1), Wanda Willcutt-Morgan, Virgil Dale Hale, the family of Gladys Irene Parker, Bobby Parker, Deneice Jackson, Pamela Ellis, Carlene Burleson, Gerald Parker, and siblings Robert, Deena, Diana, and Darla Parker for permission to use their identities and DNA test results. Parker-A01 and Parker-02 in table 7.1 remain unnamed by choice.

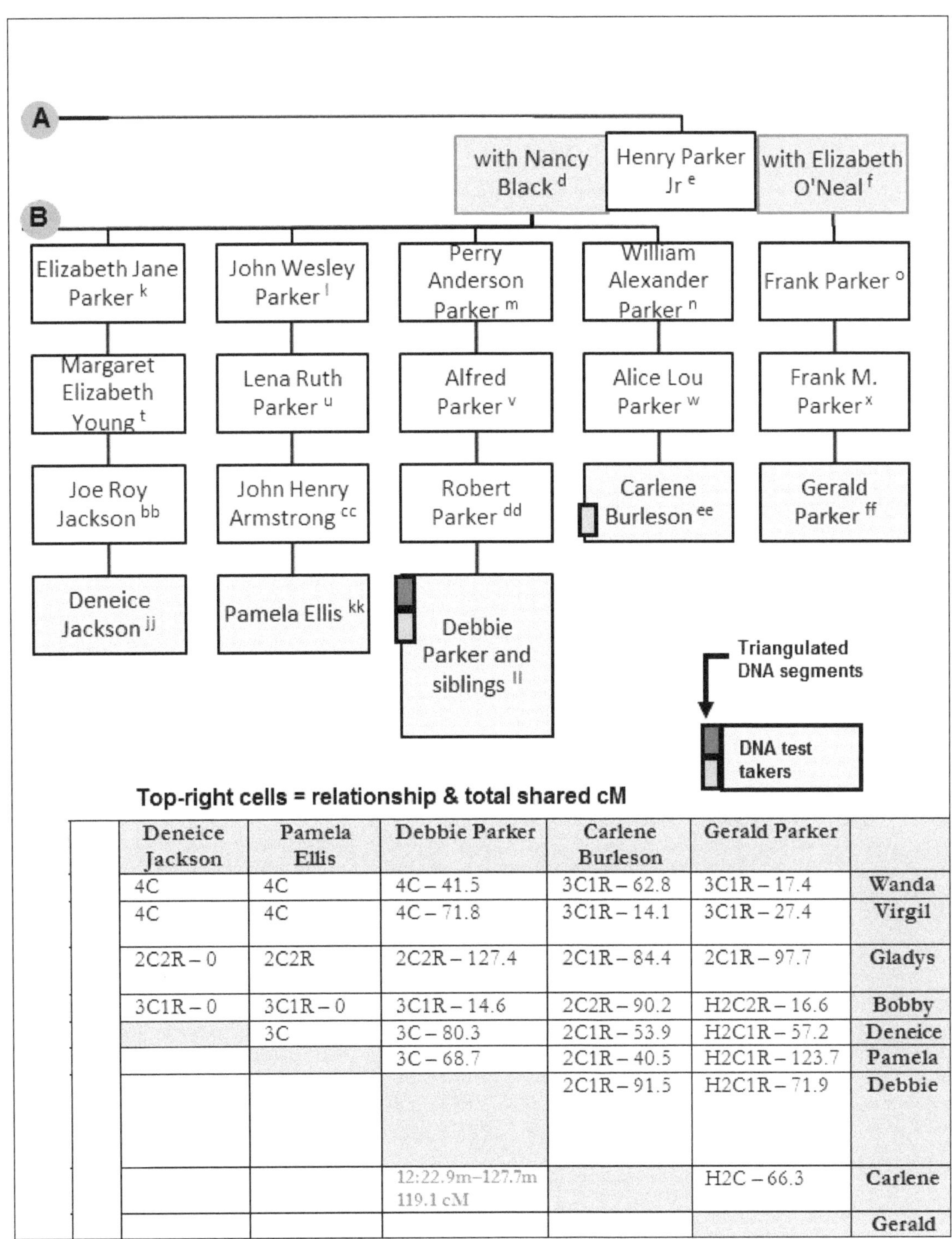

Deneice Jackson	Pamela Ellis	Debbie Parker	Carlene Burleson	Gerald Parker	
4C	4C	4C – 41.5	3C1R – 62.8	3C1R – 17.4	**Wanda**
4C	4C	4C – 71.8	3C1R – 14.1	3C1R – 27.4	**Virgil**
2C2R – 0	2C2R	2C2R – 127.4	2C1R – 84.4	2C1R – 97.7	**Gladys**
3C1R – 0	3C1R – 0	3C1R – 14.6	2C2R – 90.2	H2C2R – 16.6	**Bobby**
	3C	3C – 80.3	2C1R – 53.9	H2C1R – 57.2	**Deneice**
		3C – 68.7	2C1R – 40.5	H2C1R – 123.7	**Pamela**
			2C1R – 91.5	H2C1R – 71.9	**Debbie**
		12:22.9m–127.7m 119.1 cM		H2C – 66.3	**Carlene**
					Gerald

Figure 7.1.b. Parker descendants, atDNA test takers (extended on the facing page)

Citations for figure 7.1

Citations here are for parent-child links between the person in figure 7.1 referenced by the note identifier and the parent of that person. To save space, NARA microfilm publications are cited as Publication Number-Roll Number (such as M1650-40 or T627-4134). The test taker or account manager provided identity of parent and of test taker; no documentation is cited for living persons in most cases. Permission to use test taker names, lineages, and DNA test results for this study are in the author's files.

a. N. D. Shinn affidavit, August 1896, George W. Parker [older] claim no. 2405; Applications from the Bureau of Indian Affairs, Muskogee Area Office, Relating to Enrollment in the Five Civilized Tribes under the Act of 1896; Records of the Bureau of Indian Affairs, RG75; M1650-40. The affidavit names sons of Henry [Sr.] as George W., Edward, Ret, and Henry, and daughters of Henry [Sr.] as Winny, Sally, and Hester, other sons and daughters mentioned but not named.

b. Ibid.

c. Ibid. 1850 U.S. census, Pope County, Arkansas, pop. sch., p. 287 (stamped), dw./fam. 643, Henry Parker; M432-29.

d. Debbie Parker Wayne, "Bailing, Bigamy, Brother Love: The Family of Henry Parker and Nancy Black," Pegasus, Journal of the Dallas Genealogical Society 1 (2013): 11–20; PDF online at http://debbiewayne.com/pubs/pub_HenryParker2012_DPW.pdf.

e. N. D. Shinn affidavit, August 1896, George W. Parker claim no. 2405. Debbie Parker Wayne, "Bailing, Bigamy, Brother Love: The Family of Henry Parker and Nancy Black," Pegasus.

f. Ibid.

g. 1850 U.S. census, Pope County, Arkansas, pop. sch., p. 282 (stamped), dw./fam. 578, Jeremiah Woolcot; M432-29.

h. 1870 U.S. census, Pope County, Arkansas, pop. sch., p. 370B (stamped), dw. 856, fam. 874, Erasmus D. Ford; M593-61. 1880 U.S. census, Pope County, Arkansas, pop. sch., ED 137, p. 90B (stamped), dw. 90, fam. 91, Erasmus D. Ford; T9-54.

i. "Demurrer and Answer;" August 1896, George W. Parker claim no. 2405, Applications from the Bureau of Indian Affairs, Muskogee Area Office, Relating to Enrollment in the Five Civilized Tribes under the Act of 1896; Records of the Bureau of Indian Affairs, RG75; M1650-40; NARA, D.C.

j. Texas Department of Health, death certificate no. 10068 (1927), G. W. Parker [younger]; Bureau of Vital Statistics, Austin. Milam County, Texas, Deed Book 178:17–18, G. W. Parker affidavit, 2 May 1925; County Clerk's Office, Cameron; G. W. names his father.

k. 1860 U.S. census, Milam County, Texas, pop. sch., San Gabriel PO, p. 42A/B (stamped), dw./fam. 250, Henry Parker; M653-1301. Debbie Parker Wayne, "Bailing, Bigamy, Brother Love: The Family of Henry Parker and Nancy Black," *Pegasus.*

l. Wayne, "Bailing, Bigamy, Brother Love: The Family of Henry Parker and Nancy Black," *Pegasus.*

m. Texas Department of Health, death certificate no. 38890 (1925), Anderson Perry Parker; Bureau of Vital Statistics, Austin. 1860 U.S. census, Milam County, Texas, pop. sch., San Gabriel PO, p. 42A/B (stamped), dw./fam. 250, Henry Parker; M653-1301.

n. 1860 U.S. census, Milam County, Texas, pop. sch., San Gabriel PO, p. 42A/B (stamped), dw./fam. 250, Henry Parker; M653-1301. Wayne, "Bailing, Bigamy, Brother Love: The Family of Henry Parker and Nancy Black," *Pegasus.*

o. Texas Department of Health, death certificate no. 17910 (1942), Frank Parker; Bureau of Vital Statistics, Austin. 1900 U.S. census, Hood County, Texas, pop. sch., precinct 4, ED 91, p. 239A (stamped), dw./fam. 548, Henry Parker; T623-1645.

p. 1870 U.S. census, Pope County, Arkansas, pop. sch., Illinois Township, p. 382A (stamped), dw. 593, fam. 608, William Woolcutt; M593-61. 1880 U.S. census, Pope County, Arkansas, pop. sch., Wilson, ED 139, p. 133B (stamped), dw. 133, fam. 134, Wm. Woolcutt; T9-54.

q. 1900 U.S. census, Jefferson County, Arkansas, pop. sch., Barraque, ED 80, p. 214B (stamped), dw./fam. 151, James Roberts; T623-63. 1910 U.S. census, Polk County, Arkansas, pop. sch., Mountain View, ED 109, p. 130A (stamped), dw. 226, fam. 235, James F. Roberts; T624-61.

r. Parent also identified by test taker's nephew who administers the account.

s. Milam County, Texas, Deed Book 175:108–109, W. H. Parker et al. to G. W. Parker, 25 June 1913; County Clerk's Office, Cameron. Milam County, Texas, Deed Book 207:310–311, Dan Parker et al. to D. W. Stephens, 9 April 1928; County Clerk's Office, Cameron. Names father and links to Henry.

t. Texas Department of Health, death certificate no. 12814 (1966), Margaret Elizabeth Jackson; Bureau of Vital Statistics, Austin. 1900 U.S. census, Falls County, Texas, pop. sch., Welderville ED 27, p. 314B/315A (stamped), dw./ fam. 354, A. J. Young; T623-1632.

u. 1900 U.S. census, Chickasaw Nation, Indian Territory, pop. sch., T2N R4E, ED 133, p. 142A (stamped), dw.226, fam. 241, John Parker; T623-1847.

v. Texas Department of Health, death certificate no. 1391 (1955), Alfred Parker; Bureau of Vital Statistics, Austin. 1900 U.S. census, Lee County, Texas, pop. sch., precinct 7, ED 62, p. 124A (stamped), dw. 116, fam. 117, Perry A. Parker.

w. Taylor County, Texas, Marriage, 1:78, W. D. Parker-Annie G. Coats, 6 July 1884. 1910 U.S. census, Garvin County, Oklahoma, Simpson, ED 86, p. 15A (stamped), dw./fam. 51, Sintha A. Parker; T624-1252. "Oklahoma, County Marriage Records, 1890–1995," Aaron Burleson-Alice Parker, 1 September 1911, Garvin County; digital images, *Ancestry.com* (http://ancestry.com/ : accessed 17 July 2018); citing FHL film 1,398,971; links Alice Parker to Alice Burleson.

x. Texas Department of Health, death certificate no. 65072 (1965), Frank M. Parker; Bureau of Vital Statistics, Austin). 1930 U.S. census, Hood County, Texas, Granbury, precinct 1, ED 111-1, p. 225A/B (stamped), dw./fam. 13, Frank Parker; T626-2355.

y. Searcy County, Arkansas, delayed birth certificate no. 79132 (issued 1942), Beldon Willcutt, Arkansas Bureau of Vital Statistics, Little Rock. Arizona Department of Health Services, death certificate no. 4060 (1957), Beldon Willcutt; digital image "Arizona, Death Records, 1887-1960," *Ancestry.com* (http://www.ancestry.com : accessed 18 May 2009). Parent also identified by test taker.

z. 1930 U.S. census, Grant County, Arkansas, Simpson, ED 11, p. 265A (stamped), dw./fam. 51, Vincent Lott; T626-70. "U.S. World War II Draft Registration Cards, Young Men, 1940–1947," digital images, *Ancestry.com* (http://www.ancestry.com/ : accessed 21 January 2019), Jissie Mason Hale, serial no. 698, Order No. 135, Local Draft Board, Sheridan, Grant County, Arkansas; citing *Records of the Selective Service System, 1926–1975*, RG 147.

aa. Texas Department of Health, death certificate no. 50507 (1982), William Hugh Parker; Bureau of Vital Statistics, Austin. 1910 U.S. census, Milam County, Texas, pop. sch., precinct 4, ED 60, p. 1A, dw. 7, fam. 7, Perry Parker; T624-1578.

bb. 1930 U.S. census, Falls County, Texas, precinct 7, ED 24, p. 63A (stamped), dw. 301, fam. 309, Joe Jackson; T626-2330. 1940 U.S. census, Tom Green County, Texas, pop. sch., San Angelo, ED 226-1, p. 14A (stamped), household 311, Margaret Jackson; T627-4147.

cc. "U.S. World War II Draft Registration Cards, Young Men, 1940–1947," digital images, *Ancestry.com* (http://www.ancestry.com/ : accessed 21 January 2019), John Henry Armstrong, serial no. W351, Order No. 12,300, Local Draft Board 1, Pryor, Mayes County, Oklahoma; citing *Records of the Selective Service System, 1926–1975*, RG 147. 1930 U.S. census, Jefferson County, Oklahoma, pop. sch., Earl, ED 7, p. 124A (stamped), dw./fam. 90, William W. Armstrong. Parent also identified by test taker.

dd. Arkansas Department of Health, death certificate no. 2004.023616 (2004), Robert C. Parker, Little Rock, Arkansas. Parent also identified by test taker.

ee. Parent identified by child of deceased test taker.

ff. Parent identified by test taker.

gg. Parent identified by test taker. Use of alias requested.

hh. 1940 U.S. census, Grant County, Arkansas, pop. sch., Merry Green, ED 27-10, p. 89A (stamped), household 155, Essie M. Hale; T627-138. Parent also identified by test taker.

ii. "Texas, Birth Index, 1903–1997," database, *Ancestry.com* (http://www.ancestry.com : accessed 21 August 2018), William Randall Parker, 6 February 1939; Texas Department of State Health Services, Austin. 1940 U.S. census, San Saba County, Texas, pop. sch., Dallas, ED 206-1, p. 26A (stamped), household 564, Hugh Parker; T627-4134.

jj. Parent identified by test taker.

kk. Parent identified by test taker.

ll. Robert C. Parker, obituary, *The Mena (Arkansas) Star*, 25 November 2004, p. 2A. Parent also identified by test taker.

mm. "U.S., Obituary Collection, 1930–2018," database, *Ancestry.com* (http://www.ancestry.com : accessed 21 August 2018), William Randall Parker, 6 February 1939. Parent also identified by test taker.

Sizes and locations of shared DNA segments should be considered during analysis.[65] Locations of segments are not available from all testing companies, but third-party tools can provide the information when test takers give permission.[66]

Descendants of Henry Parker Jr. and Henry Parker Sr. share total amounts of DNA as seen in tables 7.2 and 7.3. The number of shared segments, not shown, varies from one to six. It is possible many of these segments will never triangulate because of random DNA inheritance, but some do triangulate as will be shown in the "Triangulate Segments" section.

Table 7.3. Shared atDNA in cM of Parker test takers

Ancestral Path		Henry Jr. to Perry A.	Henry Jr. to George W. (1854)	Henry Jr. to Alex. Wm.	Henry Jr. to Frank	Hester to Winnie
	Test takers	Debbie Parker	Bobby Parker	Carlene Burleson	Gerald Parker	Virgil Dale Hale
Winnifred to Wm. Henry	Wanda Willcutt	41.5 4C	10.4 4C1R	62.8 3C1R	17.4 3C1R	32.3 4C
Hester to Winnie	Virgil Dale Hale	71.8 4C	0 4C1R	14.1 3C1R	27.4 3C1R	
Henry Jr. to Frank	Gerald Parker	71.9 half-2C1R	16.6 half-2C2R	66.3 half-2C		27.4 3C1R
Henry Jr. to Alex. Wm.	Carlene Burleson	91.5 2C1R	90.2 2C2R		66.3 half-2C	14.1 3C1R
Henry Jr. to George W.	Bobby Parker	14.6 3C1R		90.2 2C2R	16.6 half-2C2R	0 4C1R
Henry Jr. to Elizabeth	Deneice Jackson	80.3 3C	0 3C1R	53.9 2C1R	57.2 half-2C1R	0 4C
Henry Jr. to John Wesley	Pamela Ellis	68.7 3C	0 3C1R	40.5 2C1R	123.7 half-2C1R	16.4 4C
George W. (1836) to Thomas	Gladys Irene Parker	127.4 2C2R	0 2C3R	84.8 2C1R	97.7 2C1R	91.2 2C2R

Refer to figure 7.1 for the test takers' full lineage to Henry Parker Sr. The ancestral path row and column lists only the child and grandchild of Henry Parker Sr. in the test taker's lineage.

65. BCG, *Genealogy Standards*, 2nd ed., 30–31, standard 52, "Analyzing DNA test results," bullet 4.

66. BCG, *Genealogy Standards*, 2nd ed., 32, standard 57, "Respect for privacy rights."

The shared segment list for the Parker descendants was compared to a list of areas where pileups and multiple matches are often found.[67] This list includes the centromeres, the HLA region, SNP-poor regions, and known excess IBD regions.[68] Genealogists differ over whether these segments should be included in analysis and how long a segment should be in these areas to be considered IBD. For this analysis, a conservative position was taken. One segment each on chromosomes 1 and 17 was eliminated from the analysis because they are part of an excess IBD region. One segment on chromosome 9 was eliminated from the analysis because it crosses the centromere and is relatively small in size (15 cM across the centromere). More investigation may show this to be an IBD segment.

In addition to problem chromosome areas that apply to all humans, each person can have individual pileup regions. Some testing companies take this into account before providing the list of DNA matches. *DNA Painter* allows a test taker to determine where these individual pileup areas occur.[69] If a chromosome segment contains hundreds or thousands of matches it is probably a pileup region. If a chromosome segment contains fewer matches and most of those can be identified as paternal or maternal matches, the shared segment is more likely to be a true segment. One segment on chromosome 6 was eliminated from analysis as a focus test taker has over 2,000 matches on that segment.

The total amounts of atDNA shared, as seen in table 7.3, by the hypothesized Parker cousins is consistent with the amounts reported in the Shared cM Project.[70]

Triangulated segments

Some cousins share DNA segments with others, but all cousins do not share the same segments. Triangulated groups, where multiple lines from a common ancestor share the same or an overlapping segment, also provide credible evidence when multiple cousins inherit the same or overlapping DNA segment.[71] Triangulated segments provide a higher level of credibility when pedigrees are complete and accurate, reducing the possibility of assigning a segment to the wrong ancestors.

With many descendants in the Parker FamGroup1 Project, the project administrator can confirm whether each of the project members shares the same or an overlapping segment with all of the others.[72] A dozen or so segments are shared, but there is not enough information to confirm triangulation. Two large triangulated segments link descendants of Henry Parker Sr. and Henry Parker Jr. Refer back to figure 7.1 for the lineage of these test takers.

67. BCG, *Genealogy Standards*, 2nd ed., 30–31, standard 52, "Analyzing DNA test results," bullet 5.

68. Sue Griffith, "Chromosome Maps Showing Centromeres, Excess IBD Regions and HLA Region," *Genealogy Junkie* (http://www.genealogyjunkie.net/blog/chromosome-maps-showing-centromeres-excess-ibd-regions-and-hla-region). See also "Identical by descent," *ISOGG Wiki* (https://isogg.org/wiki/Identical_by_descent).

69. Jonny Perl, *DNA Painter* (https://dnapainter.com/).

70. Blaine T. Bettinger, "Shared cM Project," *The Genetic Genealogist* (https://thegeneticgenealogist.com/).

71. BCG, *Genealogy Standards*, 2nd ed., 30–31, standard 52, "Analyzing DNA test results," bullets 6 and 7.

72. Debbie Parker Wayne, Project administrator, "Parker FamGroup1 Project" (https://www.familytreedna.com/groups/parker-fam-group-1/).

In figure 7.2, chromosome 11 includes a triangulated segment starting at about location 23.0–44.5 million (Mb) and ending at about location 76.3–82.7 Mb, with overlapping segment sizes between 21.1 and 48.1 cM. This segment is shared by four lines of descent from Henry Parker Sr. Of those, Wanda descends from daughter Winnifred (Parker) Woolcott/ Willcutt, Virgil descends from daughter Hester (Parker) Ford, Debbie and siblings descend from son Henry Parker Jr., and Gladys descends from son George W. Parker (born 1836). Siblings and first cousins who share this segment are counted only once as part of the group descending from one child of Henry Sr.

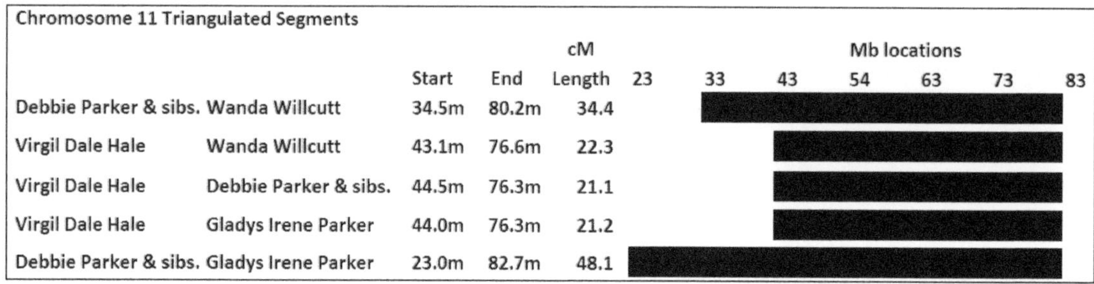

Chromosome 11 Triangulated Segments

		Start	End	cM Length	Mb locations 23 · 33 · 43 · 54 · 63 · 73 · 83
Debbie Parker & sibs.	Wanda Willcutt	34.5m	80.2m	34.4	
Virgil Dale Hale	Wanda Willcutt	43.1m	76.6m	22.3	
Virgil Dale Hale	Debbie Parker & sibs.	44.5m	76.3m	21.1	
Virgil Dale Hale	Gladys Irene Parker	44.0m	76.3m	21.2	
Debbie Parker & sibs.	Gladys Irene Parker	23.0m	82.7m	48.1	

Figure 7.2. Triangulated segment on chromosome 11

In figure 7.3, chromosome 12 includes a triangulated segment starting at about location 22.9–92.6 Mb and ending at about location 113.5–127.7 Mb, with overlapping segment sizes between 25.0 and 119.1 cM. This segment is shared by four lines of descent from Henry Parker Sr. Of those, Wanda descends from Winnifred (Parker) Woolcott, Carlene and Debbie plus her siblings descend from Henry Parker Jr., and Gladys descends from George W. Parker (1836). Siblings and first cousins who share this segment are counted only once as part of the group descending from one child of Henry Sr.

Chromosome 12 Triangulated Segments

		Start	End	cM Length	Mb locations 22 · 42 · 62 · 82 · 102 · 112 · 122
Debbie Parker & sibs.	Wanda Willcutt	92.6m	113.5m	25.0	
Carlene Burleson	Wanda Willcutt	92.2m	113.5m	25.4	
Carlene Burleson	Debbie Parker & sibs.	22.9m	127.7m	119.1	
Debbie Parker & sibs.	Gladys Irene Parker	92.6m	126.8m	51.7	
Carlene Burleson	Gladys Irene Parker	92.6m	126.8m	51.7	

Figure 7.3. Triangulated segment on chromosome 12

It is possible that additional information might refute the triangulation hypothesis. Perhaps the test takers share other yet-unknown ancestors, come from a shared population, or have recent pedigree collapse in the tree. However, the larger the segment size and the more lines from the same ancestor sharing the same segment, the less likely the hypothesis is false. Many more case studies with dozens of test takers are needed to confirm or refute the credibility of triangulated segments. So far, it appears triangulated segments are often valid, but triangulation cannot be expected in every case.

Two triangulated segments of significant size support links between hypothesized descendants of Henry Parker Jr. and Henry Parker Sr. Evidence from documents, Y-DNA, and atDNA are all consistent with the hypothesis that these families are linked. We have not yet eliminated the possibility of other common ancestors from whom this DNA might have been inherited. Mapping and clustering tools provide clues to help with this.

Chromosome mapping and Walking the Ancestor Back

The "Visual Phasing Methodology and Techniques" chapter covers a method to map chromosome segments to a grandparent when the grandparents are not available to test themselves. Additional analysis allows researchers to map chromosome segments further back based on identifying the shared ancestor. "Walking the Ancestor Back" is a confirming methodology for triangulated segments discussed in the "Lessons Learned from Triangulating a Genome" chapter.

The triangulated and shared segments just discussed trace to Henry Parker Sr. on the author's paternal grandfather's line. When compared to the chromosome map created by visual phasing, all segments shared with Parker descendants correspond to the segments inherited from the paternal grandfather. This analysis is not included here because of space limitations and this concept is covered in several other chapters.

Shared matches and In Common With

Most testing companies offer a tool to identify people who share DNA with two or more test takers. This tool is sometimes called Shared Matches, In Common With (ICW), or a similar name. Researchers have always been able to make textual lists to analyze these shared matches. Shared Matches and ICW lists do not guarantee that the DNA from all of the test takers on the list came from the same common ancestor. However, large lists where no other common ancestor can be found in complete and accurate trees increases the probability that the DNA came from one common ancestor. FamilyTreeDNA offers a matrix tool that displays the data in a chart. Genetic network graphs can also be used to analyze this data.

Graphics and genetic networks

Genetic networks offer supporting evidence for links. While it may be necessary to create, expand, or document trees for some matches, analysis of a network may take less time than needed to re-research the complete trees of *dozens* of cousins. Networks are based on shared matches, in common with, and shared locations. Genetic networks may provide evidence to support DNA conclusions, but some depend on tree accuracy. These tools include Ancestry DNA Circles,[73] *ConnectedDNA*'s graphs,[74] *Genetic Affairs* Auto Clustering,[75] and Leeds Color Clustering.[76]

73. "DNA Circles," *Ancestry* Support (https://support.ancestry.com/s/article/How-DNA-Circles-are-created-1460089695851).

74. Shelley Crawford, *Connected DNA* (https://www.connecteddna.com/).

75. Evert-Jan Blom, *Genetic Affairs* (http://geneticaffairs.com/).

76. Dana Leeds, "DNA Color Clustering: The Leeds Method for Easily Visualizing Matches," *DNA with DANA LEEDS* (http://www.danaleeds.com/).

DNA Circles

DNA Circles at AncestryDNA are formed when test takers have: *(1)* public trees, *(2)* the DNA test is linked to a person in the public tree, *(3)* two test takers match each other as second or more distant cousins, and *(4)* a common ancestor is named in both trees (with the name and facts corresponding closely enough for the algorithm to deduce that the named ancestors are the same person). DNA Circles create a genetic network linking a test taker to *(1)* those who share DNA with the test taker (a DNA match) and *(2)* those who share DNA with your DNA matches, but not with you (a DNA Match to a circle). DNA Circles change based on the contents of each tree; it is wise to save information on circles when viewing the data as it may be different when next viewed.[77]

Table 7.4 summarizes Debbie Parker Wayne's DNA Circle for Henry Parker Sr. as of April 2016. This circle includes thirteen test takers:

- three in a group descended from Henry Parker Jr.'s son Jade or J. D.
- four in a group descended from Henry Jr.'s daughter Elizabeth
- three in a group descended from Henry Jr.'s son John Wesley
- three not in a named group

This provides support for Debbie's pedigree back to Henry Parker Sr. if the trees of these test takers are complete to the Henry Parker Sr. level or further, contain no gaps to that generational depth, and contain accurate parent-child links. That is not generally a safe assumption to make.

Table 7.4 DNA Circles Genetic Network, April 2016

Family Group: Descendants of ...	Number	DNA Conf.[a]	Notes
Sr. > Henry Parker Jr. > son J. D. "Jade" Parker	3		J. D. Jade" Parker FG
Sr. > Henry Parker Jr. > daughter Elizabeth Parker Young	4		Henry Parker FG (Links to Henry Sr., Henry Jr., and Nancy Black)
Sr. > Henry Parker Jr. > son John Wesley Parker	3		John Wesley Parker FG (Links to Henry Sr., Henry Jr., and Nancy Black)
Individuals not in named group	3		
Total Members of DNA Circles for Henry Parker Sr. plus Henry Parker Jr. and Nancy Black, **April 2016**	**13**		

a. The values of the confidence prediction were not recorded in 2016.

77. Some restrictions may be placed on viewing DNA Circles for those without an *Ancestry* paid membership..

Table 7.5 summarizes Debbie Parker Wayne's DNA Circle for Henry Parker Sr. as of January 2019. This circle includes eighty-one test takers:

- nine (increased from three) in *two* strong confidence groups descended from Henry Jr.'s son Jade or J. D.
- ten (increased from four) in a strong confidence group descended from Henry Jr.'s daughter Elizabeth
- five (increased from three) in a strong confidence group descended from Henry Jr.'s son John Wesley
- nineteen (increased from three) not in a named group (rated weak to strong)
- six in a strong confidence group descended from Henry Sr.'s daughter Winnifred
- three in a weak confidence group descended from Henry Sr.'s daughter Hester
- four in a strong confidence group descended from Henry Jr.'s son William Alexander
- three in a strong confidence group descended from Henry Jr.'s son George W. [younger]
- two in a lower-level confidence group descended from Henry Jr.'s son George W. [younger] (George's son Perry Cornelius and grandson William Henry)
- two in a strong confidence group descended from Henry Jr.'s son George W. [younger] (George's son Perry Cornelius and grandson William Hugh)
- nine in a strong confidence group descended from Henry Jr.'s sons Perry Anderson
- six in a strong confidence group descended from Henry Jr.'s daughter Laura Alice "Lena"
- three in a strong confidence group with links to Milam County, Texas (where Henry Parker Jr. and Nancy Black resided)

This provides stronger support for Debbie's pedigree back to Henry Parker Sr. if the trees of these test takers are complete to the Henry Parker Sr. level or further, contain no gaps to that generational depth, and contain accurate parent-child links.

Each of these test takers is linked to other descendants of Henry Parker Sr. and Henry Parker Jr. Most are found only in the DNA Circles for the Parker family and not in any other circles. While some of the test takers have gaps in their trees, many have extensive trees. Some trees are not well-documented and do not provide credible evidence unless additional research extends or confirms the parent-child links. However, the sheer number of test takers who appear only in Parker-related circles reduces the likelihood of multiple common ancestors between the test takers. More connections and stronger connections between groups in a circle reduce the likelihood of chance connections.[78]

78. Catherine A. Ball et al., "DNA Circles™ White Paper," *AncestryDNA* (https://www.ancestry.com/corporate/sites/default/files/AncestryDNA-DNA-Circles-White-Paper.pdf), 27.

Table 7.5. DNA Circles Genetic Network, January 2019			
Family Group: Descendants of ...	**#**	**DNA Conf.[a]**	**Notes**
Henry Parker Sr. > daughter Winnifred Parker Woolcott	6	Strong	Amos Frank Anderson FG (Links to Henry Sr.)
Henry Parker Sr. > daughter Hester Jane Parker Ford	3	Weak	Erasmus W. (Ross) Ford FG (Links to Henry Sr.)
Sr. > Henry Parker Jr. > son William Alexander Parker	4	Strong	William Alexander Parker FG (Links to Henry Sr., Henry Jr., and Nancy Black)
Sr. > Henry Parker Jr. > son George W. Parker	3	Strong	George W Parker FG (Links to Henry Sr., Henry Jr., and Nancy Black)
Sr. > Henry Parker Jr. > son George W. Parker > grandson Perry Cornelius Parker > William Henry Parker	2	Some	William Henry Parker FG (Links to Henry Jr. and Nancy Black)
Sr. > Henry Parker Jr. > son George W. Parker > grandson Perry Cornelius Parker >William Hugh Parker	2	Strong	William Hugh Parker FG (Links to Henry Jr. and Nancy Black)
Sr. > Henry Parker Jr. > son J. D. "Jade" Parker	9	Strong	Alice Virginia Richard FG (Links to Henry Sr., Henry Jr., and Nancy Black) J. D. Jade" Parker FG
Sr. > Henry Parker Jr. > son Perry Anderson Parker	9	Strong	Perry Anderson Parker FG (Links to Henry Sr., Henry Jr., Nancy Black) Wade Anderson Parker FG (Links to Henry Jr., Nancy Black, and Perry Anderson)
Sr. > Henry Parker Jr. > daughter Elizabeth Parker Young	10	Strong	Henry Parker FG (Links to Henry Sr., Henry Jr., and Nancy Black)
Sr. > Henry Parker Jr. > son John Wesley Parker	5	Strong	John Wesley Parker FG (Links to Henry Sr., Henry Jr., and Nancy Black)
Sr. > Henry Parker Jr. > daughter Laura Alice "Lena" Parker	6	Strong	Earl Thomas FG (Links to Henry Sr., Henry Jr., and Nancy Black)
Milam County Parkers with tree gaps and no named link	3	Strong	Betty Joyce FG (Links to Henry Jr., Nancy Black, and Perry Anderson)
Individuals not in named group	9	Strong	
Individuals not in named group	7	Good	
Individuals not in named group	2	Some	
Individuals not in named group	1	Weak	
Total Members of DNA Circles for Henry Parker Sr. plus Henry Parker Jr. and Nancy Black, **January 2019**	**81**		
a. The hierarchy for the DNA Confidence predictions is Strong, Good, Some, Emerging, Weak. See the Ancestry "DNA Circles™ White Paper" for more information.			

The conclusion that the shared DNA was inherited from Henry Parker Sr. and his wife is supported by several factors. First, some of the trees are complete and accurate out to the level of the hypothesized common ancestor. Second, the number of members in many of the circles is high. Third, the circle members descend from multiple children of both Henry Parker Sr. and Henry Parker Jr. covering many lines back to Henry Sr.

Many Parker descendants have tested at AncestryDNA where the locations of shared segments are not indicated. AncestryDNA provides only the total amount of share DNA and the number of shared segments. The pedigrees and total amount of atDNA shared by all of these matches should be correlated in the same way as was done for the Parker cousins shown in figure 7.1 and table 7.2

The total amount of shared DNA may differ at each testing company as each uses different matching algorithms. In some instances, cousins will be listed on a match list at one company but not another because of the different algorithms. Nevertheless, when cousins have tested at multiple companies, the shared DNA may allow us to link to additional cousins.

Leeds Color Clustering

The Leeds Color Clustering Method offers a visual method to organize shared matches. Originally designed for use on AncestryDNA, the method can be used on any site that identifies shared matches.[79] This is helpful to researchers who learn visually. Clustering informs the researcher of the likelihood of multiple shared ancestors within a group.

When beginning, each person on a match list is represented in one row of a spreadsheet or table. Names of matches can be manually entered or extracted using a tool such as the *DNAGedcom* Client[80] or DNA Match Manager.[81] Ancestral lines can be simply identified as group 1, group 2, and so on as shown in figure 7.4. If each row has only one color assigned then there is no indication that multiple common ancestors are shared in recent generations by the test taker and the person on the match list. If many rows have more than one color assigned then there may be pedigree collapse or endogamy in the lineage of the test taker. This complicates analysis as described in "The Challenge of Endogamy and Pedigree Collapse" chapter.

As common ancestors are identified, ancestral surnames can be assigned to groups as shown in figure 7.5. Additional columns can be added for the predicted and actual relationships, amount of shared DNA, and other data that is helpful during the analysis. In this example, two colors are found only on a few rows. Two colors are seen when two siblings from one

79. Dana Leeds, "DNA Color Clustering: The Leeds Method for Easily Visualizing Matches," *DNA with DANA LEEDS* (http://www.danaleeds.com/). See also Roberta Estes, "The Leeds Method," *DNA-eXplained*, 26 September 2018 (https://dna-explained.com/2018/09/26/the-leeds-method/).

80. *DNAGedcom* (http://dnagedcom.com/). The *DNAGedcom* Client requires a subscription, but other tools on the site are freely available.

81. Ann Raymont, "DNA Match Manager," *DNASleuth*, 1 September 2018 (https://dnasleuth.wordpress.com/2018/09/01/dna-match-manager/).

family married two first cousins from another family giving the descendants shared ancestors on both maternal and paternal lines: dark teal and brown represent the grandparents shared by these cousins. The teal and brown lines are not related to the Parker lines. However, two third cousins share both Parker (royal blue) and Rogers (red) ancestors. This must be investigated as it may complicate the Parker analysis.

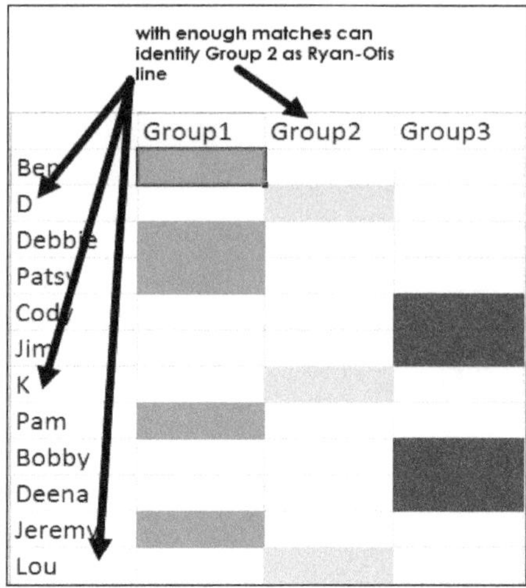

Figure 7.4. Starting with Leeds Color Clustering

Names of Matches - 2018-10-01 1st 10 pages of	Predicted Rel	Actual Rel	cM / #	Parker	Rogers	Anderson	Richards	Johnson
	1C	1C	874 / 29	tree min	tree min			
	1C	1C	847 / 37	tree 5+ gens	tree 5+ gens			
	2C	1C	623 / 28	tree bare				
	2C	1C1R	551 / 25	tree 2 gens				
	2C	2C	237 / 12			tree mine	tree mine	
	2C	2C-half	231 / 9					
	2C	2C	207 / 13			tree mine	tree mine	
	3C							
	3C	3C						
	3C							
	4C+							
	4C+							
	4C+							
	4C+							
	4C+							
	4C+							
	4C+							
	4C+							
	4C+							
	4C+							
	4C+							

Figure 7.5. Expanded Leeds Color Clustering worksheet

Genetic Affairs Autoclustering

Genetic Affairs offers automatic retrieval of DNA match updates and some analysis tools. Autoclustering organizes shared match lists graphically. The information is also provided in a spreadsheet for additional analysis of the network.[82]

ConnectedDNA

ConnectedDNA[83] graphs display the shared match lists in a visual format that is easy to explore. The web of relationships between groups and group members is easy to navigate and patterns become visible. Groups, in different colors, can be assigned to surnames or ancestral couples based on the names of matches as seen in figure 7.6.

The data in the spreadsheet provided by ConnectedDNA will be invaluable during the investigation of matches who may share more than one common ancestor. It is not apparent in this figure, but test taker names are visible when zoomed in. This graph includes 1,006 people (each represented by a colored dot on the graph), with 13,556 lines between them. A spreadsheet is provided as part of the ConnectedDNA product with clusters marked and options to filter to groups of interest in the current research phase.[84] Filtering the spreadsheet to show only the Parker cluster matches helps focus research on the matches likely to contribute to solving this problem. The spreadsheet also includes links to quickly access a match's profile and tree as analysis proceeds.

---

82. Evert-Jan Blom, *Genetic Affairs* (http://geneticaffairs.com/).
83. Shelley Crawford, *Connected DNA* (https://www.connecteddna.com/).
84. Debbie Parker Wayne, "ConnectedDNA Graphs and Clues," *Deb's Delvings in Genealogy*, 12 January 2019 (http://debsdelvings.blogspot.com/2019/01/connecteddna-graphs-and-clues.html). See also Shelley Crawford, "Visualising Ancestry DNA matches–Part 1–Getting ready," *Twigs of Yore*, 1 July 2017 (http://twigsofyore.blogspot.com/2017/07/visualising-ancestry-dna-matchespart.html).

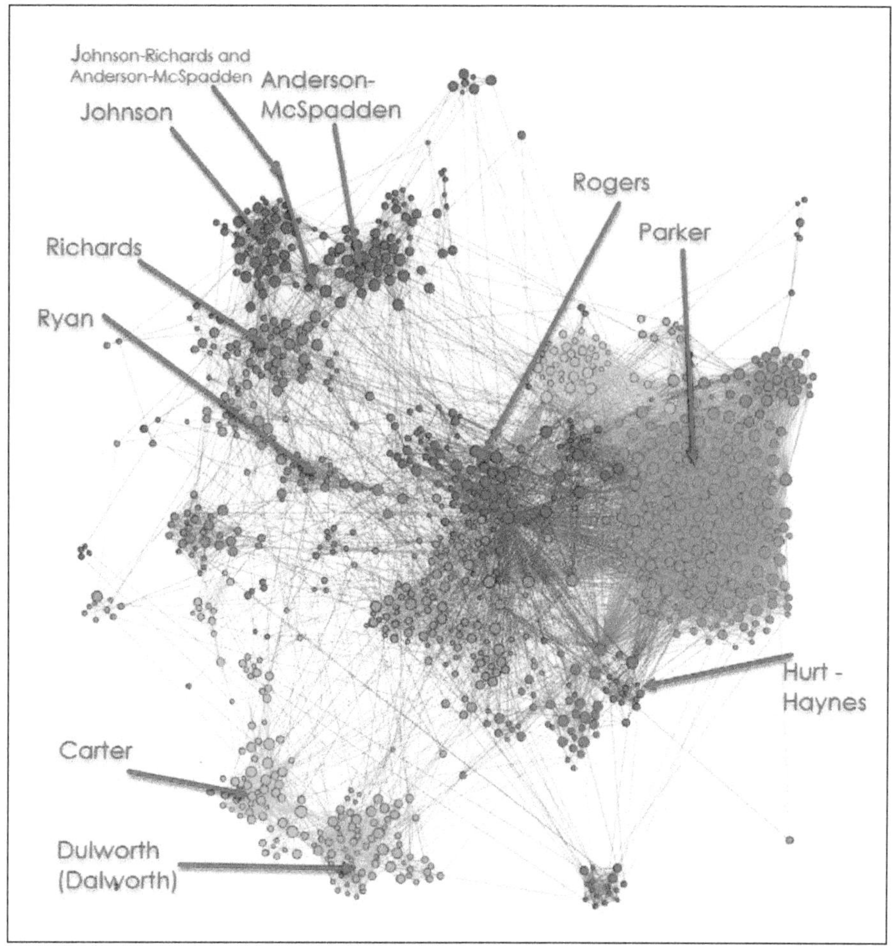

Figure 7.6. *ConnectedDNA* chart for analysis of genetic networks

The sheer number of tools available to assist genealogists with DNA analysis is mind-boggling. And the number grows daily. No one can be an expert on every tool. Pick an interesting tool, study how to use it, then analyze your family data using the tool. Practice is the only way to learn to use the tools and DNA data to solve genealogical problems.

Targeted and Selected Test Takers and Costs of Testing

Many researchers contend that DNA testing should not be required because it is expensive. Other than siblings, the author only targeted and paid for tests for one other test taker to confirm that the Henry Parker in Hood County was the same as the man in Milam County. All other test takers paid for their own tests and were selected from match lists for inclusion in this project after determining they also descended from Henry Parker Sr. or Jr. As more genealogists test, the cost for any individual researcher may only be the cost of his or her own test because hundreds or thousands of cousins will already be in the databases. This makes DNA evidence one of the best bargains in genealogy as it may help answer questions on all family lines, not just one ancestor or family named in a specific document.

Summary and Acknowledgments

When added to the documentary evidence and Y-DNA evidence, the atDNA evidence supports the hypothesis of a father-son relationship between the Henry Parker (Jr.) of Milam County, Texas, and the Henry Parker (Sr.) of Pope County, Arkansas. In addition, Gerald Parker descends from Henry Parker Jr.'s bigamous marriage to Elizabeth O'Neal. His Y-DNA and atDNA link the Henry Parker of Milam County to the Henry Parker of Hood County, Texas. The Henrys documented in the three counties must be one man for the DNA of the descendants of Henry Parker Jr. to match as it does. There is a small chance that another ancestor is shared by these cousins, but the hypothesized relationships are supported by *(1)* the sheer number of matches (many of whom are not described in this essay that was designed to teach concepts), *(2)* the amount of DNA shared by many matches, *(3)* the triangulated segments, and *(4)* the depth and accuracy of many of the trees, with no gaps in the generations under study.

Researchers must learn to accurately analyze DNA test results, but that is true for any source used to establish genealogical proof. Using DNA test results properly requires math and statistical skills, science knowledge, memorizing dozens of acronyms and abbreviations, and, above all, practice.

This analysis would be impossible without all of the Parker descendants who took DNA tests, shared results through the Parker DNA Projects, and allowed their DNA test results to be used. My thanks to all of them. Thanks also to Dennis West and Gregory Parker for organizing all of the Parker test takers into Y-DNA groups. Especially, thanks to everyone who enthusiastically sponsors and actively recruits DNA test takers. Having more test takers in the databases helps us all make new discoveries.

Would You Like Your Data Raw or Cooked?

Ann Turner, MD

Most of the genetic genealogy testing companies offer an option to download a file with raw data. "Raw" is a bit of a misnomer: the DNA in a sample has already been highly processed, split apart into single strands, chopped into small pieces, tagged with fluorescent chemicals that light up when they match a target, and passed through some interpretive steps before the ingredients are finally reduced to the simple labels A, C, G, or T. Each company cooks these ingredients into their own proprietary recipes for admixture analysis and relative matching. The latter category is the focus of this chapter.

File Structure

Many people download the raw data file without even looking at the ingredients, simply uploading it to third-party sites for use in different recipes. However, a peek at the file may reveal some interesting tidbits as well as a deeper understanding of what is happening behind the scenes in the kitchen.

Various companies use different formats, different ways of arranging the data. The contents can be read by the human eye, but the eye may glaze over at the enormity of the file: some 600,000 to 700,000 lines of data. Text editors can open the file, but spreadsheets present the data in a tidy fashion that is easy to manipulate. For example, any format can be converted to another with spreadsheet formulas.

The file will typically start with some header lines flagged with a number sign or pound sign, also called a hashtag (#). These lines describe the source of the data and define the column headers for the data. Tables 8.1 through 8.4 show the format for raw data files from five companies. The files from 23andMe, AncestryDNA, and Living DNA separate the fields with tabs and use the .txt extension. If your operating system defaults into opening a .txt file into a text editor or word processing program, use the File > Open dialog in your spreadsheet to locate and open the file. FamilyTreeDNA and MyHeritage separate the fields with commas and use the .csv extension for Comma Separated Values (CSV). Most operating systems will default into opening a CSV file in a spreadsheet. Regardless of whether the fields are separated with commas or tabs, opening the file in a spreadsheet program deposits the data into cells.

Table 8.1. 23andMe format for raw data

Column headers	# rsid	chromosome	position	genotype
Data line 1	rs4477212	1	82154	CT
Data line 2	i5005018	1	5947454	--

Table 8.2. FamilyTreeDNA and MyHeritage format for raw data

Column headers	RSID	CHROMOSOME	POSITION	RESULT
Data line 1	rs3094315	1	742429	CT
Data line 2	rs3094315	1	752566	--

Table 8.3. LivingDNA format for raw data

Column headers	# rsid	chromosome	position	genotype
Data line 1	rs9283150	1	565508	CC
Data line 2	1:726912	1	726912	--

Table 8.4. AncestryDNA format for raw data

Column headers	rsid	chromosome	position	allele1	allele2
Data line 1	rs4477212	1	82154	C	T
Data line 2	rs2051646	1	7817856	0	0

The first three columns hold identical data for these companies, with rows sorted by chromosome numbers 1 through 22, then position within the chromosome. The companies also include data for the X chromosome (called chromosome "23" by AncestryDNA to keep the field numeric). However, only 23andMe incorporates the X chromosome data into their algorithms. Some companies also include data for the Y chromosome and mitochondrial DNA (mtDNA).

The chromosome and position columns are straightforward, but the first column is more mysterious. Nevertheless, it is a springboard to many concepts and is worth exploring in some detail.

RSID and SNP selection

The abbreviation rsid stands for Reference SNP ID. A centralized database called *dbSNP* assigns numbers to new variants submitted by researchers.[1] Not all variants have been catalogued at *dbSNP*.

The rsid column shows some SNPs that are identified differently. 23andMe uses a catalogue number with the prefix "i" (internal) for some cases. This might be used where a variant described in the literature was never submitted to *dbSNP*, or where multiple probes were designed to capture a variant. Living DNA may list a variant without an rsid with just the chromosome number and position. This method depends on the build number of the human genome, which will be discussed later in this chapter.

Most of the human genome is identical across individuals. Most variants are Single (one) Nucleotide (another word for base) Polymorphisms (poly = many; morph = form), abbreviated *SNP* (pronounced "snip"). The current build of *dbSNP* has a curated list of 113,862,023 positions in the human genome where variants have been observed and confirmed.[2] The number is growing rapidly as whole genome sequencing ramps up.

Obviously, no one person has such a large number of variations. The 1000 Genomes Project compared samples collected from around the world to the human reference sequence. They found that the average number of differences between any one individual and the human reference sequence was in the 4,000,000 to 5,000,000 range.[3]

Whole genome sequencing would reveal all of those differences, but at a price of plowing through mounds of dull results. Most differences reveal nothing beyond the fact that the test taker is a fellow human being. However, sampling just a subset of known variants can be surprisingly informative. The technology for detecting DNA variations at specific locations is called a gene chip or more formally, a microarray. The nickname "chip" comes from the earliest manufacturing methods, which had some similarities to the procedure used for computer chips.

All URLs were accessed 26 January 2019 unless otherwise indicated.

1. *dbSNP: Short Genetic Variations*, National Center for Biotechnology Information (NCBI), US National Institutes of Health (https://www.ncbi.nlm.nih.gov/projects/SNP). The official name has been changed to Short Genetic Variations to reflect the fact that some entries are of a different type, such as insertions or deletions of a few bases (indels). Also, the acronym SNP has been replaced by SNV for Single Nucleotide Variant. Polymorphism has the connotation of a somewhat common variation, but frequency is no longer a distinction. However, the use of the acronym SNP is so ingrained that the database is still referred to as dbSNP.

2. Build Statistics and many interesting facts about reference sequences can be found at "RELEASE: dbSNP Build 151," *dbSNP*, NCBI (https://www.ncbi.nlm.nih.gov/projects/SNP/snp_summary.cgi).

3. The 1000 Genomes Project Consortium, "A Global Reference for Human Genetic Variation," *Nature* 526 (October 2015): 68-74 (https://doi.org/10.1038/nature15393).

Illumina is the chief provider of the chips used by the genetic genealogy companies.[4] There are trade-offs between comprehensive coverage and expense. A chip with space for 600,000 to 700,000 SNPs seems to be a sweet spot. Decisions, decisions. How are those SNPs selected? Here is where *dbSNP* comes in handy. It has much descriptive information about each SNP, helpful for filtering the possibilities down to a manageable number.

What are some desirable characteristics of SNPs to be included? Spreading the SNPs more or less evenly over the whole genome would qualify a chip test as a whole genome *scan* (an important distinction to whole genome *sequencing*). This places the SNPs about 5,000 base pairs apart on average, serving as tags for a region; the SNPs are not necessarily interesting in and of themselves, but they are like mile markers on a highway. Unlike regularly spaced mile markers, some DNA regions may have higher or lower SNP density.[5]

Another practical consideration is *allele* frequency (alternative versions of a marker). A variant that is found in one in a million people might be of great interest to the rare person carrying it, but there are a limited number of locations on the chip. Chips in this size range focus on SNPs where the *Minor Allele Frequency (MAF)* is at least 5% to make its inclusion worthwhile; that means the second most common allele value occurs in at least 5% of a given population. The ideal frequency depends on the application. Somewhat counter-intuitively, the best frequency for detecting relatives in a database is a 50-50 split where both alleles are equally frequent.

Because much of the SNP discovery has been based on samples of European origin, random selections from *dbSNP* might overlook informative SNPs in other geographic regions. The new Global Screening Array (GSA) chip, currently used by 23andMe and Living DNA, aims for more balanced and comprehensive coverage based on the 1000 Genomes Project. The GSA chip will be replacing the Illumina OmniExpress chip formerly used by several companies.

Note that these SNP selection strategies of spacing, frequency, and geographic coverage are not looking for SNPs that affect traits or medical conditions. Indeed, the only thing known about most SNPs is that a mutation occurred sometime somewhere in somebody. SNPs with known functional significance (such as mutations changing the structure of a protein) are add-ons to the basic core of tagging SNPs. Even the GSA chip, intentionally designed to include SNPs of value for clinical research, ends up with only 9% in that category.[6] *SNPedia,*

4. Living DNA introduced a new chip from Thermo Fisher Scientific / Affymetrix in late October 2018, too late for details to be included in this chapter. Their internal name for the chip is *Sirius*. See https://livingdna.com/news/we-have-a-new-chip for more information.

5. Rebekah Canada, "Ancestry V2 Chip," *Haplogroup* (https://haplogroup.org/ancestry-v2-chip). The heat map of SNP density shows low SNP density in the region of the centromeres and a very high SNP density in the HLA on chromosome 6 (Human Leukocyte Antigen; or white blood cell type). The HLA region is highly variable and important for immune systems which must distinguish self from non-self to repel invaders.

6. "Infinium™ Global Screening Array-24 v2.0," *Illumina* (https://www.illumina.com/content/dam/illumina-marketing/documents/products/datasheets/infinium-commercial-gsa-data-sheet-370-2016-016.pdf).

a retrieval system for information about SNPs, lists the number of annotated SNPs from various chips. They range from a low of 12,000 for MyHeritage to a high of 44,000 for AncestryDNA v2.[7]

Your personal data (genotype, allele, result)

The remaining columns in the raw data file pertain to your personal results. FamilyTreeDNA and MyHeritage actually use the word result, but the other companies introduce specialty terms. There are two results for every SNP, one from the paternal chromosome and one from the maternal chromosome. A *genotype* is the combination of the two results. The order is arbitrary, since the small pieces of DNA are floating around in the soup without a way to identify which chromosome they came from. Some companies maintain the order of Illumina's base-calling software, but 23andMe does some post-processing and puts the alleles in alphabetical order. AncestryDNA splits the genotype into two columns, allele1 and allele2. Again, the order is arbitrary. Some of the results in the allele1 column will derive from the paternal chromosome and some from the maternal chromosome.

Tables 8.1 through 8.4 give examples of two different categories of genotypes. A *homozygous* genotype has the same base listed twice (the Living DNA example CC in table 8.3), while a *heterozygous* genotype has two different bases listed (the 23andMe example CT in table 8.1). SNPs on a chip are virtually all bi-allelic, meaning there are only two known alternatives or variants. Theoretically, any base could be changed to any other base, but multi-allelic SNPs are relatively rare, and they are avoided in the SNP selection process.

The tables also illustrate how *no-calls* are presented: with a double dash (--) or zero (0) in both allele columns. The Illumina software evaluates signal intensity when it makes a base call. The chips contain thousands of copies of probes for every SNP. A probe is like a fishing lure, designed to attract a specific kind of fish. The attraction in the case of DNA is a short sequence that will take advantage of the base-pairing rules for forming the famous double helix, where A and T are *complementary* bases, as are C and G. The short sequence is long enough to occur uniquely at just one place in the whole genome. For example, checking *dbSNP* for rs2051646 brings up the bases that flank the position of interest (a C or a T) as shown in figure 8.1.[8] One lure for a SNP will contain a G (to attract a fish with a C in that position) and the other lure will contain an A (to attract a fish with a T in that position). When a fish takes the bait, a fluorescent molecule lights up. If not enough fish float by and take the bait, the signal intensity will be too low to read, resulting in a no-call.

7. "Testing," *SNPedia* (https://www.snpedia.com/index.php/Testing).
8. "Search results" for SNP rs2051646, *dbSNP* (https://www.ncbi.nlm.nih.gov/snp/?term=rs2051646).

...AAATTTTAAAGATTTAAAGGCTCAT[C/T]TATTTAAAACTATAAATAATTGCTA...

Figure 8.1. dbSNP Definition for RSID rs2051646

One reason for a *miscall* (where the DNA allele value is read incorrectly) is unbalanced signal intensities. A homozygous genotype for rs2051646 would have a double dose of the C or the T signal respectively, while a heterozygous genotype would have an equal amount for the C and T. The amount is seldom exactly balanced, so a heterozygous CT might be reported as CC if the T signal is weak, resulting in a miscall.

Fortunately, the error rate is quite low, as reported by various people who have tested themselves on different chips or companies. Numbers in the range of one discrepancy per 5,000 or 10,000 SNPs seem common, and some report even better results. David Pike has a handy suite of tools on his website; one looks for differently reported SNPs.[9] It found only 14 discrepancies for the 299,672 SNPs contained in both files when comparing a 23andMe version 4 (v4) kit (a custom chip) with an AncestryDNA version 1 (v1) from the same person. That is just one in 21,405 SNPs. It should be noted that some "errors" are actually nomenclature issues, when companies report the base found on the opposite strand of the double helix. Probes can be designed for either strand, but the convention is to always report the allele on the plus strand (also called the forward strand). Occasionally this conversion step is overlooked. This is a nomenclature issue and not a true mismatch in data.

No-calls and miscalls may sometimes be fixed by running the sample again with the same or different chip version. There could be some systematic reason for the failures, such as a mutation or a probe that does not perform as expected. However, the marginal benefit of rerunning a sample would be very small. If more than one kit is already available, a merge utility, described in the *GEDmatch* Genesis section, may fill in some of the blank spots.

From Genotypes to Haplotypes

The weakness of using genotypes is glaring, but current technology is not yet capable of separating paternal and maternal chromosomes at an affordable price. If paternal and maternal alleles are indistinguishable, how can they be used to find relatives on the correct side? The most accurate method is to compare a child's data SNP by SNP with the parents and deduce which allele came from which parent. This process is call *phasing.*[10] The genotype can be divided into two *haplotypes*: the set of alleles that are found on one chromosome.

9. David Pike, Autosomal Utilities (https://www.math.mun.ca/~dapike/FF23utils).

10. In casual usage, "phasing" is sometimes used to mean attribution of a match to the paternal or maternal side, based on known shared relatives. This is the sense used by FamilyTreeDNA, which is not doing true phasing, but using shared relatives to place matches in a paternal or maternal "bucket."

Phasing—using parents to phase a child's DNA data

If the genotype is homozygous, for example, AA, the data can automatically be phased. One A came from the father and the other A came from the mother. The DNA from the parent does not even need to be inspected.

If the child is heterozygous AG, and the father is AA, and the mother is GG, then the A came from the father and the G from the mother. If the child is AG, and the father is AG, and the mother is GG, then the A came from the father and the G from the mother. If the child is AG, and the father is AG, and the mother is AG, then the result is still indeterminate: the A could have come from the father and G from the mother, or vice versa.

Even phasing a child against one parent provides a useful partition into paternal and maternal alleles. *GEDmatch*[11] has a phasing utility to divide the child's alleles into paternal P1 and maternal M1 kits. The "DNA File Diagnostic Utility" gives a glimpse into the enormous improvement in the quality of matches. An unphased kit includes numerous false positive matches, which are due to weaving back and forth between paternal and maternal alleles. The trio data using a father and mother and child shows that more SNPs can be resolved, but duo data using one parent and a child is still a substantial improvement over unphased data. This is illustrated in table 8.5.

Table 8.5. Effect of phasing on matches

	Total Matches	Segments > 7cM	Hetero-zygous SNPs
Unphased	100,409	29,120	28.5%
Trio M1 or P1	16,177	8,484	5.6%
% of matches remaining	16.3%	29.3%	
Duo M1 or P1	19,721	9,236	12.5%
% of matches remaining	19.6%	31.7%	

Statistical phasing

If no parents are available, an alternative approach is statistical phasing. This compares the child to a reference population that has already been phased. The child will not completely match any one of the reference population, but short sections, perhaps 500 SNPs or so, may be used to create a dictionary with the frequency of various haplotypes. The computer

11. *GEDmatch* (https://gedmatch.com/). This a database containing raw data files uploaded by customers tested at many different companies. Because several *GEDmatch* tools allow comparisons between kits on a SNP-by-SNP basis, it is perhaps the closest thing to working directly with raw data files. Many examples in this chapter are based on *GEDmatch* utilities. At the time of this writing, *GEDmatch* "Classic" is in the process of switching to *GEDmatch* Genesis; the names and locations of some tools may change.

algorithm will experiment with different ways of dividing the two alleles into two different chromosomes and pick the best option.

This process is not fool-proof, and sometimes an erroneous conclusion will send the algorithm down the wrong path. This is known as a **phase switch error**. As an exercise, I used the Sanger phasing service for my son's data from 23andMe version 3. Sanger uses data from the Haplotype Reference Consortium for the dictionary which includes about 64,000 haplotypes.[12] The service returns a file with genotypes, where nominally the first allele is always on one chromosome and the second allele is on the alternate chromosome. The paternal and maternal sides would not necessarily remain the same for all chromosomes, but they should be consistent within a chromosome if all goes well.

The results were eye-opening when I uploaded them to *GEDmatch* Genesis. Comparing his alleles from the first column (in the file resulting from the Sanger phasing) with me at the default level of 7 cM, chromosome 1 had many gaps. Using a 1 cM threshold, chromosome 1 was broken up into about 30 segments (color coded on the bottom above the blue bar) as shown in figure 8.2. The gaps between the segments show where the phasing switched; my son should match me but he does not. Comparing his alleles from the second column, chromosome 1 was also broken up into about 30 segments, largely filling in the gaps as shown in figure 8.3. About 99% of the genome was covered by switching back and forth.

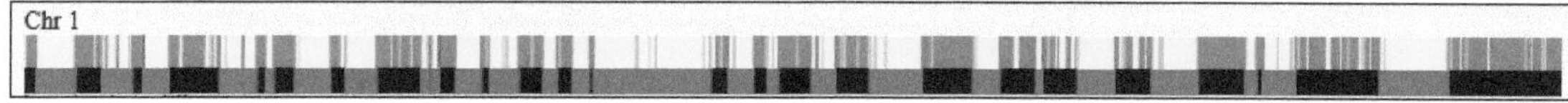

Figure 8.2. Chromosome comparison between mother and child with Sanger Phasing A

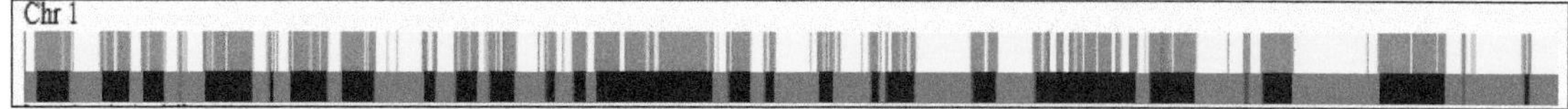

Figure 8.3. Chromosome comparison between mother and child with Sanger Phasing B

Dropping down to the 1 cM level was essential for filling in the gaps. At 2 cM, only 90% was covered. The median segment size was 2.9 cM. This indicates that statistical phasing by itself can operate only over relatively short distances.

23andMe, AncestryDNA, and MyHeritage all use a form of statistical phasing for some of their tools. In addition to having larger dictionaries of haplotypes, they each take additional steps to smooth things out. For instance, MyHeritage mentions stitching segments together.[13] AncestryDNA has a very detailed white paper covering its approach.[14] The company starts a comparison with a haplotype from the dictionary, but then continues using genotype

12. Sanger Imputation Service (https://imputation.sanger.ac.uk/).

13. Yael, "Major Updates and Improvements to MyHeritage DNA Matching," *MyHeritage*, 11 January 2018 (https://blog.myheritage.com/2018/01/major-updates-and-improvements-to-myheritage-dna-matching).

14. Catherine A. Ball et al., "AncestryDNA Matching White Paper," *AncestryDNA*, 31 March 2016 (https://www.ancestry.com/corporate/sites/default/files/AncestryDNA-Matching-White-Paper.pdf).

data to bypass the phase switch problem. 23andMe uses statistical phasing for its "Ancestry Composition" chromosome paintings by assigning ancestry to relatively short sections at a time. Statistical phasing is not used for matching test takers. Each chromosome in the diagram refers to a single parent, but the "top" and "bottom" copies, as seen in figure 8.4, could be paternal and maternal, or vice versa. However, if a parent has tested, actual values are used. In that case, the chromosome painting will display the maternal side on top and the paternal side on the bottom.[15]

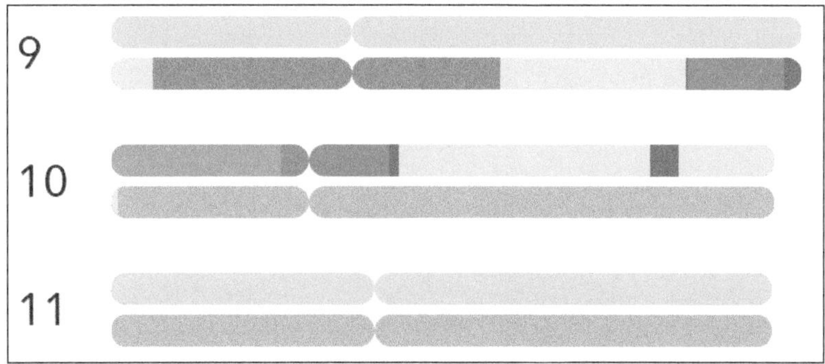

Figure 8.4. 23andMe Ancestry Composition showing two chromosome copies

Child phasing

Not everyone has a parent to test, and statistical phasing is best left to professional chefs with industrial equipment. However, many of the benefits of phasing can be garnered by testing a child, then using the child's DNA data to phase a parent's data. A child will not reveal any genuine new matches, but the overall quality of the matches will be improved by reducing false matches.

GEDmatch offers a unique tool called "'My Evil Twin' Phasing." Regular *GEDmatch* phasing results in a kit containing just the alleles inherited from one parent, with the suffix M1 or P1. The "evil twin" kit contains all the alleles *not* inherited from that parent, with the suffix M2 or P2. This is rather like a photographic negative.

The child's M1 and M2 kits amount to "quasi-phased" kits for the mother, where "quasi" is used in the sense of "almost" or "nearly." As the example in figure 8.5 shows, the mother's results are not phased across the entire length of a chromosome, but long stretches shared with a match will be effectively phased. The exceptions will be a small number of segments split by a recombination point. Chr1, Chr4, and Chr9 are abbreviations representing chromosome 1, 4, and 9, respectively. See the Chr9 segment in figure 8.5 for an illustration of how matching segments would be assigned to M1 and M2 kits.

15. "Phasing and Inheritance," Technical Details section, *23andMe* Customer Care (https://customercare.23andme.com/hc/en-us/articles/212193688#details).

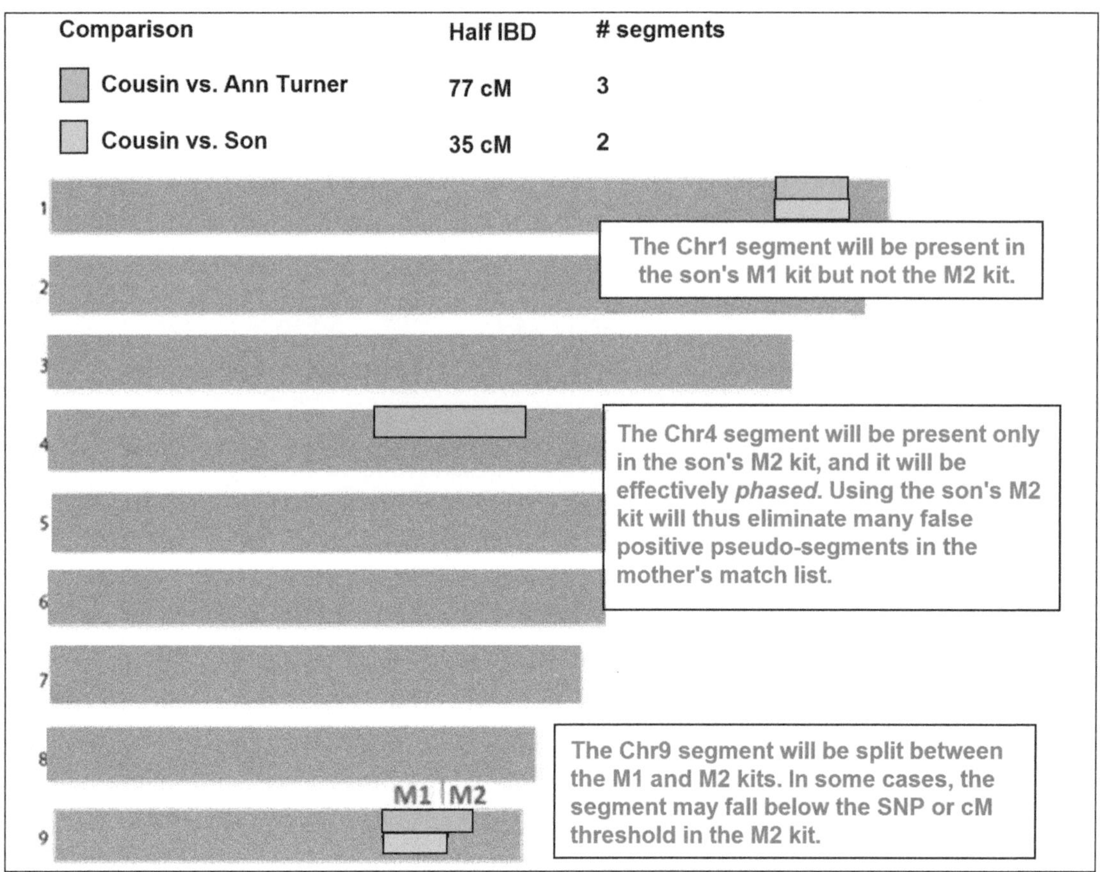

Figure 8.5. *GEDmatch* "My Evil Twin" Phasing

Phased data eliminates ambiguous segments. When lacking phased data from any method described above, one option available to everyone is to increase the segment size under consideration. Use segments large enough in size to be confident that the match is not a false positive match.

Segments

A segment may sound like a concrete object, but it is actually a conclusion based on analysis of the raw data. The recipe for determining how one matches, or does not match, another person begins with a search for the right ingredients: a long consecutive run of SNPs that are at least half identical. For example, if both people are homozygous AA, or if one person is AA and the other is heterozygous AG, or both people are AG, then at least one allele in one party matches at least one allele in the other party. An acronym HIR was coined for a half identical region. Designating a region as HIR is an agnostic, neutral observation about the DNA, not a conclusion.

That leaves a lot of wiggle room, and it is very likely that any two random people will have a half identical match for many individual SNPs scattered over the whole genome. In fact, just by coincidence, they may have quite a few SNPs in a row that are half identical. For a concrete example see "Anatomy of an IBS segment" on Jim Bartlett's *Segment-ology* blog.[16] Although there is a run of 850 half-identical SNPs in that example, phasing and triangulation disprove the segment—or, phrased differently, prove the segment was not inherited from the common ancestor originally identified. Ironically, the match broke down a brick wall; this is an object lesson for the fact that the paper trail can appear to "prove" a DNA segment when the segment was not inherited via the presumed lineage path.

At a certain point, though, the best explanation for sharing many SNPs in a row is inheritance from a common ancestor, also called Identical by Descent (IBD). That certain point is, however, a judgment call about when the ingredients are fully cooked. 23andMe's rule of thumb requires 700 consecutive SNPs and must include at least 400 homozygous SNPs in one part. FamilyTreeDNA requires 500 SNPs. AncestryDNA and MyHeritage do not provide details. In all cases, the goal is to maximize the number of true positives (with a lenient threshold) and minimize the number of false positives (with a strict threshold).

Boundaries

The absolute boundaries for a segment are SNPs that are opposite homozygotes. If one party is AA and the other party is GG, there is no way they could trace that result back to the same common ancestor. The actual boundaries may be fuzzier. A segment may be extended by some number of SNPs that are half identical, just by chance, when one or both parties are heterozygous for a SNP.

Intuitively, it might seem that matching on an allele with a low minor allele frequency (MAF) is significant, but the flip side is that the genotype is more likely to be heterozygous. With only two variants per SNP, both variants are represented in any person with a heterozygous value. A heterozygous SNP is a universal match: at least one allele in one party matches at least one allele in the other party. The natural tendency is seeking to verify an assumption, but falsifying a matching segment is the best approach to a robust conclusion. There is a 0.0625 chance of observing opposite homozygotes when the MAF is 0.5, as seen in table 8.6. This is almost eight times as likely as the 0.0081 chance when the MAF is 0.1, as seen in table 8.7. Fortunately, the chips are designed to include SNPs with a wide range of MAF making it easier to accurately define segment boundaries.

16. Jim Bartlett, "Anatomy of an IBS Segment," *segment-ology*, 2 October 2015 (https://segmentology.org/2015/10/02/anatomy-of-an-ibs-segment/). This website is a treasure trove of information about segments and triangulation. A failure to triangulate can reveal a large proportion of false positive pseudo-segments.

Table 8.6. Chance of observing a mismatch (AA vs GG) for MAF = 0.50 is 0.125					
	Person 2	AA	AG	GA	GG
Person 1	chance of genotype	0.25	0.25	0.25	0.25
AA	0.25	0.0625	0.0625	0.0625	***0.0625***
AG	0.25	0.0625	0.0625	0.0625	0.0625
GA	0.25	0.0625	0.0625	0.0625	0.0625
GG	0.25	***0.0625***	0.0625	0.0625	0.0625

Table 8.7. Chance of observing a mismatch (AA vs GG) for MAF = 0.10 is 0.0162					
	Person 2	AA	AG	GA	GG
Person 1	chance of genotype	0.81	0.09	0.09	0.01
AA	0.81	0.6561	0.0729	0.0729	***0.0081***
AG	0.09	0.0729	0.0081	0.0081	0.0009
GA	0.09	0.0729	0.0081	0.0081	0.0009
GG	0.01	***0.0081***	0.0009	0.0009	0.0001

Builds

The celebration of the "completion" of the human genome in 2000 was premature. The reference sequence has been revised many times since then. Gaps have been filled in and some sections have been shuffled to a different location. Segment boundaries are marked by their base position on a chromosome diagram. Different *builds* of the human genome reference sequence may place a SNP at a different location. But a SNP is still recognized by the flanking bases that make up the probe as shown in the discussion around figure 8.1 earlier in this chapter.

Currently FamilyTreeDNA and *GEDmatch* Classic use Build 36 for mapping segments on a chromosome diagram. 23andMe, AncestryDNA, Living DNA, and MyHeritage all use Build 37. *GEDmatch* Genesis offers an option to display boundaries in Build 36, Build 37, or even Build 38 coordinates (released in 2013 but not yet used by the genetic genealogy companies). Hendrik Wendland has developed a utility "MapS Converter" to convert between Build 36 and Build 37.[17] This is handy for tools such as *DNA Painter*[18] when including data from different sources. It also has a feature to calculate the *genetic distance* between two points, for example, the size of the overlap in two matching segments.

17. *MapS Phasing* (https://www.maps-phasing.com/).
18. *DNA Painter* (https://dnapainter.com/).

Genetic distance

Genetic distance, as used in autosomal DNA discussions, is a logical measurement, contrasting with the physical distance in base pairs used to diagram the location of segments. Genetic distance is based on observations of a specific study.[19] In figure 8.6, notice how there is not a straight line comparing megabases (Mb) and centimorgans (cM); thus the cM unit is measuring something different than Mb. Sometimes the curve flattens out, especially noticeable at about 120–140 Mb on chromosome 1. This is the position of the centromere, where little recombination takes place. A segment mapped in this location may appear very long visually, yet the cM value is very small. Also, notice that the lines are somewhat different for the two measurements.[20]

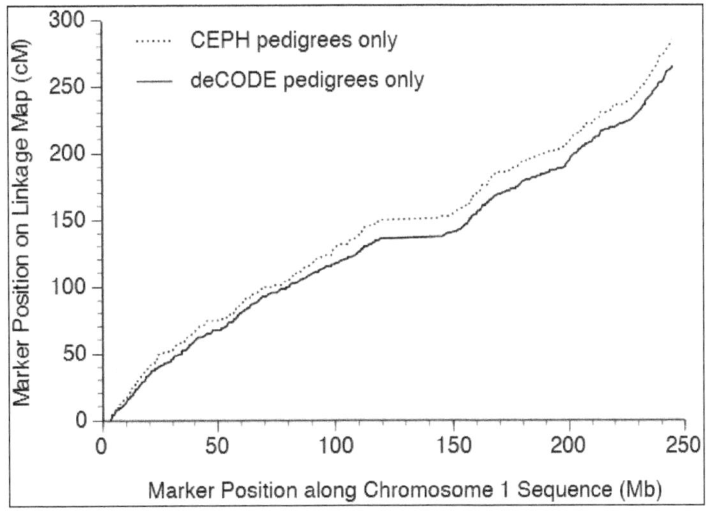

Figure 8.6. Megbase-centimorgan comparison

Just what is this mysterious, abstract unit called the centimorgan (cM)? A cM represents a 1% chance that the SNPs at two locations will be separated by recombination. It measures "effective" distance. As an analogy, a mile is a fixed quantity (5,280 feet), and so are megabases (million bases, or Mb). However, the probability that a person can walk a mile in twenty minutes is more fluid. If the terrain is very rough, the "effective" distance of a literal mile might be more like two miles. AncestryDNA even doubles down on the notion of effective distance: if a segment is very frequent in the general population, it could have come from many ancestral sources. AncestryDNA will then downweight the cM value with its Timber algorithm.[21]

19. Archived copy of Rutgers University Map Interpolator described in TC Matise et al., "A second-generation combined linkage physical map of the human genome," *Genome Res* (17): 1783–6 (http://web.archive.org/web/20070113005025if_/http://compgen.rutgers.edu:80/maps/compare.pdf).

20. *Centre d'Etude du Polymorphisme Humain* (CEPH) (http://www.cephb.fr/en/familles_CEPH.php) established family reference panels in 1984 to build and lead the first international collaboration for human genome mapping. *deCODE genetics* (https://www.decode.com/) is a global leader in analyzing and understanding the human genome

21. Catherine A. Ball et al., "AncestryDNA Matching White Paper."

In any event, the probability that a segment will be passed on intact is more informative than its size in Mb. A 40 cM segment will be split apart more frequently than a 10 cM segment, so it is more likely to come from a recent common ancestor *on the average*. Individual cases can vary widely. Simulations from Speed and Balding show that segments between 10 and 20 cM can be found in siblings (see orange bar for G=1 in figure 8.7). Yet about 50% of such segments will be fourteen or more generations back. At the same time, segments between 40 and 50 cM can sometimes persist for more than 10 generations (see green bars designated by arrows in figure 8.7).[22] It is easy to pay lip service to the randomness of inheritance, but examples like this can still be surprising.

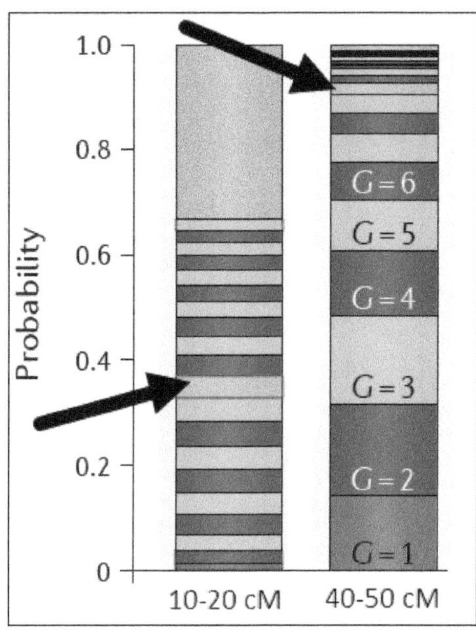

Figure 8.7. Extracted portion of IBD region length, Speed and Balding

Number of segments

A frequent question takes the form "Which is better, a single segment of X cM or multiple segments adding up to X cM?" where X is some number. 23andMe developed its algorithm by simulating a large number of pedigrees using real genotype data and recombination rates. The scattergram in figure 8.8, from 23andMe's paper, demonstrates that any given relationship can include a large number of combinations for number of segments and total cM.[23]

22. D. Speed and DJ Balding, "Relatedness in the post-genomic era: is it still useful?" *Nature Reviews Genetics* (January 2015): 33-44 (https://doi.org/10.1038/nrg3821 is behind a paywall; author's copy online at http://dougspeed.com/wp-content/uploads/nrg_relatedness.pdf); extracted from figure 2B.

23. B.M. Henn et al., "Cryptic Distant Relatives Are Common in Both Isolated and Cosmopolitan Genetic Samples," *PLOS ONE* 7 (2012) (https://doi.org/10.1371/journal.pone.0034267).

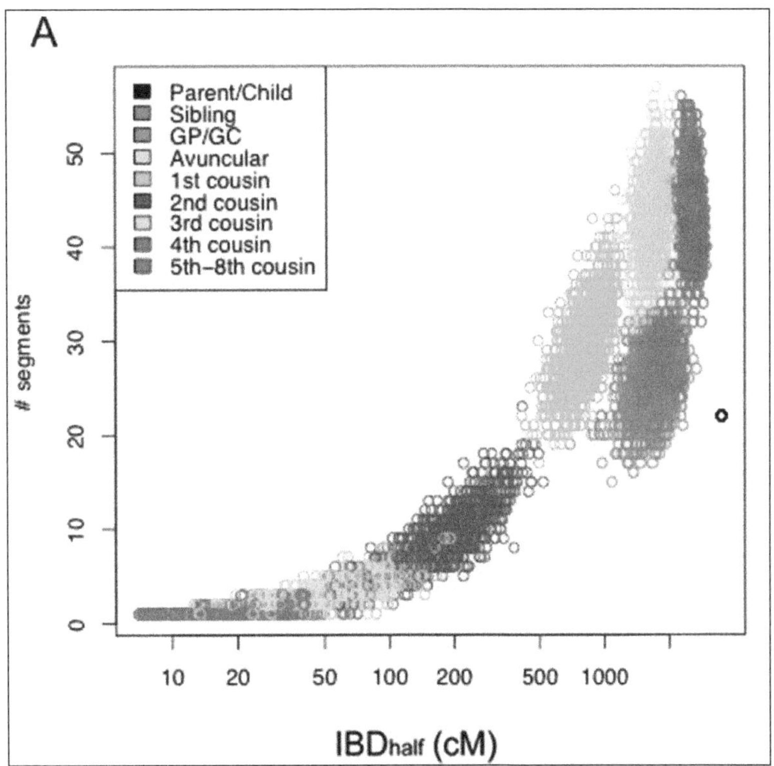

Figure 8.8. 23andMe number of segments and Total cM for different relationships

In fact, there can be a dramatic difference in the distribution of segments in close cousins, depending on the number of male or female transmissions. The recombination rate in females is 1.7 times higher than males.[24] For instance, a paternal first cousin once removed, the child of a male first cousin, shares 491 cM in 10 segments, the longest being 196 cM, with his cousin Ann. Another paternal first cousin once removed, the child of a female first cousin, shares 472 cM in 21 segments, the longest being 78 cM, with the same cousin Ann. For more distant relationships, the male and female transmissions will begin to average out. The genetic genealogy companies use sex-averaged rates in their algorithms.

24. Reshmi Chowdhury et al., "Genetic Analysis of Variation in Human Meiotic Recombination," *PLOS Genetics* 5 (September 2009) (https://doi.org/10.1371/journal.pgen.1000648). This paper has other interesting tidbits. For example, the recombination varies across a chromosome in "jungles" and "deserts," and the rate is itself a genetic trait.

SNPs

What is the role of SNPs? Why do lists of matching segments even include the number of SNPs? Strictly speaking, it is not necessary for the genealogist to have that data. AncestryDNA does not supply it. MyHeritage includes a large number of imputed SNPs, predicted values for SNPs not directly tested. Imputation is discussed later in this chapter.

SNPs are sampling points. The goal is to include enough SNPs to expose any invalid or false segments. There are diminishing returns to adding more SNPs beyond that point. Chips with more SNPs usually add ones with a lower Minor Allele Frequency, which means that more people will end up matching each other for the common allele. That is easy pickings and less informative. Additional SNPs will also tend to be closer to each other and more likely to be inherited together as a "package." The packaged SNPs are effectively saying the same thing twice, which does not make it any truer.

The various testing companies each have determined that there are sufficient SNPs available for comparisons within their own proprietary databases. However, a problem arises when doing cross-platform comparisons at *GEDmatch* or with any database that includes test takers from multiple companies or tested with multiple versions of the chip.

In the best of all possible worlds, we could get the same list of matches no matter which company did the testing. This was virtually true in the olden days, when the overlap between companies was in the vicinity of 660,000 to 680,000 SNPs. That overlap was seen with 23andMe v3, AncestryDNA v1, and FamilyTreeDNA (FTDNA), and as seen in table 8.8.[25]

Then 23andMe (v4) and AncestryDNA (v2) each developed their own custom chips, and the SNP overlap dropped to the 300,000 to 400,000 range. The concordance, the agreement, of match lists was still largely true, albeit with some drop-off in the number of matches at 300,000 SNPs overlap compared to 400,000 SNPs. The discrepancy between match lists may be partly due to segments that fall below a SNP threshold calibrated back when there were close to 700,000 SNPs available for comparison. The "quality" of the matches for each kit was high; they would not disappear with phased kits.

---

25. "SNP Comparison Chart," *ISOGG Wiki* (https://isogg.org/wiki/Autosomal_SNP_comparison_chart). The full chart has additional columns for 23andMe v2, *Genes for Good*, and *Genographic* which is now tested by *Helix*. Data for the *Sirius* chip will be added when it becomes available. One of the design considerations for the Sirius chip was improving SNP overlap.

Table 8.8. Autosomal SNP overlap								
	Ancestry v1	Ancestry v2	FTDNA	23andMe v3	23andMe v4	23andMe v5	Living DNA	My Heritage
Ancestry v1	**682,549**	407,968	672,031	659,830	300,404	160,803	161,681	676,049
Ancestry v2	407,968	**641,908**	410,572	436,948	307,950	149,394	149,078	411,075
FTDNA	672,031	410,572	**698,179**	673,166	304,208	162,602	161,862	702,442
23andMe v3	659,830	436,948	673,166	**930,434**	510,793	178,994	148,715	677,159
23andMe v4	300,404	307,950	304,208	510,793	**585,974**	109,518	103,865	304,059
23andMe v5	160,803	149,394	162,602	178,994	109,518	**630,132**	584,753	162,602
Living DNA	161,681	149,078	161,862	148,715	103,865	584,753	**603,129**	163,280
My Heritage	676,049	411,075	702,442	677,159	304,059	162,602	163,280	**702,442**

Table 8.9 summarizes details about the match lists for 23andMe v4 and AncestryDNA v1 and v2. Only matches with 15 or more total cM and in the top 2,000 matches to the primary kit were considered. The diagonal shows the number of matches for each kit. The numbers above the diagonal are the count of matches found in both lists. The numbers below the diagonal convert the count to percentages. The "quality" column is the percentage of matches found in the quasi-phased M1 or M2 kits or both, which eliminate many false positives based on genotype data.

Table 8.9. Overlap of match lists for kits in *GEDmatch* Classic				
	Ancestry v1	Ancestry v2	23andMe v4	"Quality"
Ancestry v1	***1,553***	1,367	1,254	93.2%
Ancestry v2	88.0%	***1,403***	1,226	95.2%
23andMe v4	81.7%	87.4%	***1,269***	95.4%

GEDmatch Genesis

The *GEDmatch* developers began experimenting with different approaches in their Genesis beta test site.[26] The goal was to make the match lists even more consistent across platforms. Then the new Global Screening Array (GSA) chip came along, used first by Living DNA in 2016 and then by 23andMe v5 in 2017. The chip itself has about 600,000 SNPs, but the SNP overlap with earlier platforms dropped to 160,000.

26. *GEDmatch* Genesis (https://genesis.gedmatch.com/); may move to https://gedmatch.com/.

Is that enough sampling points to expose any contradictions to a match? One way to approach this question is to merge data files from different chips to achieve the maximum possible number of SNPs available for comparison with any other kit. Wilhelm HO's "DNA Kit Studio" has a module to perform just that function.[27] A merged kit was created using data from AncestryDNA v1, AncestryDNA v2, 23andMe v4, and 23andMe v5 (GSA).

The *GEDmatch* Genesis database contains a mixture of kits from GSA chips (not acceptable at *GEDmatch* Classic) and older platforms. The overlap of two GSA kits will be reported as about 300,000 SNPs, lower than the values seen in table 8.8. This is due to a change in the way SNPs are filtered to retain just the most informative SNPs. Those with a low minor allele frequency (easy pickings) and heterozygous SNPs (universal matches) are omitted.

Table 8.10 shows a massive disconnect between GSA kits compared to each other versus a GSA kit compared to an older platform. The merged kit contains all SNPs from all platforms, so logically, a match from a GSA kit should also be found in the merged kit, even if the SNP overlap is low. If so, it has passed a more stringent test using more SNPs. If not, then the merged kit has exposed some mismatches and the GSA kit match has failed.

Table 8.10. GSA kit match list compared to merged kit based on SNP overlap

Overlap	Fail	Pass	Total	% Pass
<300,000	1,274	216	1,490	14.5%
>=300,000	34	1,476	1,510	97.7%
Total	1,308	1,692	3,000	56.4%

Inspection of a failure with the *GEDmatch* one-to-one compare tool shows a matching segment (color-coded blue on the bottom bar and surrounded by the box in figure 8.9) was based on just 276 SNPs. Even this segment shows a mismatch (color-coded with a red bar). Occasional mismatches are tolerated in order to accommodate rare genotyping errors, but the error rate is so low that the allowance for mismatches is probably too generous. Even a single red bar in a small segment should be grounds for suspicion.

27. Wilhelm HO, "DNA Kit Studio," *DNA Ancestry Tools* (https://wilhelmhgenealogy.wordpress.com/dna-kit-studio).

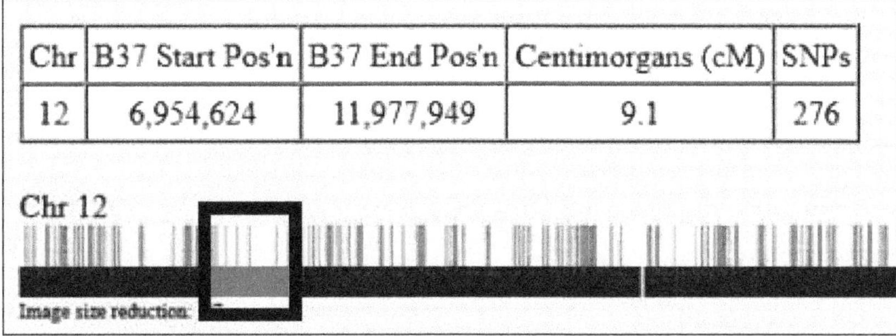

Chr	B37 Start Pos'n	B37 End Pos'n	Centimorgans (cM)	SNPs
12	6,954,624	11,977,949	9.1	276

Figure 8.9. *GEDmatch* one-to-one comparison with 276 SNPs

The same region in the merged kit (shown at full resolution) has approximately 1,150 SNPs, but there lots of red bars as seen in figure 8.10. It is not a true segment.

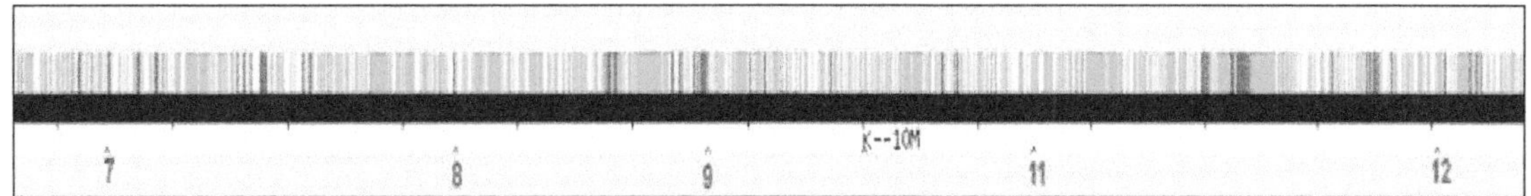

Figure 8.10. *GEDmatch* one-to-one comparison with 1,150 SNPs

Wilhelm HO's tool will be of most benefit for those who have tested with a GSA chip and one of the older platforms. In fact, it might be worthwhile to test on both generations of chips if one's DNA budget permits and the older chips are still available.

Failing that, segment length comes to the rescue again. In this situation, the longest segment is the one to use. Total cM may be inflated by including small segments that would not stand up to further inspection with more SNPs. A longest segment of 15 cM gives a reasonable chance for a repeatable match with any chip, even if the SNP overlap is low. This is shown in table 8.11. Even so, phasing might disprove some segments.

Table 8.11. GSA kit match list compared to merged kit based on longest segment

Longest	Fail	Pass	Total	% pass
<15 cM	1,298	1,577	2,875	54.9%
>=15 cM	10	115	125	92.0%
Total	1,308	1,692	3,000	56.4%

Imputation

The low SNP overlap between various chips is an increasing frustration as more and more people test on the new GSA chip. Can computational methods increase the number of SNPs for comparison?

Imputation is a statistical method for predicting the results for a SNP that has not been directly measured, one that is not tested on the microarray. It compares a sample to a reference set of people with more comprehensive data, such as whole genome sequencing data from the 1000 Genomes Project.

Suppose there are two SNPs about 5,000 bases apart with possible alleles

SNP1 A or G
SNP2 C or T

The reference set has a SNP positioned between SNP1 and SNP2, and it turns out that the possible values for this SNP are A and C. It also turns out that everyone who has an A for SNP1 and a T for SNP2 has a C for the SNP in between. Then it is likely that a person who has the same results for SNP1 and SNP2 with a direct test would also have a C for the SNP in between: the C would be imputed. The C is in *linkage disequilibrium* with SNP1 and SNP2. In other words, the C is *not* equally likely to be found with various combinations of the alleles for SNP1 and SNP2.[28]

In practice, things can be fuzzier—perhaps the connection between SNP1 and SNP2 holds for only 87% of the reference set samples. The farther apart those two SNPs are located, the greater the chance that recombination will have split the segment into two parts that are no longer firmly linked. Faulty conclusions are a result, especially when the minor allele is uncommon and the reference set does not have many exemplars.

Effect of errors

As an exercise, I uploaded 23andMe kits from v4 and v5 (GSA) to *DNA.Land*.[29] One of *DNA.Land*'s services is imputation based on the 1000 Genomes Project data. They supply a VCF (Variant Call Format) file with data for about 39,000,000 SNPs. This is a very large file, hard to manipulate without additional tools to convert the data into a format similar to that used by the genetic genealogy companies. Wilhelm HO has created a utility to extract a subset of SNPs corresponding to chips in common use.[30] I requested an output file with 23andMe v3 SNPs. This is the chip with the largest number of SNPs, approaching one million.

I then compared the imputed file with results from a different chip, AncestryDNA v1. As shown in table 8.12, the source files with genotyped data had a very good concordance rate (agreement) across the chips; only 0.004% to 0.005% of the SNPs differed. The concordance rate with the data imputed from a v4 file is noticeably lower, 0.620%, but still decent in

28. For more on linkage disequilibrium and other advanced topics see Joel T. Dudley and Konrad J. Karczewski, *Exploring Personal Genomics* (Oxford, UK: Oxford University Press, 2013).

29. *DNA.Land* (http://dna.land/).

30. Wilhelm HO, "Convert 23andme V5 RAW to Gedmatch classic and other companies valid format," *DNA Ancestry Tools* (https://wilhelmhgenealogy.wordpress.com/convert-23andme-to-myheritage-and-gedmatch-classic).

absolute terms. The error rate increases sharply to 2.517% for the GSA chip. To determine whether another imputation method might improve the outcome, I repeated the exercise using the Sanger imputation service. Sanger has an option to use a larger reference set, the Haplotype Reference Consortium of 64,976 haplotypes.[31] The error rate did not change much with the larger reference sample.

Table 8.12. Comparison of AncestryDNA v1 kit with 23andMe genotyped and imputed kits				
	v4 genotype	v5 genotype	v4 -> v3 imputed	v5 -> v3 imputed
SNP overlap	299,675	159,815	659,815	659,865
% SNPs different	0.005%	0.004%	0.620%	2.517%

Imputation is still far, far better than chance, and it might be quite satisfactory for admixture analysis or Genome Wide Association Studies (GWAS), where genotypes for many people are pooled into a dataset. But imputation may not be good enough to use for finding cousins, where minute differences between two specific individuals are enough to disqualify a matching segment.

Table 8.13 gives some statistics for an uncle and me with kits compared at *GEDmatch* Classic and *GEDmatch* Genesis. They show that imputation results in a decreased probability that he would be classified in the uncle range, 1,349 to 2,175 cM as per the "Shared cM Project."[32] In fact, even the v5 kits with a sparse number of genotyped SNPs available for comparison outperforms the imputed file with a larger number SNPs. The v5 kit in *GEDmatch* Classic is severely impacted by the SNP threshold; *GEDmatch* Genesis allows segments with a smaller number of SNPs and consequently includes more segments.

31. Sanger Imputation Service (https://imputation.sanger.ac.uk/). The mechanics of setting up an account for data transfer are somewhat more complex than using *DNA.Land.*

32. Blaine T. Bettinger and Jonny Perl, "The Shared cM Project 3.0 tool v4," at *DNA Painter* (https://dnapainter.com/tools/sharedcmv4).

Table 8.13. *GEDmatch* Comparison of Uncle and Niece (with Uncle tested on v2 and Niece tested on two chips used for imputation)						
	Classic				Genesis	
	v4 genotype	v5 genotype	v4 -> v3 imputed	v5 -> v3 imputed	v5 genotype	v5 -> v3 imputed
Total cM	1,451.9	1,201.9	1,395.8	782.7	1,384.9	931.7
# segments	50	31	52	61	56	70
SNP overlap	449,588	98,123	532,994	521,341	97,531	513,299
Probability	94.72%	16.41%	90.23%	0.00%	89.35%	0.00%

Overcoming the low SNP overlap

These results in table 8.13 are rather dismal, even for a close relative such as an uncle. The problems are magnified for more distant relationships with fewer and shorter segments. In fact, for *GEDmatch* Genesis, a file with a sparse SNP overlap outperforms an imputed file with more SNPs.

Can we extrapolate these findings to the companies that are now employing imputation? Fortunately, they seem to have some tricks up their sleeves to overcome the error rate. Conceptually, they could employ various strategies: a better reference set based on their own internal data, a method to filter out the SNPs with the lowest confidence, a greater tolerance for an occasional mismatch, and perhaps other means.

23andMe is the first company to actually employ imputation when comparing its v5 kits with previous versions. More than 90% of my match lists for v4 and v5 overlap. Since 23andMe has a cap on the number of relatives reported, some of those discrepancies might disappear if all matches were listed. On the other hand, more discrepancies might crop up if smaller matches were compared. In any event, the aforesaid uncle matched me for 1,534 cM (47 segments) in v5 and 1,494 cM (53 segments) in v4 (including the X). The reduced number of segments was a result of stitching together two segments with a small gap between them, an indirect sign that a newly calibrated error tolerance is one of 23andMe's strategies.

Living DNA was the first company to employ the GSA chip for its product; it initially offers only admixture analysis, plus mtDNA and Y haplogroups. Cousin matching is on the horizon at the time of this writing, and Living DNA has been hard at work fine-tuning the imputation process. David Nicholson indicates that the company dedicates a large amount of staff time to selecting the most reliable SNPs, an ongoing process. The company uses the Haplotype Reference Consortium along with a custom set of SNPs.[33]

33. David Nicholson, to Ann Turner, email, 7 September 2018; privately held by Turner.

MyHeritage has always employed imputation for uploads it accepted from the start (23andMe prior to v5, AncestryDNA v1 and v2, and FamilyTreeDNA). The SNP overlap for these kits appears to be adequate, based on informal anecdotal reports. As for the GSA chip, MyHeritage wrote on 11 January 2018 that "We have support for DNA data from GSA chips working in our lab; it is in fairly good shape, but still not perfect, so we decided to exclude it from this release until it is perfected. This will be added in the next few months."[34]

This was not a trivial task, but on 12 September 2018 this came to pass.[35] I uploaded my 23andMe v5 kit for comparison with my previously uploaded AncestryDNA v1 kit. My aforesaid uncle matched the v5 kit at 1,445 cM (46 segments) and the v1 kit at 1,429 cM (45 segments), both squarely in the expected range for an uncle.

Table 8.14 shows how the match lists compare for a broader range of shared amounts of DNA. It appears that 23andMe v5 matches that make it through the process are present in a high percentage of AncestryDNA v1 matches, but many of the smaller AncestryDNA v1 matches do not show up in the 23andMe v5 match list. I do not regard that as a tragedy.

Table 8.14. Comparison of Ancestry v1 and 23andMe v5 kits uploaded to MyHeritage

Ancestry v1 upload	in 23andMe v5	total	%		23andMe v5 upload	in Anc v1	total	%
All	3,508	6,610	53.1%		**All**	3,508	4,970	70.6%
Sum >= 10 cM	2,579	4,486	57.5%		**Sum >= 10 cM**	2,975	4,040	73.6%
Sum >= 20 cM	571	812	70.3%		**Sum >= 20 cM**	1,014	1,215	83.5%
Sum >= 30 cM	152	194	78.4%		**Sum >= 30 cM**	288	311	92.6%
Sum >= 40 cM	66	75	88.0%		**Sum >= 40 cM**	89	94	94.7%

Other Tools for Raw Data

Many of the preceding examples involve investigating one raw data file in the context of a database comparison at *GEDmatch*. Access to more than one raw data file opens up an opportunity to explore more facets of the file.

David Pike has developed a suite of tools:[36]

- Search for Runs of Homozygosity (ROHs)
- Search for Heterozygous Sequences
- Search for Shared DNA Segments in Two Raw Data Files
- Inspect a Shared DNA Segment in Two Raw Data Files

34. Yael, "Major Updates and Improvements to MyHeritage DNA Matching," *MyHeritage*, 11 January 2018 (https://blog.myheritage.com/2018/01/major-updates-and-improvements-to-myheritage-dna-matching).

35. Esther, "New: MyHeritage supports 23andMe v5 and Living DNA uploads," *MyHeritage*, 12 September 2018 (https://blog.myheritage.com/2018/09/new-myheritage-supports-23andme-v5-and-living-dna-uploads).

36. David Pike, Autosomal Utilities (https://www.math.mun.ca/~dapike/FF23utils).

- Inspect Shared DNA Segments in a Trio of Raw Data Files
- Search for Discordant SNPs in Parent-Child Raw Data Files
- Search for Discordant SNPs when given data for child and both parents
- Search for Differently Reported SNPs
- Phase a Child when given data for child and both parents
- Phase Siblings with Data from Both Parents

Some of these tools have functional equivalents at *GEDmatch* but give more detail at the SNP level. For example, "Runs of Homozygosity" is the basis for the *GEDmatch* tool "Are Your Parents Related?" but optionally includes shorter ROHs. A "Search for Differently Reported SNPs" will let you determine which specific SNPs are represented by a yellow bar at *GEDmatch* with a one-to-one comparison of two kits for the same person. Pike's phasing routines give the same results as *GEDmatch* but include the data necessary to create a stand-alone file. However, the output requires some post-processing to massage it into a format suitable for uploading to other sites. Hendrik Wendland's "MapS Phasing" creates a file in one of the formats used by the genetic genealogy companies. [37]

The "Search for Discordant SNPs" lists the specific SNPs represented by red bars when comparing a parent and child at *GEDmatch*. This information may lead to the discovery of microdeletions.[38]

Another site offers a novelty tool that creates a QR code with just enough SNPs to distinguish you from any other person on the planet. Take a guess at how many SNPs that takes before checking out the link.[39] It is a surprisingly low number, provided the SNPs are carefully selected to be out of linkage disequilibrium (that is, they are inherited independently of each other) and the alleles have a high Minor Allele Frequency, reinforcing two concepts discussed in this chapter. My QRC is in Figure 8.11. Do you match me if you run your QRC through the tool? If so, we will need to inform the authors!

Figure 8.11. Ann Turner DNA QR Code

37. *MapS Phasing* (https://www.maps-phasing.com/).

38. Ann Turner, "Generation Gaps: A Sign of Microdeletions?" Satiable Curiosity column, *Journal of Genetic Genealogy* 8 (2016): 1–6 (http://jogg.info/pages/vol8/sc/generation-gaps.pdf). *GEDmatch* Genesis introduces artificial breaks in regions of low SNP density and MyHeritage and 23andMe may stitch segments together so parts of this column may become outdated.

39. Yonghong Du et al., "A SNP panel and online tool for checking genotype concordance through comparing QR codes," *PLOS ONE* 12 (September 2017) (https://doi.org/10.1371/journal .pone.0182438). The tool mentioned in the article has been moved to http://qrc.its.unc.edu/.

What's Next?

So, do you want your data raw or cooked? This chapter should leave you with the impression that the genetic genealogy companies have done an admirable job of making the raw data more tasty and digestible. The notion of cooking data can have a negative connotation, and there is a whiff of this in the compromises the chefs make to whip imperfect ingredients into shape. False positives and false negatives are a fact of life when dealing with genotype data.

Will whole genome sequencing (WGS) bring relief? It will almost certainly help with cross-platform problems. A core set of SNPs can be extracted for comparisons with existing databases. Much of the extra data in a WGS will be redundant, but there will be many variants with low Minor Allele Frequency. Each child may harbor some seventy or so mutations compared to the parents.[40] Some new variants can become familial or clan markers if they persist over multiple generations. Every mutation occurs on a specific haplotype background. Perhaps short matching sections of a genome can be identified using common variants, then inspected to see if this haplotype block harbors a distinctive variant pointing to a more recent Most Recent Common Ancestor.

40. Jared C. Roach et al., "Analysis of Genetic Inheritance in a Family Quartet by Whole-Genome Sequencing," *Science* 328 (April 2010): 636-639 (https://doi.org/10.1126/science.1186802).

Drowning in DNA? The Genealogical Proof Standard Tosses a Lifeline

Karen Stanbary, MA, CG

Many excellent sources for documentary research exist, but few are available for DNA analysis. This chapter uses a hypothetical case to illustrate the essential research, analysis, and reasoning employed when using genetic sources. The reader is asked to believe reasonably exhaustive research in documentary sources is complete as summarized in "The Starting Point" section. Of course, since this is fictional, actual documentary research is impossible. References to documentary research are included solely to demonstrate that genealogical conclusions about biological relationships must correlate research in both documentary and genetic sources. All names, dates, relationships, pedigrees, user IDs, and *GEDmatch* kit numbers are fictional. For examples correlating documentary and DNA evidence in real-life case studies see the "Y-DNA Analysis for a Family Study," "Parker Study: Combining atDNA & Y-DNA," and "Correlating Documentary and DNA Evidence to Identify an Unknown Ancestor" chapters.

DNA testing entered the genealogy world in 2000 as a quiet wave washing over our toes whispering the possibilities of Y-DNA-surname and mitochondrial DNA (mtDNA) matrilineal studies. Nine years later, ripples caressed our ankles as autosomal DNA (atDNA) testing confused us with distracting ethnicity pie charts. Seers peeked behind the swells and glimpsed the promise of cousin matching. White caps formed as each new testing company launched its products. Today, DNA testing possibilities hit us with tsunami-like force. Some in our community dive deep into the water and use it skillfully in daily work. Some prefer to wade knee-deep into the water, curiously dabbling in their own results. Others stand at the shore, with an erroneous belief that DNA will not reveal anything new. Some, stand paralyzed, intimidated by the unknown.

Good news! The Genealogical Proof Standard (GPS), offers both a compass and a lifeline to help navigate the stormy waters. The GPS includes five interdependent components:

1. Reasonably exhaustive research
2. Complete and accurate source citations
3. Tests of analysis and correlation of sources, information, and evidence
4. Resolution of conflicts among evidence items
5. A soundly reasoned, coherently written conclusion[1]

Genealogists use the GPS to "measure the credibility of conclusions about ancestral identities, relationships, and life events" and "assess the credibility of a genealogical conclusion."[2] The GPS underlies the product and process standards described in *Genealogy Standards*. This small publication is a valuable guide for successful and efficient genealogical research, reasoning, and communication of the results.

The Board for Certification of Genealogists (BCG) adopted modifications to four existing standards and seven new standards specific to using DNA Evidence. These are published in a second edition of *Genealogy Standards* to be released in early 2019. There is a new sub-section "Using DNA Evidence" added to Chapter 3—Standards for Researching. It introduces the use of DNA evidence.

> Meeting the Genealogical Proof Standard requires using all available and relevant types of evidence. DNA evidence both differs from and shares commonalities with documentary evidence. Like other types of evidence, DNA evidence is not always available, relevant, or usable for a specific problem, is not used alone, and involves planning, analyzing, drawing conclusions, and reporting. Unlike other types of evidence, DNA evidence usually comes from people now living.[3]

Dr. Thomas W. Jones, in the essential work *Mastering Genealogical Proof*, provides in-depth descriptions of the methodology and reasoning required to meet the GPS. He discusses the use of DNA test results within the context of the GPS.

> DNA testing gives family historians access to biological data via DNA records and reports. We must interpret these documents in the same way we interpret other kinds of complex sources (land records, for example.) DNA samples that do—or do not—match are genealogically significant, but without documentary data, DNA reports cannot support or disprove any conclusion of relationship or non-relationship. The GPS is as important in contexts using DNA results as it is in contexts without them.[4]

1. Board for Certification of Genealogists (BCG), *Genealogy Standards*, 2nd ed. (Nashville: Ancestry.com, 2019), 1–2.

2. BCG, *Genealogy Standards*, 2nd ed., 1.

3. BCG, *Genealogy Standards*, 2nd ed., 29.

4. Thomas W. Jones, *Mastering Genealogical Proof* (Falls Church, Va.: National Genealogical Society, 2013), 4–5. Reprinted by permission of the author who holds the copyright.

The concepts and strategies we learn from studying the GPS, *Genealogy Standards*, and *Mastering Genealogical Proof* can be applied to the use of genetic evidence in genealogical conclusions. We are familiar and comfortable with these concepts as applied to documentary research:

- Planning for reasonably exhaustive research, including the careful crafting of a research question
- Providing detailed source citations for DNA information
- Analyzing, correlating and grouping evidence, and considering the independence in evidentiary items
- Resolving conflicts within the evidence
- Developing a complex, documented argument explaining the rationale used to consider a conclusion proved

Successful incorporation of DNA test results into the body of evidence to achieve genealogical proof employs the same methodologies, principles, and research strategies. A hypothetical case example incorporating autosomal DNA evidence makes the point.

The Starting Point

Our research problem seeks unknown information about the biological parents of a unique individual. A fictitious grandmother, Florence May (Carpenter) Forrest, born about 1882, ends her biological ancestral line and blocks further advancement on this line. Family lore says she was adopted as an infant. Thorough research in court records does not suggest the adoption passed through legal channels. Documentary research for the rest of her life reveals she consistently named her adoptive parents as her only parents. The town orphanage records burned in a fire. Florence and her adoptive parents consistently reported Florence's birthplace as Illinois across the multiple records created throughout her life. Information on her biological parents is needed to advance this line further.

We begin all research projects with an analysis of the accuracy of the starting-point information to assure the soundness of our foundation.[5] In this hypothetical case, previous documentary research provides a source-cited timeline for our subject. In the real case that this hypothetical study is based on, reports of the findings from documentary research focused on the mined evidence about Florence's parentage and the circumstances of her birth. The parameters of negative findings for court records are explained. We conclude documentary research does not provide information to identify her biological family. Yet DNA survives to tell the forgotten tale.

5. BCG, *Genealogy Standards*, 2nd ed., 12, standard 11, "Sound basis."

Research Questions

An effective research question is both sufficiently broad to be answerable and sufficiently focused to yield hypothetical answers when using tests of analysis and correlation. Research questions clearly define the subject as a unique person and the unknown information we seek.[6] Researchers using DNA sources seek answers about biological relationships as well as the identity of an unknown biological ancestor. These types of problems are typically answered with a series of related and nested research questions.

The Principal Research Question

Our hypothetical case begins with the principal focused question

Who was *one* of the *biological* parents of Florence May (Carpenter), born about 1882 and who married Clyde Forrest on 3 January 1904 in LaSalle County, Illinois?

The research question includes two italicized words. If both biological parents are unknown, search for *biological* parents *one* at a time. Every individual inherits paired chromosomes—one from the mother and one from the father. A shared atDNA segment with a match could be from either the mother or father for both the focal person and the match. Therefore, initially, atDNA evidence can lead to either biological parent and their ancestors. Since both of Florence's biological parents are unknown, the research question must be broad enough to encompass both parental possibilities. To seek the identity of only one specific parent (mother or father) would be too narrow and hinder our ability to search for meaningful evidence. Of course, once one biological parent is identified, we can pose a new question about the identity of the remaining parent. Elimination of matches associated with the first parent culls the match list for the second parent, making the second phase easier.

We seek a *biological* parent. We do not simply say "parent." A parent could be of any type: biological, adoptive (both formal and informal), foster, legal, to name a few. Distinction between familial relationship types is a hallmark when working in the new age of DNA.[7] We can no longer assume the nature of the connecting relationship. The research question distinguishes the type of familial relationship, in this hypothetical case, as biological.

The principal research question informs the development of the initial research plan. In Florence's case, several distinct research phases will seek answers to intermediate questions using DNA sources, documentary sources, or a combination of both. Each intermediate question and its conclusion develops the subsequent intermediate research question. For example, figure 9.1 previews the principal research question with six supporting research questions for six interrelated research phases:

---

6. BCG, *Genealogy Standards*, 2nd ed., 11–12, standard 10, "Effective research questions." Also, Jones, "Research Questions," *Mastering Genealogical Proof*, 7–8.

7. BCG, *Genealogy Standards*, 2nd ed., 28–29, standard 50, "Assembling conclusions from evidence."

- Supporting research question 1 seeks information about biological relationships using DNA test results to create a focus study group.
- Supporting research question 2 seeks information about the identity of hypothetical biological great-grandparents of the focus person. (Florence's great-grandparents are the common ancestors between Florence's grandchildren and their fourth cousins.)
- Supporting research question 3 seeks information about the identity of candidates for one of Florence's biological parents.
- Supporting research question 4 seeks information useful to eliminate candidates from consideration.
- Supporting research question 5 seeks living and willing DNA test takers.
- Supporting research question 6 seeks information in genetic sources to test a hypothesis.

While these research phases are fully described later in this chapter, focused on the DNA analysis, figure 9.1 depicts the relationship of intermediate research questions to each other. Genealogical works discussing DNA use abbreviations such as 2C for second cousin and 3C1R for third cousin once removed. This convention is followed here.

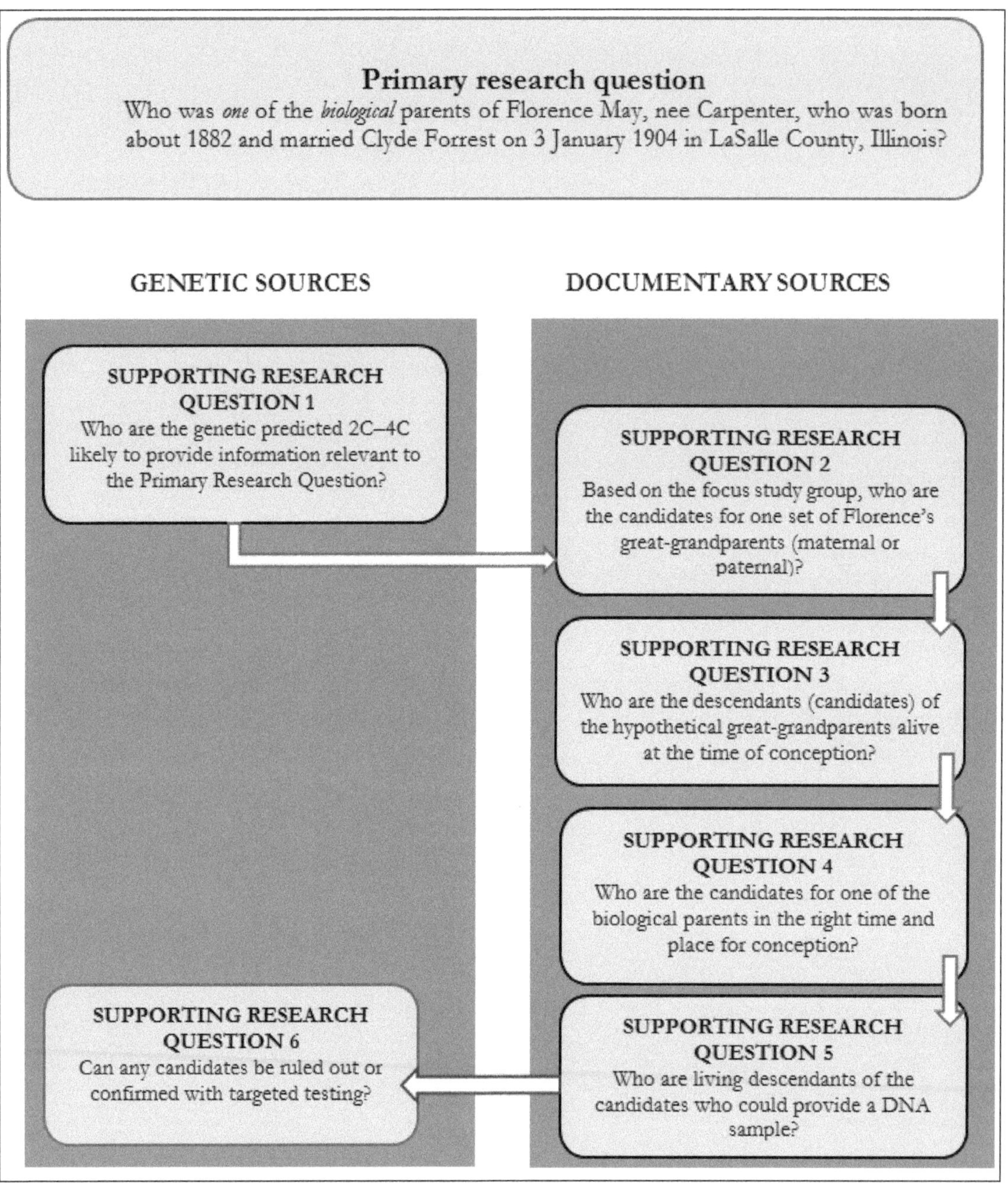

Figure 9.1. Primary and supporting research questions

The Initial Targeted Testing Plan

The documentary path is bleak, but the possibilities of DNA evidence offer a ray of hope. The next stage is the development of a research plan in sources likely to provide independent and high-quality evidence. Think of a targeted testing plan as a research plan using DNA sources. Familiar with the qualities of an efficient and flexible research plan, we know to look for "sources likely to provide or lead to evidence that helps meet a plan's objective." We prioritize our plans to facilitate "efficient discovery of useful evidence." We recognize and embrace the flexibility of dynamic research plans. Proof is our goal. We plan for reasonably exhaustive research with the understanding the plan will continue to be refined as research progresses. We seek to "gather all reliable information potentially relevant to the research question."[8] An effective plan for DNA testing is both selective and targeted. Similarly, a targeted testing plan specifies three parameters of the DNA sources likely to provide meaningful evidence: test takers, types of DNA tests, and testing companies.[9]

Florence bore three daughters and one son, all now deceased. Yet, their surviving children (Florence's grandchildren) are willing to test. Just as we seek the most efficient path in documentary research, we carefully consider and choose the *best* type of DNA test to provide the *best* evidence to answer the research question.

In this case, neither Y-DNA nor mtDNA is an appropriate test choice. Since Florence did not inherit Y-DNA, that test would be useless. Florence inherited mtDNA and passed it on to each of her four children. However, only Florence's daughters passed mtDNA to their children. Also, mtDNA mutates rarely, so even a perfect match could share a common ancestor as much as 500 years back.[10] Mitochondrial DNA testing might be useful in a later phase of the research.

The *best* choice for the type of DNA to test for this research question is autosomal, which includes X-DNA. The atDNA test provides a list of "cousin matches" (people who share identical significant segments of DNA) on all recent ancestral lines.[11] Autosomal DNA testing provides the added benefit of possible shared X-DNA segments.[12] Remember eighth grade biology—every individual inherits two sex chromosomes. A female inherits two X chromosomes (XX). A male inherits only one X chromosome and a Y chromosome (XY). The unique inheritance pattern of the X may prove useful to determine the predicted common ancestral couple.[13]

---

8. BCG, *Genealogy Standards,* 2nd ed., 12, standard 13, "Source-based content"; 13, standard 15, "Efficient sequence"; 14, standard 16, "Flexibility"; and 14, standard 17, "Extent."

9. BCG, *Genealogy Standards,* 2nd ed., 29–30, standard 51, "Planning DNA tests."

10. Blaine T. Bettinger and Debbie Parker Wayne, *Genetic Genealogy in Practice* (Falls Church, Va.: National Genealogy Society, 2016), 24–25, 47–48.

11. Bettinger and Wayne, *Genetic Genealogy in Practice*, 67–68.

12. Bettinger and Wayne, *Genetic Genealogy in Practice*, 101.

13. Bettinger and Wayne, *Genetic Genealogy in Practice,* 102–3.

Because of the random nature of autosomal DNA recombination,[14] each of Florence's children inherited segments of her genome; some segments are the same and some differ between the children. Testing grandchildren from each of Florence's children is efficient—neither redundant nor wasteful. It provides the opportunity to capture more of their grandmother's genome, thereby increasing the chance of meaningful cousin matching. Additionally, extending the testing plan to four grandchildren provides opportunities for analysis and correlation that one or even two tests would not.[15]

Testing a daughter of Florence's son provides unique evidence. That daughter received X-DNA from her father that he inherited only from his mother, the focus of this study. Florence passed to her son either one non-recombined X chromosome or a recombination of her two X chromosomes (one inherited from her mother and one from her father, which he inherited from his own mother). Florence's son in turn passes on his single X chromosome intact to his daughter. The X-DNA segments passed from a father to a daughter only include his own mother's X-DNA. X-DNA passed from mother to a daughter will include a recombination of X-DNA from both maternal grandparents. In this case, the fewer X-DNA matches from a son in the line of descent are useful for the elimination of candidates from consideration.

Genealogy Standards reminds us to seek independent information when collecting data.[16] Testing a descendant of each of Florence's children creates independent information items. By contrast, the results of DNA testing of four grandchildren, all siblings who inherited DNA from only one of Florence's children, would need to be grouped as related items into one unit. That unit would be assigned credibility no greater than the weight of the strongest item.

The testing plan takes advantage of the four willing grandchildren—each from a different child of Florence May (Carpenter) Forrest. Spreading the DNA samples among the four primary autosomal DNA testing companies[17] maximizes the possibility of finding significant matches. Purchasing more than four tests may not be necessary. Depending on the matches in the database, the problem might even be solved if all the grandchildren tested at the same company. In this case, we use all companies. With the test takers' informed consent,[18] the researcher can upload raw data to *GEDmatch*. Doing so permits the researcher to see many of each test taker's matches from the other companies, helping to achieve reasonably exhaustive research. Uploading to *GEDmatch* provides yet another benefit. AncestryDNA's website does not provide information about X chromosome matching even though the raw data includes X-DNA data. Uploading the AncestryDNA test taker's raw data to *GEDmatch*

14. For explanation of autosomal DNA recombination, see Bettinger and Wayne, *Genetic Genealogy in Practice,* 70–72.

15. BCG, *Genealogy Standards,* 2nd ed., 31, standard 53, "Extent of DNA evidence."

16. BCG, *Genealogy Standards,* 2nd ed., 27, standard 46, "Evidence independence."

17. *23andMe* (https://www.23andme.com/), *AncestryDNA* (https://www.ancestry.com/dna/); *FamilyTree DNA* (https://www.familytreedna.com/), and *MyHeritageDNA* (https://www.myheritage.com/dna).

18. BCG, *Genealogy Standards,* 2nd ed., 32, standard 57, "Respect for privacy rights."

provides both X-DNA and atDNA data for analysis.[19] Additionally, uploading to *GEDmatch* provides sufficient detail and identifying information for others to verify or dispute the final conclusion.[20] Table 9.1 plans for transfer of the raw data from AncestryDNA and 23andMe into FamilyTreeDNA (FTDNA) and creates more possibilities for meaningful information.

Table 9.1. Initial targeted testing plan					
Florence's child	**Grandchild**	**Type of test**	**Company**	**Transfer**	**Written consent for upload to *GEDmatch***
Fannie	James	atDNA	AncestryDNA	FTDNA, MyHeritage (MH)	yes
Mary	Doris	atDNA	FTDNA	MH	yes
Elizabeth	Thomas	atDNA	23andMe	FTDNA, MH	yes
George	Marlene	atDNA[21]	AncestryDNA	FTDNA, MH	yes

The most recent common ancestral *couple* for the four grandchildren is Florence *and* her husband Clyde. Any matches who share atDNA with the grandchildren could be related through either Florence's or Clyde's ancestral lines. Supplemental targeted testing, if needed, may resolve the questions of whether DNA matches are in Florence's family or Clyde's. Testing a descendant of one of Clyde's siblings helps to eliminate matches related on Clyde's ancestral lines from consideration. See table 9.2.

Table 9.2. Supplemental targeted testing plan					
Clyde's sibling	**Grandchild of Clyde's sibling**	**Type of test**	**Company**	**Transfer**	**Written consent for upload to *GEDmatch***
Joan	Carol	atDNA	AncestryDNA	FTDNA, MH	yes

Documentation of Parent-Child Links

Each parent-child link on a relevant ancestral line must be documented. We use specific source citations to document each conclusion we establish.[22] Ancestral lines of descent are genealogical conclusions. Figure 9.2 illustrates the test takers' connections to Florence and Clyde and Clyde's parents (George Forrest and Susan Wood). Note references indicate a

19. To register at *GEDmatch* (https://www.gedmatch.com/register.php). Registration will create a personalized home page and subsequently, provide instructions to upload raw data.

20. BCG, *Genealogy Standards,* 2nd ed., 31, standard 54, "Sufficient verifiable data."

21. Marlene is the daughter of George, the only son of Florence. The X-DNA Marlene inherited from George is inherited from only Florence's ancestral lines (Marlene's paternal great-grandparents). Analysis of Marlene's paternal X-DNA is automatically phased to Florence. By contrast, the other three grandchildren all descend from Florence's daughters. They inherited X-DNA from both Florence and Clyde's mother.

22. BCG, *Genealogy Standards,* 2nd ed., 6, standard 2, "Specificity."

requirement of documentation to tie each individual to the one parent or parents in the relevant ancestral line.

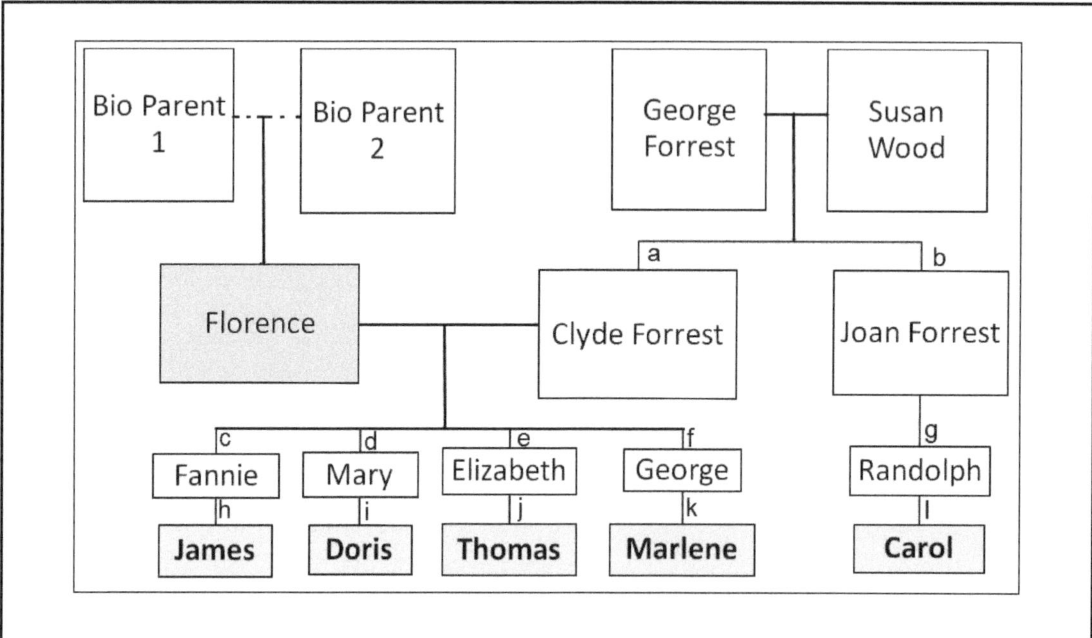

This is a hypothetical case. In an actual case, the sources supporting each parent-child relationship would likely be detailed here. Alphabetical note references in the chart, as well as sources linking Florence to the hypothetical biological parents, each require source citations. For real-life examples see figure 7.1 in the "Parker Study: Combining atDNA and Y-DNA" and tables 10.1 and 10.2 in the "Correlating Documentary and DNA Evidence" chapters.

Figure 9.2. Parent-child links (bold-face font indicates DNA test takers)

Planning for Research

We value a sufficiently broad research scope, seeking all potentially relevant information to the research question.[23] Doing so reduces the chance of overlooking significant information. Yet, a broad research scope is tempered by planning for efficient sequence. Sequencing prioritizes efficient discovery of useful evidence."[24] DNA test results include much information that is not useful evidence for a specific research question.

Relationships among the four grandchildren are known. James, Doris, Thomas, and Marlene are all first cousins to each other. They are second cousins to Carol. In isolation, their DNA test results do not provide meaningful information to answer our question. Their test results simply provide the base to collect data relevant to the research question. The populated list of their genetic matches and details about the shared atDNA is the buried treasure to seek.

23. BCG, *Genealogy Standards,* 2nd ed., 16, standard 19, "Data-collection scope."
24. BCG, *Genealogy Standards,* 2nd ed., 13, standard 15, "Efficient sequence."

Assuming no pedigree collapse and no misattributed parentage, each of Florence's grandchildren inherited about 25% of Florence's genome, but also inherited approximately 25% from Florence's husband Clyde Forrest and each of their other two grandparents. We do not want to mine for evidence from all their genetic matches. We employ evidence discrimination to exclude information items irrelevant to the research question. We justify our reasoning to include and exclude sources and information items for data collection.[25] About 75% of the grandchildren's matches will be related on each test taker's other three grandparent lines and thus, are irrelevant to the research question. Efficient research planning, including the supplemental targeted test on Clyde's ancestral lines, winnows the grandchildren's matches to only those who are likely related on Florence's ancestral lines—a focus study group. Figure 9.3 illustrates this concept for one of Florence's grandchildren James. A similar chart could be created for the remaining three test taking grandchildren.

Recombination may result in cousins that do not share enough detectable amounts of DNA to be listed as matches. Be aware that some matches that are not listed as in-common-with or shared matches could still be related through Florence's line. Those cousins are not useful at this step of the process but may provide helpful information later.

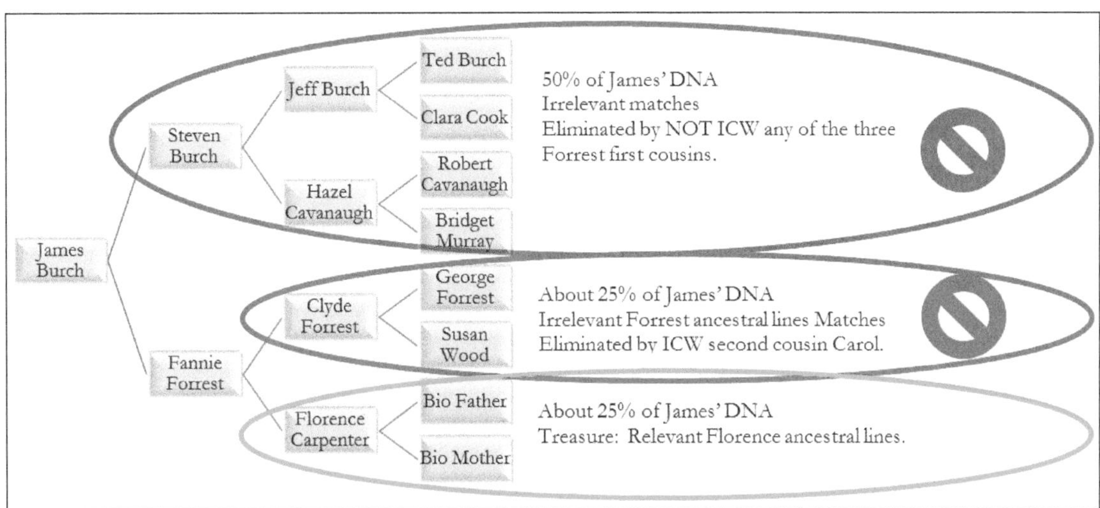

Figure 9.3. Grandchild inheritance-elimination of irrelevant matches

Creation of a focus study group requires we understand the meanings of concepts associated with DNA analysis and interpretation. Just as we understand the meanings when we undertake documentary research, we educate ourselves to understand the meanings of fundamental concepts associated with DNA.[26] This knowledge helps us to discriminate among the many matches, eliminating the irrelevant.[27]

25. BCG, *Genealogy Standards,* 2nd ed., 25, standard 42, "Evidence discrimination."

26. BCG, *Genealogy Standards,* 2nd ed., 17, standard 24, "Understanding meanings" and 30–31, standard 52, "Analyzing DNA test results."

27. BCG, *Genealogy Standards,* 2nd ed., 25, standard 42, "Evidence discrimination."

Two essential skills needed at this stage are the ability to: 1) analyze using "in common with" (ICW), also known as shared matching; and 2) predict potential relationships based on the amounts of shared atDNA. We use company and third-party tools to help with identification and analysis of ICW.[28] We use statistical references, such as data gathered from citizen science projects or scientific studies, to help with prediction of potential relationships.[29]

An ICW match is someone who shares atDNA with the test taker and with another person on the test taker's match list. The shared DNA can be on any chromosome. A group of ICW matches may or may not share atDNA on the same segment. A group of ICW matches may or may not all descend from the same common ancestor. They appear on each other's match lists, but they may not all share one common ancestor. A list of ICW matches is a good place to begin the process of selecting members of a focus study group. Our initial targeted testing plan intentionally included four independent grandchildren to meet the goal of sufficiently extensive research.[30] If we had tested only one grandchild, we would not have enough data to begin a focus study group and would not achieve thorough research. We also tested a second cousin on grandfather Clyde's side to help eliminate relatives on his line.

Using the match lists for each grandchild, a multi-stage process of elimination aids in the creation of the meaningful focus study group. Figure 9.4 illustrates efficient pruning of James's match list:

1. First elimination phase: Ideally, we seek predicted second to fourth cousin matches. Predicted first cousins share more genealogical lines that must first be eliminated to provide data relevant to this research question. Second cousins almost always share atDNA and their common ancestors are great-grandparents (the unknown ancestors we seek). Third and fourth cousins are not guaranteed to all share DNA, but the odds are high we can collect meaningful data. By contrast, fifth cousins share fourth great-grandparents. Assuming no pedigree collapse, an individual has 128 fourth great-grandparents. A genealogical conclusion requires that all evidence points to just one answer.[31] A conclusion identifying the common ancestor between fifth cousins requires elimination of 127 ancestors from consideration—a daunting task. Each ancestor, not eliminated, represents a competing hypothesis. More distant cousins may share atDNA inherited from a common ancestor, but single small segments could be false segments, from a common ancestor far outside of a genealogical time frame, or a frequently-occurring segment within a population, known as a pileup region.[32] None of these are useful for solving genealogical problems.

28. For a description of "In Common With" and "Shared Matches," see Blaine T. Bettinger, *The Family Tree Guide to DNA Testing and Genetic Genealogy* (Cincinnati, Ohio: Family Tree Books, 2016), 114–19.

29. Blaine T. Bettinger, "The Shared cM Project," version 3.0, *The Genetic Genealogist*, August 2017 (https://www.thegeneticgenealogist.com/). Check for the most current version.

30. BCG, *Genealogy Standards*, 2nd ed., 14–15, standard 17, "Extent" and 31, standard 53, "Extent of DNA evidence."

31. BCG, *Genealogy Standards*, 2nd ed., 28–29, standard 50, "Assembling conclusions from evidence."

32. Blaine T. Bettinger, "The Danger of Distant Matches," National Genealogical Society, *Paths to Your Past*:

Therefore, for this research question, we prune each grandchild's match list leaving only matches who are predicted second to fourth cousins.

James tested at Ancestry. The targeted testing plan includes transfer of his raw data into FamilyTreeDNA and MyHeritageDNA. This will identify matches who tested at those three companies. It also includes uploading his raw data to *GEDmatch,* which identifies matches who tested at other companies and some that may not meet the company thresholds for matching. First, we create a comprehensive list of all predicted second to fourth cousin matches from each database.

2. Second elimination phase: Identify and eliminate matches from the remaining group who are likely related through the test taker's other grandparent lines (except the Forrest line). We identify only those matches who are ICW at least two of the other three grandchildren. If a match is related to at least two grandchildren, then that match will not likely be related to the grandchildren's other two grandparents (the grandparents other than Florence and Clyde). Because we chose four independent grandchildren, we can limit our search to their shared matches. Matches related to Florence and Clyde will be the only matches the four grandchildren will share in common (unless their parents are also related).

Each of the four grandchildren are spread across multiple databases and included in *GEDmatch.* We use third-party and company tools to cull the match list to only those who are ICW James and at least two of Florence's test taking grandchildren.

3. Third elimination phase: James's match list now includes matches likely related to either Florence's biological lines or Clyde Forrest's ancestral lines. The documentary descendant tree is source-cited (figure 9.2), creating a sound basis to establish the paper-trail relationships.[33] We examine the match's ICW to seek parallels, families in common, as an aid to group the matches.[34] We sort them into two buckets: "Likely Clyde Matches" and "Likely Florence Matches." Because Clyde's parents are known and documented, and the sibling relationship between Clyde and his sister Joan is documented, matches who match both Florence's grandchildren and Carol (Joan's grandchild) are part of the "Likely Clyde Matches" bucket. This brings Clyde's bucket into focus, justifying the choice to eliminate Forrest ancestral line matches from consideration.[35]

2018 Program Syllabus (Grand Rapids, Mich., 2018), 427–428. Also, audio recording of the same, available at *Playback NGS* (https://www.playbackngs.com/). Also, Bettinger, "Evaluating Genealogical Conclusions Using DNA," National Genealogical Society, *Paths to Your Past: 2018 Program Syllabus,* 651–2. Also, audio recording of the same. Also, "Excess IBD Sharing," International Society of Genetic Genealogy (ISOGG) *Wiki* (https://isogg.org/wiki/Identical_by_descent).

33. BCG, *Genealogy Standards,* 2nd ed., 12, standard 11, "Sound basis."

34. BCG, *Genealogy Standards,* 2nd ed., 27, standard 47, "Evidence correlation."

35. BCG, *Genealogy Standards,* 2nd ed., 25, standard 42, "Evidence discrimination."

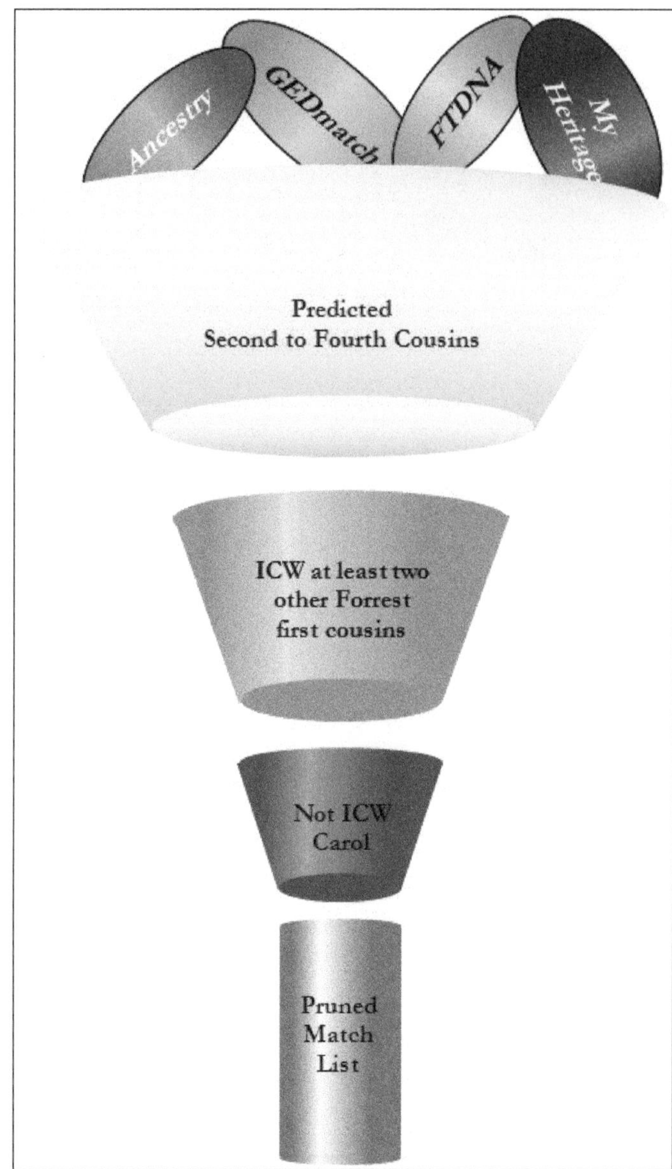

Figure 9.4. Pruning James's match lists

The Focus Study Group

Targeted testing justifies both the inclusion and elimination of matches in the focus study group. Initially, included matches are those who match at least three of the four grandchildren; eliminated are those who match cousin Carol (Clyde's line) and those who do not match at least three of Florence's grandchildren. Once we complete the pruning for all four grandchildren, we have four lists of likely Florence ancestral line matches. We merge the four lists and eliminate duplicates. Like planning for research in documentary sources,[36] focus study groups are flexible, not fixed. As we analyze and correlate test results, we delete some people from the list and add others. See table 9.3 for a sample Focus Study Group

36. BCG, *Genealogy Standards,* 2nd ed., 14, standard 16, "Flexibility."

Tracking template. At the end of this step, the first four columns will be complete; the last column will be completed in the next step.

Grouping the Matches (Again)

The remaining matches in table 9.3 are grouped based on shared DNA (the ICW column). Genealogists regularly group evidence. This is no different. Grouping individual matches into subgroups facilitates analysis. Our pruned focus study group is now limited to likely genetic cousins related to Florence's grandchildren on only Florence's ancestral lines. Yet, each member of the focus study group could be related to the grandchildren on *either* Florence's paternal side *or* Florence's maternal side. At this stage, we do not know which of Florence's parents is the center of each of the two groups. We only know Group 1 and Group 2 are different. Again, we use our understanding of ICW to sort the people into likely Group 1 matches and likely Group 2 matches. Some focus group members may not match anyone else at this point; this is indicated by "undetermined" as the group designation for Anna Kross. She could be related to Florence's grandchildren in one way, perhaps through Florence's husband's side, and related to focal study group member James through an ancestor not related to Florence.

The probability that second cousins and closer matches share DNA is about 99%. Any matches more distant may be genealogically related but may not share atDNA segments.[37] Therefore, we cannot infer lack of a genealogical relationship for matches more distant than second cousins who do not share any atDNA segments.[38] See table 9.4 for sample Focus Study Group Shared DNA template.

We choose the group with the closest matches for the next research phase. In this case, Group 1 is a good place to begin. Group 1 is shaded yellow in table 9.4. Analysis is facilitated by selecting the closest matches in this group, the matches who share the most DNA. In this case, we select the match Peter S. and then see if correlation can tie him with the other Group 1 matches.

Correlation

Now comes the fun part—correlation of the evidence. The beloved third element of the Genealogical Proof Standard, "tests of analysis and correlation—of all sources, information items, and evidence contributing to an answer to a genealogical question or problem"[39] leads the way. Each ancestral line in a match's tree provides a competing hypothesis of the source of the shared DNA between each match and Florence's grandchildren. Inferential reasoning allows us to eliminate competing hypotheses until the evidence points to only one answer to the question.[40]

37. Bettinger and Wayne, *Genetic Genealogy in Practice,* 74, Table 4, "Likelihood of detectable amounts of shared DNA with a genealogical relative (%)."

38. BCG, *Genealogy Standards,* 2nd ed., 37, standard 65, "Content," third bullet point.

39. BCG, *Genealogy Standards,* 2nd ed., 1–2.

40 BCG, *Genealogy Standards,* 2nd ed., 28–29, standard 50, "Assembling conclusions from evidence."

Table 9.3. Focus study group tracking ICW[a]				
Match Name	**Identifiers**	**Matching Grand-children**	**ICW**	**Group**
Peter S.	Psdna (A) A555667 (G)	All four	Kent J., Charles J., Sandra J.	1
Charles J.	Cjdna (F) T124765 (G)	All four	Peter S., Kent J., Sandra J.	1
Kent J.	Kjdna (A) A444379 (G)	All four	Peter S., Sandra J.	1
Clifford W.	Cwdna (23) M877543 (G)	All four	none	2
Sandra J.	Sjdna (F) T789003 (G)	All four	Charles J., Kent J., James B.	1
James B.	JBDNA (A)	James, Thomas, Marlene	Charles J., Sandra J.	1
Anna K.	AnnaKross (A)	James	none	undetermined

a. Match name is the full name, if possible. Identifiers include the user names by test company and the *GEDmatch* kit number (the first letter of each company or site name is in parentheses). Group 1 or Group 2 refers to Florence's parents. We cannot know yet if it is the mother or the father, so we label Group 1 or Group 2.

Table 9.4. Focus study group tracking amount of shared atDNA						
Match Name	**James**	**Doris**	**Thomas**	**Marlene**	**Aver-age**	**Predicted Relationships to Grandchildren[a]**
Peter S.	485 cM	380 cM	365 cM	398 cM	407 cM	Cluster 4, but close to cluster 5
Charles J.	74 cM	56 cM	82 cM	101 cM	78 cM	Clusters 6–9
Kent J.	55 cM	87 cM	64 cM	63 cM	67cM	Clusters 6–10
Clifford W.	225 cM	190 cM	187 cM	122 cM	181 cM	Clusters 5–6
Sandra J.	56 cM	79 cM	69 cM	112 cM	79 cM	Clusters 6–9
James B.	44 cM	0	56 cM	89 cM	47cM	Clusters 6–10
Anna K.	45 cM	0	0	0	n/a	Clusters 6–10

a. Shared cM is the amount of total shared cM for segments over 7cM in size with each grandchild. Cluster numbers in the "Predicted Relationships to Grandchildren" column are derived from the reference work by Blaine T. Bettinger, "The Shared cM Project — Version 3.0 (August 2017)," PDF report (https://thegeneticgenealogist.com/wp-content/uploads/2017/08/Shared_cM_Project_2017.pdf). The instructions state to select clusters for the 95th percentile range. For more on cluster numbers see figure 5.4 in the "Unknown and Misattributed Parentage Research" chapter.

This is not as difficult as it sounds. Genealogists use this strategy in daily work. Imagine five different Thomas Woodses of similar ages at the same time in Frederick County, Virginia, any of whom might be the father of John Woods, our hypothetical focus person. Each Thomas Woods must be considered as a competing hypothesis until the researcher eliminates four of them, leaving only one potential father. If the researcher cannot, the case cannot achieve genealogical proof. Similarly, each recent ancestral line in the focus study group's pedigree provides a competing hypothesis *until* the evidence and logic rules it out. If the researcher cannot rule out competing lines, the case cannot achieve genealogical proof.

Have you noticed, thus far, there is no mention of online pedigrees of matches? While some working on recent unknown parentage cases may use trees before starting analysis, researchers with some known family tree information are better off waiting. Immediately looking at trees in search of a common ancestor can lead to a hasty mistake in atDNA analysis. If the same ancestor appears in the trees of the matches, it must be correct, right? Many of us say, *WRONG!* We would not do this with our documentary research. We should not do this with our DNA-based research. Genealogists prefer to reason from original records.[41] DNA test results, properly interpreted, are akin to original records. An unsourced, online tree is an unsourced, derivative record. Skilled genealogists use online trees as hints, appreciating that overdependence on derivative records can lead to faulty conclusions. Quickly consulting the online trees can sway to a preconceived conclusion and potential research bias. Disciplined research builds your case.

Just as in documentary research, we seek to accurately place individuals into their families.[42] Let the evidence speak. We know how to group evidence. This is no different. Create potential subgroups from the members of the focus study group based on the amount of shared DNA between and among focus study group members. Grouping individual matches into subgroups facilitates analysis. Sort through the pruned focus study group. Group 1 should represent an ancestral line with multiple descendant branches. The line begins with a common ancestral couple as the DNA could be inherited from either the father or the mother at the top. We predict and place each focus study group member into the ancestral line using the amount of shared DNA between and among the focus study group members.

Just as *Black's Law Dictionary* helps us derive meaning from archaic words in legal documents, "The Shared cM Project" reference[43] helps us derive meaning for the amounts of shared DNA. "The Shared cM Project" provides statistical relationship parameters from data gathered from over 25,000 relationships. Interpreting the data allows us to distinguish between degrees of cousins and, ultimately, potential placement in the line of descent from a common ancestral couple.[44] Part of "The Shared cM Project" includes categorization of total shared cM into relationship clusters.[45] Identify the cluster (or range of clusters) that include the specific amount of shared cM range in the 95th percentile range. Table 9.5 is a matrix illustrating the concept and is used to compare relationships among the focus study group members. The amount of shared DNA among focus study group members in Group

41. BCG, *Genealogy Standards,* 2nd ed., 23–24, standard 38, "Source preference."
42. BCG, *Genealogy Standards,* 2nd ed., 28–29, standard 50, "Assembling conclusions from evidence."
43. Bettinger, "Shared cM Project," version 3.
44. BCG, *Genealogy Standards,* 2nd ed., 30–31, standard 52, "Analyzing DNA test results."
45. See figure 5.4 in the "Unknown and Misattributed Parentage Research" chapter.

1 (green shaded boxes) provides a range of potential relationships based on the associated clusters (yellow shaded boxes).[46]

Table 9.5. Shared DNA among focus study group members

	Peter S.	Charles J.	Kent J.	Sandra J.	James B.
Peter S.		125 cM	101 cM	124 cM	66 cM
Charles J.	Clusters 5–7		82 cM	78 cM	77 cM
Kent J.	Clusters 5–8	Clusters 6–9		231 cM	127cM
Sandra J.	Clusters 5–7	Clusters 6–9	Clusters 5–6		3487 cM
James B.	Clusters 6–10	Clusters 6–9	Clusters 5–7	Parent-child	

James B. shares 3,487 cM with Sandra J. suggesting a parent-child relationship. Correlating further, there is a marked difference between the amount of shared DNA between Sandra with Peter and Kent as compared to James with Peter and Kent. Sandra shares more DNA, suggesting Sandra is the mother of James.

Discipline counts. Now, and only now, is the time to evaluate the matches' trees. The tree linked to James B.'s kit reports his mother is Sandra (Jones) Bailey; this is consistent with the shared amount of DNA between Sandra and James. The next closest match is between Sandra and Kent with shared DNA of 231 cM. Our next goal is to hypothesize how Sandra and Kent are related to each other.

We begin with an analysis of the pedigrees for each. We consider and accommodate for the accuracy, gaps, and depth of the compared pedigrees.[47] We may improve the tree's accuracy by documenting the connections between generations in the tree. Many strategies are available to accommodate for gaps in trees. The easiest is to do more documentary research. Some gaps require inferential reasoning, perhaps based on locations or ethnic origins, to eliminate them as competing ancestral lines. Assure the trees are sufficiently deep to include the potential common ancestral couple for all members. Correlate to identify each subgroup's likely common ancestral couple. Figure 9.5 illustrates the trees of Kent and Sandra.

Sandra's pedigree is partially source cited. There are no gaps in the tree back to the level of second great-grandparents. Kent's tree is undocumented but extends to his great-grandparents. There is one gap in Kent's pedigree—an unknown paternal grandmother and by

46. BCG, *Genealogy Standards,* 2nd ed., 37, standard 65, "Content," third bullet point.

47. BCG, *Genealogy Standards,* 2nd ed., 30–31, standard 52, "Analyzing DNA test results," first bullet point.

extension, her parents. A gap could be the source of the shared DNA between Sandra and Kent. However, as we would see when analyzing shared DNA segments, Kent and Sandra share a significant segment of X-DNA of 55 cM. The gap on Kent's line does not represent ancestors in his X-DNA inheritance line. While Kent and Sandra may be related on more than one line, including the gap line, we begin with their X-DNA inheritance lines. Since we have multiple members in our focus study group, the future body of evidence may help to eliminate Kent's gap line from consideration. Disciplined researchers record observations and assumptions, distinguishing between those that are valid and those that may be unsound. Unsound assumptions may be valid but require additional supporting evidence.[48] When analyzing trees, we may not make *assumptions* exactly like the ones named in *Genealogy Standards*, but instead we observe the *reliability and credibility* of the trees and consider this as we proceed with analysis and correlation.

Reliable credibility observation

- Both pedigrees do not illustrate any pedigree collapse.

Unreliable credibility observations

- Sandra's pedigree is partially source cited. Further documentary research may be needed.
- Kent's pedigree is unsourced. Further documentary research may be needed.
- Kent's pedigree has a gap—an unknown paternal grandmother. This grandmother is not on Kent's X-DNA inheritance line. Further documentary research and/or analysis is needed.
- Sandra and Kent's pedigrees appear sufficiently deep to correlate with each other. However, another generation is required to correlate with more distant Group 1 matches.

Assumptions and observations recorded, we correlate the two pedigrees to look for common ancestors. Given the X-DNA matching segments, we focus on the X-DNA inheritance lines. We find a hypothesized common ancestral couple—Charles Jefferson and Mary Brooks as shown in figure 9.5.

48. BCG, *Genealogy Standards*, 2nd ed., 26–27, standard 45, "Assumptions."

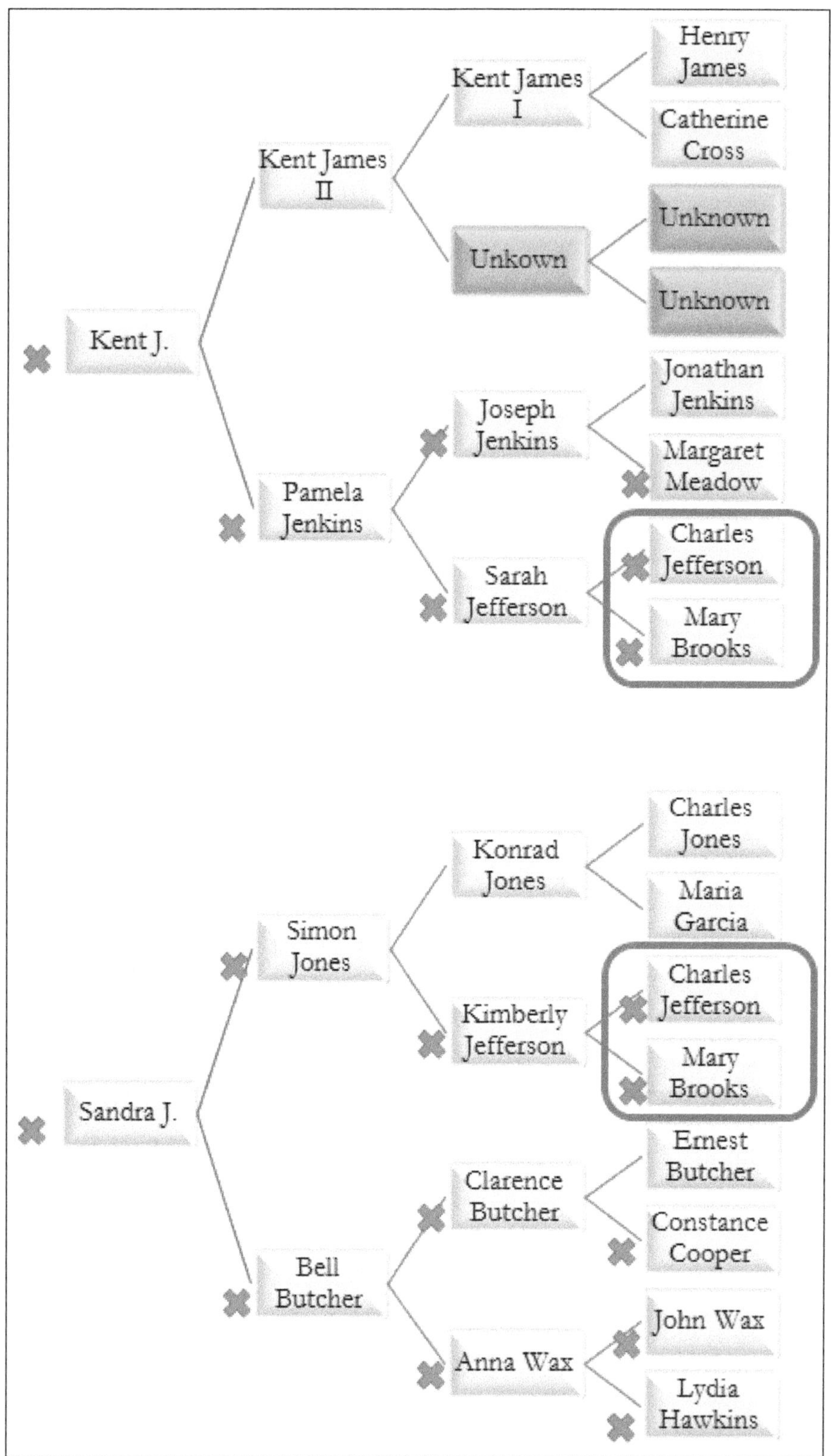

Figure 9.5. Pedigrees of Kent and Sandra

A Theoretical Tree

A theoretical tree as shown in figure 9.6 helps to visualize and correlate the multiple layers of evidence.[49] The objective is to use evidence analysis and correlation from both documentary and genetic sources to incorporate all Group 1 focus study group members into an extended descendant tree. The creation of the theoretical tree builds by generations. While Peter shares the most DNA with Florence's grandchildren, we defer exploring this relationship until we create a firm foundation. The first objective is to place each of the Group 1 focus study group members into one descendant family. We previously identified Sandra as the mother of James B. Next, we want to fit in the closest genetic match among our focus study group to Sandra—in this case, Kent. Using documentary sources, we hypothesize Sandra and Kent are second cousins, each descending from their common ancestral couple Charles Jefferson and Mary Brooks (their great-grandparents).

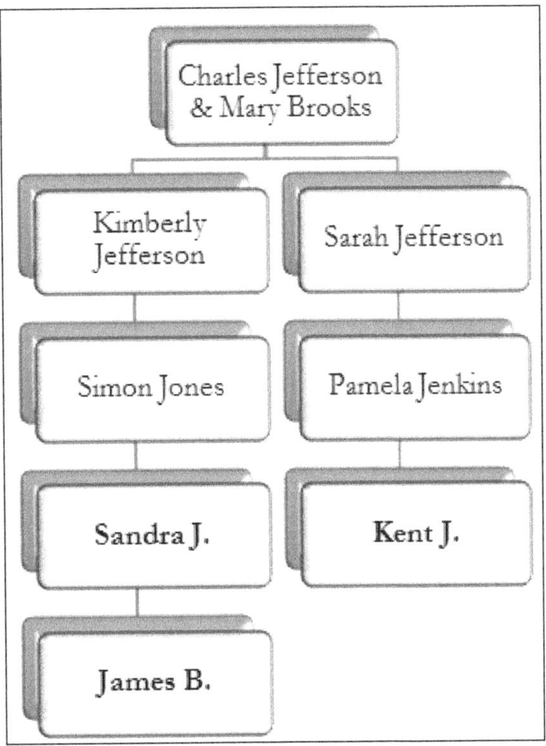

Figure 9.6. Theoretical tree—first stage

Documentary Research

The goal of the next stage of research is to extend the hypothesized most recent common ancestral couple back one generation. Given that matching segments could be inherited from either Charles Jefferson or Mary Brooks, this phase of research includes the parents of each. Initially, we use documentary sources to accomplish this.

49. BCG, *Genealogy Standards*, 2nd ed., 37, standard 65, "Content," second bullet point.

Documentary research identifies the parents of both Charles Jefferson and Mary Brooks. We add them to the theoretical tree and then delve into the documentary sources to identify all the children of the couple, eliminating any who died before the ability to conceive children. This is illustrated in figure 9.7.

Extension of the pedigree back one generation, allows for the incorporation of Group 1 matches Charles and Peter into the tree. The most recent common ancestral couple for Peter, Charles, Kent, Sandra and James are *Frank Jefferson and Kathleen Lowe*. Additionally, we discover another son of Carolyn Jefferson—Timothy, now deceased. Timothy has a living son—Robert, but he cannot be identified in any DNA database.

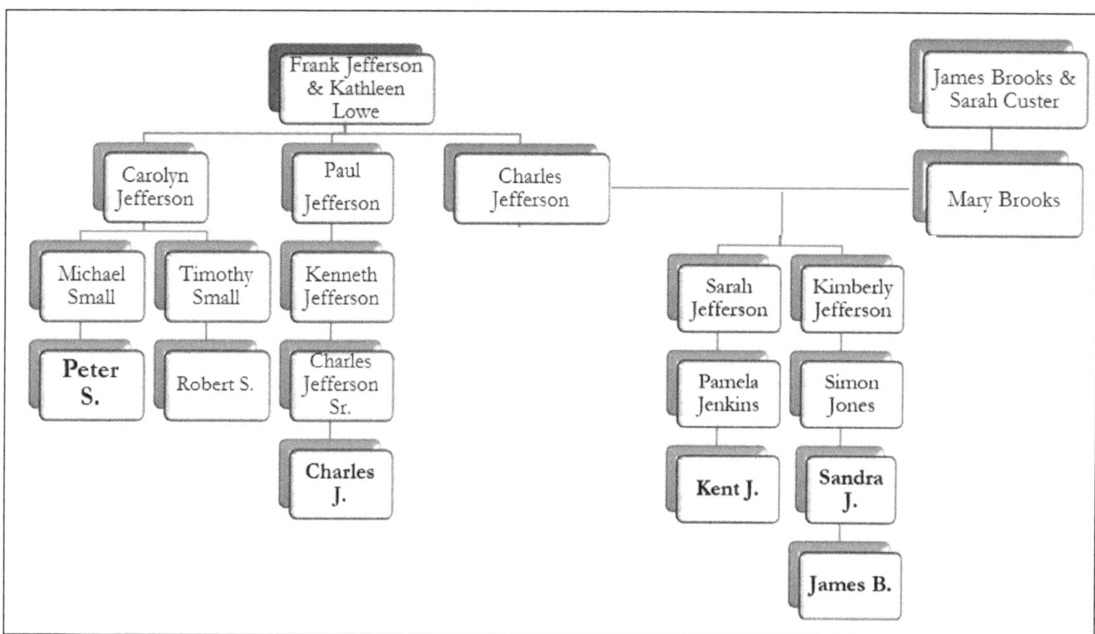

Figure 9.7. Theoretical tree—second stage

Using the previous matrix (table 9.5), we create a similar matrix, exchanging the clusters for the traced relationships. The amount of shared DNA between the Group 1 members of the focus study group (green shaded boxes) is correlated with the traced relationships (yellow shaded boxes) in table 9.6. We assure the traced relationships (documentary sources) are consistent with the amount of shared DNA (genetic sources). The two independent sources (genetic and documentary) provide consistent information to correctly place each into one descendant tree.[50]

50. BCG, *Genealogy Standards*, 2nd ed., 32, standard 55, "Integrating DNA and documentary evidence."

Table 9.6. Correlation of shared DNA with traced relationships					
	Peter S.	Charles J.	Kent F.	Sandra J.	James B.
Peter S.		125 cM	101 cM	124 cM	66 cM
Charles J.	2C1R		82 cM	78 cM	77 cM
Kent F.	2C1R	3C		231 cM	127cM
Sandra J.	2C1R	3C	2C		3487 cM
James B.	2C2R	3C1R	2C1R	Parent-child	

Formulation of a Hypothesis

The theoretical tree illustrates the relationships between members of the focus study group. The focus study group is sufficiently extensive to suggest one of Florence's biological parents is a grandchild of Frank Jefferson and Kathleen Lowe.

Previously we deferred analysis of the potential relationship of Peter to Florence's grandchildren until we had a solid foundation (theoretical tree) of the Group 1 focus study group members. Looking at table 9.7, we examine the Group 1 member who shares the most DNA with Florence's grandchildren. Peter shares more DNA with the grandchildren. Correlation of the amount of shared DNA between the traced descendants of Frank Jefferson and Kathleen Lowe with Florence's grandchildren suggest Peter is the most closely related to Florence's grandchildren of the tested descendants of Frank Jefferson and Kathleen Lowe.

We propose a reasonable hypothesis—one of Florence's biological parents is a child of Carolyn Jefferson.

Table 9.7. Correlation of shared DNA between focus study group members and Florence's grandchildren									
	Peter S.	Charles J.	Kent F.	Sandra J.	James B.	James	Doris	Thomas	Marlene
Peter S.		125 cM	101 cM	124 cM	66 cM	485 cM	380 cM	365 cM	398 cM
Charles J.	2C1R		82 cM	78 cM	77 cM	74 cM	56 cM	82 cM	101 cM
Kent F.	2C1R	3C		231 cM	127cM	55 cM	87 cM	64cM	63 cM
Sandra J.	2C1R	3C	2C		3487 cM	56 cM	79 cM	69 cM	112 cM
James B.	2C2R	3C1R	2C1R	Parent-child		44 cM	0	56 cM	89 cM
James	Cluster 4	Clusters 6–9	Clusters 6–10	Clusters 6–10	Clusters 6–10				
Doris	Cluster 4	Clusters 6–10	Clusters 6–8	Clusters 6–9	none				
Thomas	Cluster 4	Clusters 6–9	Clusters 6–10	Clusters 6–9	none				
Marlene	Cluster 4	Clusters 6–8	Clusters 6–10	Clusters 5–8	none				

Hypothesis Tests

Carolyn Jefferson bore two known children, sons Michael Small and Timothy Small, both deceased. Each has a living son. Michael's son Peter is already in the DNA databases. Documentary research confirms both Michael and Timothy lived in Chicago, Illinois, in 1882, when Florence was conceived.

It is tempting to simply conclude Michael is the biological father of Florence, based on the shared DNA between his son Peter and Florence's grandchildren. Yet, the average amount of shared DNA between Peter and the grandchildren is 407 cM, right on the border between clusters four and five. The histograms included in "The Shared cM Project" provide clues to the *most likely* relationship between two people. The peak of each histogram represents the *most common amount* of shared DNA for that specific relationship, surrounded by the "shoulder" ranges on either side of the peak, above and below the most common amount but still within the range for that relationship. Unfortunately, there is insufficient data to provide a supporting histogram for the specific relationship of half-great-uncle.[51]

51. Blaine T. Bettinger, "The Shared cM Project — Version 3.0 (August 2017)," PDF report (https://thegeneticgenealogist.com/wp-content/uploads/2017/08/Shared_cM_Project_2017.pdf), 11.

Genealogists try as hard to disprove a hypothesis as we try to prove it. Michael's brother Timothy provides a competing hypothesis. The evidence could point to either as the biological father of Florence.

Timothy is deceased and his living son Robert has not provided a DNA sample. *We expand our research to seek genetic information about Timothy's son to compare to Florence's grandchildren.* We contact Robert to explain the study and ask if Robert would be willing to provide a DNA sample. He agrees.

If Michael were the biological father of Florence, then son Peter would be a half-great-uncle to Florence's grandchildren and Robert would be a first cousin twice removed. If Timothy were the biological father then the relationships would be reversed. A half-great-uncle relationship would share, on average, 440 cM, falling within cluster 4. First cousins twice removed share, on average, 232 cM.

Once Robert's test results are received, his data suggests Michael and Timothy Small were full siblings. The data is also consistent with the hypothesis that Michael Small is the biological father of Florence. Timothy is eliminated from consideration. All the evidence points to one conclusion. Florence's grandchildren and each significant match all fit into one expanded descendant tree shown in figure 9.8.[52]

Table 9.8 illustrates the amount of DNA shared by the grandchildren of Florence with Peter and Robert. For sample DNA hypothesis charts, see figure 11.5 in the "Writing about DNA Evidence" chapter.

52. BCG, *Genealogy Standards*, 2nd ed., 37, standard 65, "Content," bullet point 2.

Table 9.8. Correlation of shared DNA between two hypothetical candidates with Florence's grandchildren							
	Peter S.	**Robert**	**James**	**Doris**	**Thomas**	**Marlene**	**Average**
Peter S.		890 cM	485 cM	380 cM	365 cM	398 cM	407 cM
Robert M.	1C		230 cM	244cM	298 cM	199 cM	243 cM
James	Cluster 4	Clusters 5–6					
Doris	Cluster 4	Clusters 4–5					
Thomas	Cluster 4	Clusters 4–5					
Marlene	Cluster 4	Clusters 5–6					

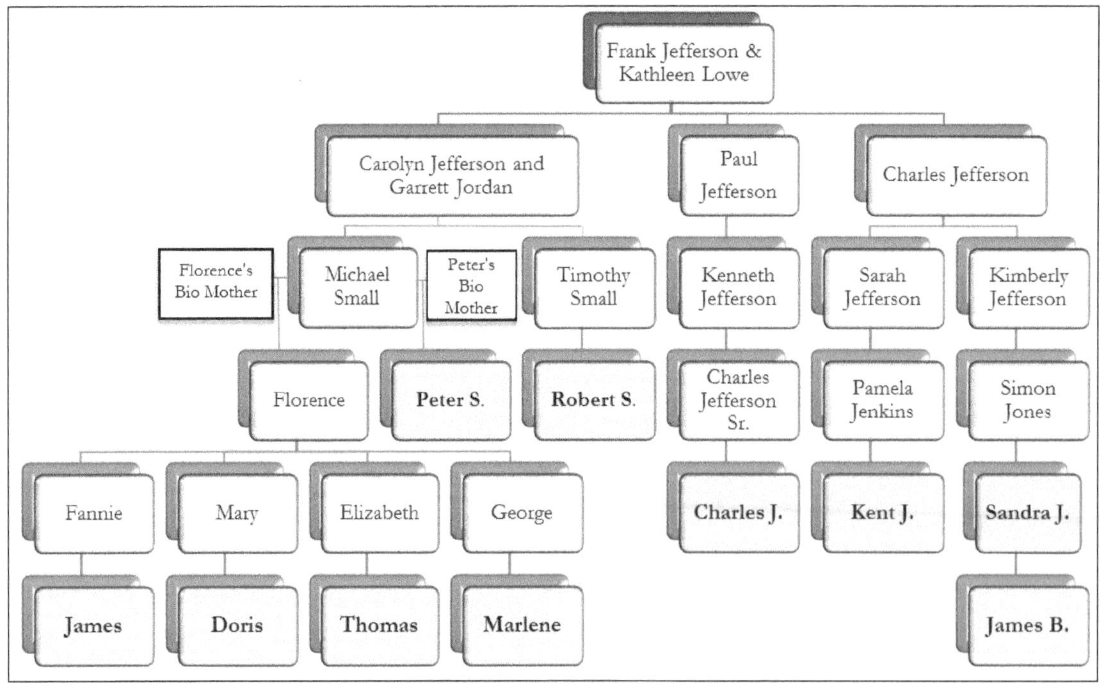

Figure 9.8. Theoretical tree—final

Documentary Research, Again

DNA alone cannot prove any biological relationship.[53] The GPS requires we source-cite each traced relationship (using documentary records) included in the relationship conclusion.[54] We previously provided source citations from each grandchild through their parents to Florence.

An analysis of the pedigrees of Peter and Robert back to the parents of Frank Jefferson and Kathleen Lowe does not reveal any gaps nor pedigree collapse in either pedigree.[55] If it did, we must factor that into our analysis and accommodate the gap through research or other means of eliminating it as a potential conflict. We conduct documentary research to prove each parent-child relationship connecting Peter S., Robert S., Charles J., Kent J., and Sandra J. to their most recent common ancestral couple, Frank Jefferson and Kathleen Lowe. Inclusion of Charles J., Kent J., and Sandra J. contribute to the intermediate conclusion that Florence's grandchildren are not related on any other ancestral lines of the focus study group.

Written Conclusion

A proof argument details the evidence (both genetic and documentary) and reasoning supporting the conclusion. The argument "eliminates the possibility that the conclusion is based on bias, preconception, or inadequate appreciation of the evidence."[56]

- Genetic evidence establishes that Florence's grandchildren are each a child of four distinct full siblings, children of Florence. We demonstrate this by the analysis of each parent-child link from the grandchild to Florence and her husband (documentary) *and* analysis of the amount of shared DNA among each grandchild (genetic) which is consistent with the relationship of first cousins.

- Florence's grandchildren share significant amounts of atDNA with five descendants (grandchildren and great-grandchildren) of Frank Jefferson and Kathleen Lowe. We demonstrate this by examining and filtering the match lists of each grandchild and the ICW for the relevant matches.

- These matches are from three different children of Frank Jefferson and Kathleen Lowe. We demonstrate this by analysis and confirmation of each parent-child link on the focus study group 1 members and the incorporation of each into a theoretical descendant tree (documentary).

- The amount of shared DNA points to only one of the descendants as a candidate for the biological father of Florence. We demonstrate this by comprehensive correlation

53. Karen Stanbary, "Can a DNA Relationship be Proved by DNA Alone?" *BCG Springboard,* 17 January 2018 (https://bcgcertification.org/can-a-genetic-relationship-be-proved-by-dna-alone/). BCG, *Genealogy Standards,* 2nd ed., 32, standard 56, "Conclusions about genetic relationship."

54. BCG, *Genealogy Standards,* 2nd ed., 6, standard 2, "Specificity."

55. BCG, *Genealogy Standards,* 2nd ed., 30–31, standard 52, "Analyzing DNA test results."

56. BCG, *Genealogy Standards,* 2nd ed., 3.

of the specific amounts of shared DNA among the grandchildren and the focus study group 1 members. We test the hypothesis with an additional targeted test.

- Documentary research confirms each parent-child link in each studied line of descent. We demonstrate this by the source citations in each of the pedigrees for each grandchild and each relevant match.
- Pedigree analysis confirms there are not multiple common ancestral lines. We demonstrate this by building the tree to the depth required by the amount of shared DNA.
- Pedigree analysis documents the accuracy of the studied lines of descent. We demonstrate this by the source citations linked to each parent-child relationship.
- Pedigree analysis confirms there are no gaps in the pedigrees, eliminating the possibility of a competing hypothesis.
- Documentary research is consistent with Illinois as the likely place of Florence's conception about 1882.

We achieve genealogical proof—even though not all of the documentary work is included here for this hypothetical case. We no longer need to stand on the shore intimidated by the murky waters of DNA test results. Genealogists use the GPS and *Genealogy Standards* every day. Come on in—the water is just fine.

Now onto Florence's biological *mother*, likely related to the grandchildren's genetic match Clifford W. While Clifford matches all four grandchildren of Florence, he does not match any test takers in focus study group 1; therefore, he is likely related on Florence's maternal side and is an initial clue in the search for Florence's mother.

Correlating Documentary and DNA Evidence to Identify an Unknown Ancestor

Identifying the Mother of Mary Elizabeth (Overbeck) Lee, born in Jefferson County, Pennsylvania

Patricia Lee Hobbs, CG

Introduction

Research on any genealogical question always begins with the traditional documentary sources. DNA test results alone can never identify an ancestor. They need to be supported by thorough documentary research—by knowing as much as possible about our ancestral families. DNA-identified ancestors will often fit into known holes in our trees. However, we will not know how and where without thorough research before, during, and after DNA testing.

The Overbeck and Lee families left few records. Nevertheless, thorough documentary research led to a birth location for Mary Elizabeth (Overbeck) Lee, which then led to autosomal-DNA matches from the same location. (The approach worked in this case because the family did not have many ancestral lines in the area to confuse matters.) Her newly-identified maternal ancestors then had to be confirmed by further documentary research.

Once the DNA results point toward newly-identified ancestors, more documentary research is always necessary. In this case, the DNA evidence led to the paternal ancestry of the unknown mother. As that DNA evidence was collected, it became apparent that some DNA connections were also made with the unknown mother's maternal family, too, requiring documentary research into two previously unknown families.[1]

This back-and-forth process is normal and necessary. Documentary and DNA research tag-team each other. An advance on one side is followed by research on the other side. The Overbeck and Lee families follow this pattern.

1. The author thanks all test takers for permission to use their identities and DNA test results in this work.

The Life of Mary Elizabeth (Overbeck) Lee

Mary Elizabeth Overbeck was born 29 May 1874 in Jefferson County, Pennsylvania.[2] She and Ira Newell Lee were both living in Mount Jewett, McKean County, Pennsylvania, 8 September 1892 when they applied to marry in Limestone, Cattaraugus County, New York, immediately to the north.[3] They lived in Campbelltown, adjacent to Mount Jewett, from then until 1903.[4] (This portion of Mary Elizabeth's life was the most critical in identifying her mother.) The Lees were in Elk County and then Forest County before returning to McKean County in 1912.[5] Mary Elizabeth lived there until two months before her death in Punxsutawney, Jefferson County, 7 May 1956.[6]

Mary Elizabeth Overbeck's Parents

Mary Elizabeth (Overbeck) Lee's death certificate lists only her father's name: Fred Overbeck. The mother's maiden name field says that the information was not ascertainable.[7] Ira Lee and Mary Elizabeth (Overbeck) Lee were not party to any deeds that might indicate they inherited land from Mary's parents.[8]

2. Commonwealth of Pennsylvania, Department of Health, Certificate of Death, File No. 44567 (1956), Mary E. Lee; Division of Vital Statistics, New Castle gives birth date. Her full name of Mary Elizabeth is given in the marriage record cited for Lee-Overbeck, 8 September 1892. Death certificate of a daughter gives more specific place of birth of Sigel, which is in Jefferson County: "Pennsylvania, Death Certificates, 1906-1964," *Ancestry* (http://search.ancestry.com/search/db.aspx?dbid=5164 : accessed 29 September 2017) > 1943 > Certificate Number Range 55401-58100, image 724, File No. 55925, Clara Barber. Marriage records of another daughter gives her mother's birthplace as Brookville, Pa which is also in Jefferson County: *FamilySearch,* access through McKean County, Pennsylvania, location search in catalog (https://www.familysearch.org/catalog/search : accessed 29 September 2017) > Vital records > Marriage license applications, 1885-1916 > image 492, Application for Marriage License, No. 6375, Cecil Baughman and Orpha Lee, 21 June 1915.

3. State of New York, Department of Health, Return of a Marriage, Lee-Overbeck, 8 September 1892; Bureau of Vital Statistics, Albany. Fields for parents' names of both bride and groom were left blank.

4. For location of Campbelltown, see *Mount Jewett Quadrangle*, US Topo [map] (Washington D.C.: U.S. Geological Survey, 2016). The Lee's residence in Campbelltown is documented by yearly appearances in McKean County, Pennsylvania, Seated Assessments in Hamlin Township, McKean County, arranged alphabetically by first letter of last name, from 1893 through 1903; Tax Assessor's office, Smethport.

5. Ira and Mary Elizabeth Lee's son, Ira, was born in 1904 in Elk County: Elk County, Pennsylvania, certified copy of birth, dated 19 September 1940 for Ira Lee, 12 November 1904; copy held by Patricia Lee Hobbs, Clever, Missouri. The Lee family was enumerated in 1910 U.S. census, Forest County, Pennsylvania, population schedule, Jenks Township, Enumeration District 59, Sheet 6 A, dwelling and family no. 53, Ira N. Lee household; citing NARA microfilm publication T623, roll 1439.

6. For date of return to McKean County: "Ira N. Lee, former Chief of Police, Dies Suddenly at his Home," *The Kane Republican*, 12 December 1940, p. 1, c. 1; State Library of Pennsylvania microfilmed newspaper collection. Mary Elizabeth lived with her daughter, Orpha: Robert Lee interview at Clever, Missouri, April 2007, transcribed by Patricia Lee Hobbs. Death date, place, and length of stay from Commonwealth of Pennsylvania, Department of Health, Certificate of Death, File No. 44567 (1956), Mary E. Lee.

7. Commonwealth of Pennsylvania, Department of Health, Certificate of Death, File No. 44567 (1956), Mary E. Lee.

8. Searches in deed indexes in the following places: Jefferson County, Pennsylvania, deed indexes available on computer, electronically searched in all indexed deeds ever recorded; Recorder of Deeds office, Brookville. Elk County, Pennsylvania, computerized deed indexes, electronically searchable; Recorder of Deeds office,

Since Mary Elizabeth Overbeck was born in 1874 and married in 1892, there is only one census when she might be listed with her birth family. The 1880 census reveals two Overbeck girls with similar names and ages:

- Mary E. Overbeck, age 7, daughter of Frank Overbeck in Rose Township, Jefferson County.[9]
- Elizabeth M. Overbeck, age 6, granddaughter of Constantine Overbeck in Rose Township, Jefferson County.[10]

Frank's daughter Mary E. cannot be Mary Elizabeth (Overbeck) Lee:

- Frank's daughter was born 5 November 1872, while Mary Lee was born 29 May 1874.[11]
- Frank's daughter grew up to marry a McKelvey, and in 1900 was reported born November 1872.[12]
- Frank was born in Belgium,[13] while Mary Lee's father, Fred, was born in Pennsylvania.[14]
- Constantine was the apparent father of Ferdinand/Fred in 1860 and 1870, so the presence of six-year-old Eliza M. as Constantine's stated granddaughter makes sense.

Ridgway. McKean County, Pennsylvania, Grantee Index L (1842-1852) and Grantor Index L under I, M, and E in given name sections for Ira, Mary, and Elizabeth; Recorder of Deeds office, Smethport.

9. 1880 U.S. census, Jefferson County, Pennsylvania, population schedule, Rose Township, Enumeration District 198, p. 62 (penned), p. 227 (stamped), dwelling 263, family 273, Mary E. Overbeck; citing NARA microfilm publication T9, roll 1136.

10. 1880 U.S. census, Jefferson County, Pennsylvania, population schedule, Brookville, Enumeration District 186, p. 43 (penned), p. 23 (stamped), "Const." Overbeck family; NARA microfilm publication T9, roll 1136.

11. For Frank's daughter's birth date, see Pensioner questionnaire, 4 October 1898, Franklin Overbeck, (Co. K, 8th Pennsylvania Infantry), pension certificate no. 837864, Case Files of Approved Pension Applications …, 1861-1934; Civil War and later Pension Files; Record Group 15: Records of the Department of Veterans Affairs; National Archives, Washington, D.C. For Mary Lee's birthdate, see Commonwealth of Pennsylvania, Department of Health, Certificate of Death, File No. 44567 (1956), Mary E. Lee.

12. Mrs. W. F. McKelvey of Pittsburgh is named as a surviving daughter in "Death of Frank Overbeck," *Brookville Republican* (Pennsylvania), 14 September 1911, p. 1, c.5. Name, place of residence, birth month and year correspond to 1900 U.S. census, Allegheny County, Pennsylvania, population schedule, Pittsburgh, Ward 38, Enumeration District 332, Dwelling 207, family 222, Mollie "McKeloey"; citing NARA microfilm publication T623, roll 1365.

13. See Table 10.1.

14. Mary Lee's father's birthplace as Pennsylvania in 1900 U.S. census, McKean County, Pennsylvania, population schedule, Hamlin Township, Enumeration District 110, Sheet No. 40 A, dwelling no. 724, family no. 763, Elizabeth Lee; citing NARA microfilm publication T623, roll not given; and also 1910 U.S. census, Forest County, Pennsylvania, population schedule, Jenks Township, Enumeration District 59, Sheet 6 A, dwelling and family no. 53, Elizabeth Lee; citing NARA microfilm publication T623, roll 1439. For Ferdinand/Fred's birthplace as Pennsylvania, see Table 10.2.

Both fathers were called Ferdinand in 1860. The older Ferdinand (later Frank) was born in Belgium about 1841, apparent son of Charles and Emily. The younger Ferdinand (later Fred) was born about 1855 in Pennsylvania, apparent son of Constantine and Angeline.

Table 10.1. Ferdinand/Frank Overbeck families in censuses

1850[a]	1860[b]		1870[c]	1880[d]
Charles Vanhaverbeke, age 47, born in Belgium	Charles Overbeck, age 60, born in Belgium		**Frank Vanoverbeck**, age 28, born in Belgium	**Frank Overbeck**, age 40, born in Belgium
Amelia, age 47, born in Belgium	"Emly" Overbeck, age 60		Margaret, age 30	Margaret, age 39
Constant, age 24, born in Belgium			Emma, age 7	Emma, age 17
Harriet, age 21, born in Belgium			Harriet, age 5	Harriet, age 15
Cordula, age 17, born in Belgium			Caroline, age 4	Carline, age 13
Marie, age 12, born in Belgium	Maria L., age 22, born in Belgium		Jeromus, age 2	Jeromes, age 11
Fernand, age 9, born in Belgium	**Ferdenand**, age 18, born in Belgium		Rebecca, age 5/12	Rebecca, age 9
Philomena, age 5, born in Belgium				Mary E., age 7 Francis A, age 5 Maggie, age 4 John, age 2 Flora, age 1

a. 1850 U.S. census, Elk County, Pennsylvania, population schedule, Jones Township, p. 620 (penned) dwelling and family nos. 41, "Fernand" Vanhaverbeke household; digital image; citing NARA microfilm publication M432, roll 776.

b. 1860 U.S. census, Jefferson County, Pennsylvania, population schedule, Brookville, p. 33 (penned), p. 83 (stamped), dwelling & household no. 241, "Ferdenand" Overbeck; citing NARA microfilm publication M653, roll 1118.

c. 1870 U.S. census, Jefferson County, Pennsylvania, population schedule, Rose Township, p. 22, dwelling no. 146, family no. 152, Frank Vanoverbeck; citing NARA microfilm publication M593, roll 1352.

d. 1880 U.S. census, Jefferson County, Pennsylvania, population schedule, Rose Township, Enumeration District 198, p. 62 (penned), p. 227 (stamped), dwelling no. 263, family no. 273, Frank Overbeck; citing NARA microfilm publication T9, roll 1136.

Table 10.2. Ferdinand/Fred Overbeck families in censuses

1860[a]	1870[b]	1880[c]	1900[d]	1910[e]
Constantine Overbeck, age 36, born in Belgium	Constantine Overbeck, age 44	**Fred Overbeck**, age 25, born in Pennsylvania	**Ferdinand Overbeck**, Apr 1856, age 44, married 11 years, born in Pennsylvania	
Angeline, age 40	Angeline, age 50	Eliza J., age 24	Elizabeth, b. 1866, age 33	Elizabeth Overbeck, age 42, widow
Charles, age 7	Charles, age 17	Ida M., age 8	Harry, b. 1890, age 9	Harry, age 20
Ferdinand, age 5, born in Pennsylvania	**Ferdinand**, age 15, born in Pennsylvania		Blanche, b. 1892, age 8	
John, age 2	John age 13		David, b. 1895, age 5	Frank, age 15[f]
Bernarga, age 3/12	Barnett age 11			
	Elizabeth age 4			Mamie age 9
				Leroy, age 5

a. 1860 U.S. census, Jefferson County, Pennsylvania, population schedule, Brookville, p. 34 (penned), p. 84 (stamped), dwelling no. 246, family no. 246, Constantine Overbeck; citing NARA microfilm publication M653, roll 1118.

b. 1870 U.S. census, Jefferson County, Pennsylvania, Rose Township, p. 6, dwelling 38, family 40, Constantine Overbeck household.

c. 1880 U.S. census, Jefferson County, Pennsylvania, population schedule, Brookville, Enumeration District 186, p. 41, dwelling and household nos. 351, Fred Overbeck household; citing NARA microfilm publication T9, roll 1136.

d. 1900 U.S. census, McKean County, Pennsylvania, population schedule, Sergeant, Enumeration District 123, Sheet 6 B, dwelling no. 114, family no. 116, Ferdinand Overbeck household; citing NARA microfilm publication T623, roll 1439.

e. 1910 U.S. census, Elk County, Pennsylvania, population schedule, Johnsonburg, Enumeration District 42, Sheet 17 A, dwelling no. 292, family no. 330, Elizabeth Overbeck household; citing NARA microfilm publication T624, roll 1341.

f. *FamilySearch* (https://www.familysearch.org/search/film/004838933: accessed 2 July 2018), digital film 004838933 > image 368, Elk County, Pennsylvania Marriage License Docket 13:321, David Franklin Overbeck to Ethel May Bedford, 3 July 1915 shows names in both censuses. His father's name is given as Ferdinand Overbeck and states that he is dead. The mother's name is given as Elizabeth Haines and her residence Johnsonburg, corresponding to the 1910 census.

Ferdinand/Fred Overbeck born in Pennsylvania to Constantine Overbeck must be Mary Elizabeth (Overbeck) Lee's father. His children that can be identified by the censuses are six in number: one from the 1880 census, and five in the 1900 and 1910 censuses by a later wife. A 1908 newspaper notice states, "Fred Overbeck, aged 53 years, died this morning [25 September 1908] at his home in this place of a complication of diseases. He is survived by his wife and five children. Interment will be made Sunday in the cemetary [*sic*] at Brookville."[15] Although no names are given for his family of a wife and five children, the 1900, and 1910 census households fit this information. Further supporting that Fred and Ferdinand were the same person, a corresponding obituary appeared in the *Jeffersonian Democrat:*

> "The body of Ferdinand Overbeck, who died at Johnsonburg, Elk County, was brought to Brookville ... He was about 53 years of age and was a son of Constantine Overbeck."[16]

Although there is some reason to add Elizabeth M. of Constantine's 1880 household to this child list, no evidence has been found stating that Fred Overbeck had any other children than the six seen in 1880, 1900, and 1910.

The 1880 Fred Overbeck household shows only one child, eight-year-old Ida, and wife Eliza J., age twenty-four (born about 1856). This is a different wife than the Elizabeth of 1900:

- In 1900 Ferdinand and Elizabeth had been married for eleven years (since about 1889), and their children's ages are consistent with that date.
- With a birth year of 1867, 1900 wife Elizabeth[17] would have been too young to be the mother of eight-year-old Ida in the 1880 Fred Overbeck household, or of six-year-old Elizabeth M. of the 1880 Constantine Overbeck household.

Eliza of the 1880 census household of Fred Overbeck is not the same person as his later wife Elizabeth. The large gap between the birth years of Ida and Mary Elizabeth and 1880, in which there are no children, may indicate Fred and Eliza married later and that she might not be the mother of the girls. But there is a better explanation for the absence of additional children after 1874. Fred/Ferdinand served almost three years in the Pennsylvania State Penitentiary, where he was prisoner number 5276, called "Ferdinand" in the population index and "Fred" on arrival 22 September 1876. Fred was married and had two living children, thus further supporting the contention that the grandchild of Constantine Overbeck in the 1880 household was Ferdinand's child, not Charles's—Ferdinand's probable brother.[18]

15. Fred Overbeck death notice, *Johnsonburg Press*, 25 September 1908, p. 5, c. 3; State Library of Pennsylvania microfilmed newspaper collection.

16. Ferdinand Overbeck death notice, *Jeffersonian Democrat*, 1 October 1908, p. 8, c.2; microfilmed newspaper collection, Jefferson County History Center, Brookville.

17. "Pennsylvania, Death Certificates, 1906-1966," *Ancestry* (https://search.ancestry.com/search/db.aspx?dbid=5164 : accessed 3 February 2019) > 1922 > Certificate Number Range 072451-075000 > image 385, File No. 72698, Elizabeth Haines Overbeck (1947).

18. Descriptive Lists, 1876-1880, MM B 128, Western State Penitentiary Prison Population Records, RG-15, Pennsylvania State Archives, Harrisburg; microfilm #9.

Research Turns to DNA

To identify Mary Elizabeth (Overbeck) Lee's mother, DNA match lists on *Ancestry* were searched to identify test-takers who had ancestors from Mary Elizabeth (Overbeck) Lee's 1874 birthplace, Jefferson County, Pennsylvania.[19]

This technique is useful in this case because no other family lines are known to have origins in Jefferson County. The Overbecks are distinctive since they came from Belgium in 1849.[20] Their descendants can pinpoint their connections easily, and those matches can be set aside in the process of identifying DNA connections to Mary Elizabeth Overbeck's mother's family. The remaining matches in Jefferson County would be strong candidates for the family of Mary Elizabeth (Overbeck) Lee's mother.

Among the Jefferson County matches of both Patricia Lee Hobbs and her paternal aunt, Mary (Lee) Moore, on AncestryDNA, the descendants of James McManigle (born ca. 1782 in Ireland) and his wife Susan Baecker are well represented. See figure 10.2 for relationships of Mary Elizabeth (Overbeck) Lee descendants.

The following children of James and Susan have descendants who match Hobbs and Moore:

- John McManigle, born 1817, married to Sarah Cobaugh
- Robert McManigle, born 1820, married to Margaret Steele
- Alexander McManigle, born 1823, married to Juliette Graham
- Christina Catherine McManigle, born ca. 1826, married to Stephen Oaks
- Susanna McManigle, born 1828, married to Harvey Bruner
- Solomon McManigle, born 1830, married to Mary Jane Mason[21]

---

19. "AncestryDNA Results" for Patricia Hobbs and Mary Elizabeth Moore (https://www.ancestry.com/dna). Death certificate of daughter gives the specific birthplace of Sigel, which is in Jefferson County: *Ancestry* (http://search.ancestry.com/search/db.aspx?dbid=5164 : accessed 29 September 2017) > "Pennsylvania, Death Certificates, 1906-1964" > 1943 > Certificate Number Range 55401-58100, image 724, File No. 55925, Clara Barber. The marriage record of a daughter gives her mother's birthplace as Brookville, Pennsylvania, which is also in Jefferson County: *FamilySearch,* access through McKean County, Pennsylvania, location search in catalog (https://www.familysearch.org/catalog/search : accessed 29 September 2017) > Vital records > Marriage license applications, 1885-1916 > image 492, Application for Marriage License, No. 6375, Cecil Baughman and Orpha Lee, 21 June 1915.

20. Origin from Belgium and date of arrival from Jefferson County, Pennsylvania, petition for naturalization, arranged chronologically by court term, 14 September 1870, Constantine Overbeck; Office of the Prothonotary microfilmed records, Brookville.

21. *Ancestry* account access was provided to the editor with links to the matches for their verification. All other associated family members have tested only with FamilyTreeDNA. Since entering the ancestor as identified in this narrative, the author now has a DNA circle of 30 members representing seven different descendant lines.

In the absence of any other McManigle family in Jefferson County, a connection to this family seems certain.[22] The ages of the children suggest that Mary Elizabeth Overbeck's mother could be a grandchild of James and Susan McManigle. Significantly more DNA is shared as a whole with Solomon and Mary Jane (Mason) McManigle descendants than with descendants of the other children, making them more likely to be ancestors, but more DNA evidence was sought. The greater amount of DNA shared with Solomon descendants was not as extreme for Beverly Pierce, a descendant of Mary Elizabeth Overbeck's daughter Orpha, as it was for the descendants of Mary Elizabeth's son Ira Lee.

Table 10.3. Shared amounts of DNA with McManigle Descendants on *GEDmatch*[a]

Mary Elizabeth (Overbeck) Lee Descendant		# of matching donors	Range of shared DNA amounts	Avg amount of shared DNA
Beverly Pierce	Solomon	2 of 3	0 – 30.3 cM	**15 cM**
	Other McManigles	1 of 7	0 – 11.9 cM	**1.7 cM**
Patricia Hobbs	Solomon	3 of 3	84.8 – 114.7 cM	**95.7 cM**
	Other McManigles	4 of 7	0 – 39.0 cM	**16 cM**
Carol Polk	Solomon	3 of 3	49.8 – 73.6 cM	**58.1 cM**
	Other McManigles	4 of 7	7.9 – 47.2 cM	**12.5 cM**
Deborah Lee	Solomon	3 of 3	54.4 – 80.9 cM	**65.0 cM**
	Other McManigles	3 of 7	0 – 24.9 cM	**7.1 cM**
Mary Moore	Solomon	3 of 3	12.1 – 59.6 cM	**29.3 cM**
	Other McManigles	6 of 7	0 – 42.7 cM	**19.4 cM**

a. Autosomal DNA analysis, on-request reports, *GEDmatch: Tools for DNA & Genealogy Research*, "one-to-one" comparisons among kits A714224 (Carlson), T522679 (Cross), A798456 (Heasley), A373279 (Warmbrodt), A844758 (Sherry), A255016 (Barnes), M888331 (Barr), T615126 (Pierce), T514993 (Hobbs), T664909 (Polk), T302206 (Lee), T639731 (Moore), A476189 (Flegal), A585788 (Walburn), A612761 (Hunter). Ancestry supplied kits A714224, A798456, A373279, A844758, A255016, A476189, A585788, and A612761; FamilyTreeDNA supplied kits T522679, T615126, T514993, T664909, T302206, T639731; 23andMe supplied kit M888331.

Although Mary (Lee) Moore is a generation closer to James and Susan McManigle than her nieces, it is apparent that her brother Robert Lee (the father of Patricia Hobbs, Carol Polk, and Debbie Lee) inherited more of the DNA that matches these McManigle descendants than she did.

The AncestryDNA tool "Shared Matches" identifies Patricia Hobbs and Mary Moore matches, whose trees have no McManigle connections but who claim Nancy Ellen Mason as an ancestor. Nancy was a sister of Mary Jane Mason who was married to Solomon McManigle. Nancy Ellen Mason was married twice, and each of these matches descends from different

22. See the many records searched throughout this essay.

husbands.[23] This strongly suggests that Mary Elizabeth Overbeck's mother was a child of Solomon McManigle and Mary Jane Mason.

Back to Documents

Solomon McManigle married Jane Mason in Jefferson County 13 March 1857. As of 4 July 1898 their living children were

Esther Ann, born 2 February 1858
William James, born 4 October 1859
Solomon Elsworth, born 6 June 1861
Walter, born 7 October 1864
Mary Jane, born 15 September 1866
Nancy Elizabeth, born 9 April 1868
David Franklin, born 24 November 1870
John Henry, born 4 January 1872
Sarah Matilda, born 9 August 1875[24]

Of these daughters, the only one old enough to be the mother of Mary Elizabeth (Overbeck) Lee, born in 1874, was Esther Ann McManigle.

After the identification of a probable family through DNA, an onsite visit to a church parish revealed a record that identifies a previously unknown wife of Ferdinand Overbeck. The baptismal record of a Maria Anna Overbeck "*uxor Ferdinandi sub conditione baptizata conversa e Methodsim[o]*" identifies a wife who was the daughter of "Salomon McManigle and Maria Johanna Mason" and born 1 February 1858. The record indicates that Maria Anna Overbeck, the wife of Ferdinand Overbeck, was baptized on 1 April 1877 in Brookville, because of her conversion from the Methodist faith.[25] This *cannot* be a marriage for Ferdinand/ Frank Overbeck because he had been continuously married to the same woman, Margaret Baughman, since 1861.[26] The birth date given for Maria Anna Overbeck is just one day off

23. For Nancy's marriages, and proof argument for her relationship, see Patricia Lee Hobbs, "Hypothesized Descents from Jacob Mason," *Quotidian Genealogy* (https://quotidiangenealogy.com/files/masonlineages.pdf). Although descent from Jacob is not as assured and proof is not attempted in this case study, greater confidence can be placed in the closer relationship between Mary Jane (Mason) McManigle and Nancy Elizabeth (Mason) Preston Joyner.

24. Pension questionnaire, 4 July 1898, Solomon McManigle, pension application no. 966267, certificate no. 740572 (Private, Co. B, 105th Pennsylvania Infantry, Civil War); Case Files of Approved Pension Applications . . ., 1861–1934; Civil War and Later Pension Files; Record Group 15: Records of the department of Veterans Affairs; National Archives, Washington, D.C.

25. Immaculate Conception Catholic Church (Brookville, Pennsylvania), "1877-1905 Baptisms," arranged chronologically, 1 April 1877, Maria Anna Overbeck; Parish rectory, Brookville. Thanks to Julie Michutka for translation assistance. There are no baptismal entries for either Ida Overbeck or Mary Elizabeth Overbeck searching in the baptismal book up through 1895. Another book containing marriages and baptisms 1863-1876 was searched by the parish secretary.

26. Pension questionnaire, 4 October 1898, Franklin Overbeck, pension certificate no. 675,338, Civil War, RG 15, NA–Washington.

from that given in Solomon McManigle's pension file for Esther Ann. There are variations in the names, but not enough to make them unrecognizable. "Maria Johanna" is an obvious variant of Mary Jane. Although less compelling, "Maria Anna" appears to be the variant given to Esther Ann. Giving the names of Christian saints at baptism is a common Catholic practice.[27]

Although this record indicates Esther McManigle was Ferdinand's wife in 1877, it does not prove that she was the mother of Mary Elizabeth (Overbeck) Lee. Furthermore, Esther Ann was living with her parents in 1880 with her name reported as "McNigle," and her marital status as single.[28] The court records of Jefferson, Clarion, Clearfield, and McKean counties were searched with unfruitful results to find evidence of a divorce.[29]

More concrete evidence was sought by viewing chromosomal segment data on *GEDmatch*.

Triangulated DNA Segments

Two recognized means of validating common ancestors through DNA are triangulation and by identifying multiple matching segments among descendants of a common ancestral couple. In both methods, the DNA must be in the same or overlapping location on the same chromosome in order to count as a "matching segment." The *Ancestry* matches imply multiple matching segments, but do not tell whether they are triangulated or not, nor do they give the matching locations.

There is no way to identify our DNA segments and from whence they came except by matching with other descendants from the shared ancestral couple.

Triangulation identifies at least three people who are not closely related but who all match each other on the same segment.[30] Many people have unidentified lines in their ancestry or

27. "Inclusive Language and Baptismal Validity," *Canon Law Made Easy* (http://canonlawmadeeasy.com/2008/09/11/inclusive-language-and-baptismal-validity/: posted on 11 September 2008, viewed 8 January 2018). Although "Esther" is a Biblical name, it would not be considered Christian since the Biblical Esther makes her appearance in the Old Testament.

28. 1880 U.S. census, Jefferson County, Pennsylvania, population schedule, Washington Township, Enumeration District No. 200, p. 23 (penned), p. 257 (stamped), dwelling no. 199, family no. 211, Soloman "McNigle" household; citing NARA microfilm publication T9, roll 1136.

29. No index of court records is known to exist by Jefferson County court staff other than those beginning in 1900. The author searched page by page in Jefferson County, Pennsylvania, Continuance Docket 29 through Continuance Docket 33 covering dates from May 1877 court term through February 1881 court term; Office of the Prothonotary, Brookville. Other counties were subsequently searched: Clearfield County, Pennsylvania, Divorce Index 1 1828-1942; Office of the Prothonotary, Clearfield and Clarion County, Pennsylvania, Continuance Dockets 24 through 27 (each are indexed) covering August 1875 through August 1881; Office of the Prothonotary, Clarion. The McKean County clerk in Smethport also searched their records, but did not allow viewing the records. She reported that the index covered divorces in 1880 through 1895. The author also wrote to the Catholic Diocese in Erie, which incorporates Jefferson County, to ask about an annulment, with no response.

30. Jim Bartlett, "Does Triangulation Work?" *Segment-ology*, (https://segmentology.org/2015/10/19/does-triangulation-work/: accessed 10 May 2016), See Triangulation Criteria section.

may match through more than one common ancestral couple. Triangulation helps to minimize the risk of attributing a segment to the wrong ancestor.

The figures on following pages show the matching segments. Each set of matching segments is identified by chromosome number and the start and end points of the matching segments by megabase pairs (Mbp, where 1 Mbp=1,000,000 bp). The identification and sequence of the bases is reported in DNA testing. Matching individuals have the same bases in the same sequence. The centimorgan (cM) value given implies the length of the segment. Centimorgan is not strictly a measure of length because statistical probabilities of recombination are factored in. Some of the matching segments are offset in that two people match a third in overlapping segments, but without the exact same beginning and endpoints. Because centimorgan is not a strict measure of length, we cannot add and subtract cM values to derive a new "length" in cM. Therefore, any segment centimorgan value which is actually longer than the greatest shared matching value is appended with a + symbol to show that the segment is longer.

Because this study's purpose is to demonstrate the relationship of Mary Elizabeth Overbeck's descendants to Mary's mother Esther A. McManigle, every triangulated segment includes at least one of Mary's descendants. Some segments are shared with descendants of only one other McManigle descendant. But there are triangulated segments that include Mary Elizabeth's descendants and two or more descendants of the other McManigle siblings.

Closely related people are considered to be only one of the three matches triangulated (for more, see the "Lessons Learned from Triangulating a Genome" chapter). The only closely related people involved in this project, other than the descendants of Mary Elizabeth Overbeck, are Sandy Sherry and William Warmbrodt who are first cousins. Each name entered in the chromosome maps is followed by the name of the McManigle ancestor of the test taker.[31] The lineages of the McManigle descendants sharing DNA with Mary Elizabeth (Overbeck) Lee's descendants are shown in figures 10.1 and 10.2. The documentary proof of the lineages of all the test takers to the common ancestor is available online.[32] Chromosome maps follow.

Figures 10.3 through 10.8 illustrate triangulated segments shared by three or more descendants; triangulation was verified using *GEDmatch*. Figures 10.9 and 10.10 illustrate segments shared by fewer than three descendants or with insufficient overlap to triangulate. The segment shown for each person is the ancestral segment carried by the DNA test taker. The an-

31. Autosomal DNA analysis, on-request reports, *GEDmatch: Tools for DNA & Genealogy Research*, "one-to-one" comparisons among kits A714224 (Carlson), T522679 (Cross), A798456 (Heasley), A373279 (Warmbrodt), A844758 (Sherry), A255016 (Barnes), M888331 (Barr), T615126 (Pierce), T514993 (Hobbs), T664909 (Polk), T302206 (Lee), T639731 (Moore), A476189 (Flegal), A585788 (Walburn), A612761 (Hunter). Ancestry supplied kits A714224, A798456, A373279, A844758, A255016, A476189, A585788, and A612761; FamilyTreeDNA supplied kits T522679, T615126, T514993, T664909, T302206, T639731; 23andMe supplied kit M888331.

32. Patricia Lee Hobbs, "Hypothesized Descents from Jacob Mason," *Quotidian Genealogy* (https://quotidiangenealogy.com/files/masonlineages.pdf).

cestral segment is determined by the start location on the chromosome with the first person matching and has an end location determined by the matching end location of the last person. Strictly speaking, triangulated segments are only the part of the segment that all three (or more) share. However, staggered segments are most common, and can also be traced to the common ancestor. It is statistically improbable that recombination events occurring in both persons would cause adjacent segments to match when those segments come from different ancestors. All matching segment locations and triangulations were verified by "one-to-one" comparisons among all test takers matching in the same region.

Figure 10.1. Selected descendants from Solomon McManigle & Susan Baecker (on facing page)

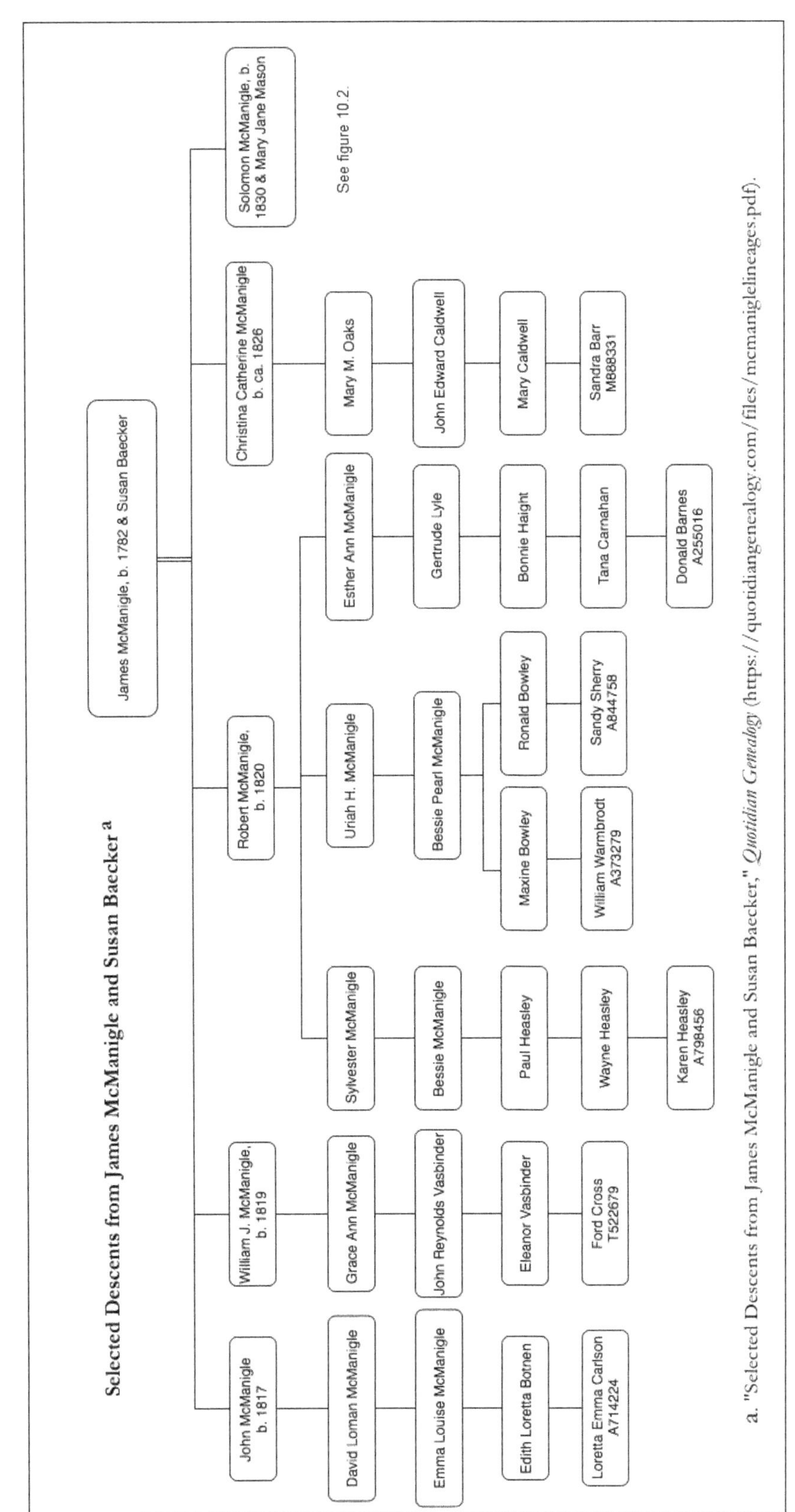

a. "Selected Descents from James McManigle and Susan Baecker," *Quotidian Genealogy* (https://quotidiangenealogy.com/files/mcmaniglelineages.pdf).

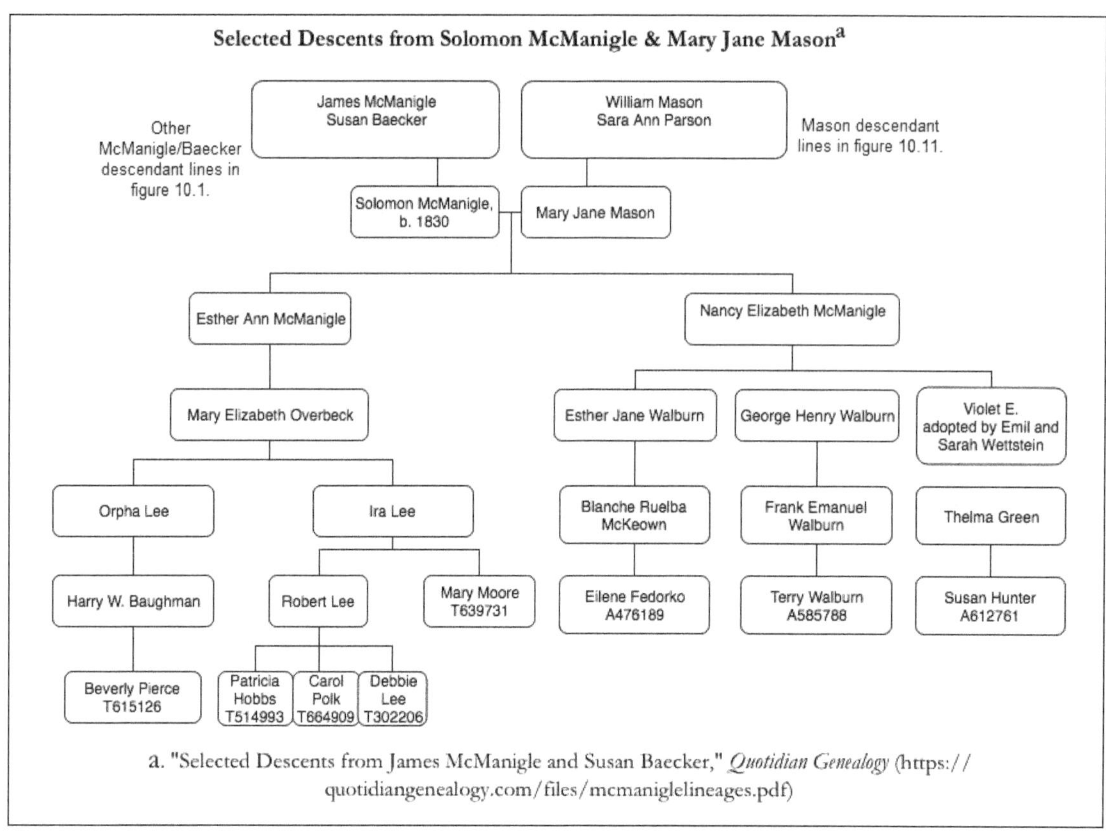

Figure 10.2. Selected descendants from Solomon McManigle & Mary Jane Mason

In the examples provided, the smallest segment is 11.8 cM.[33] Though not shown in the figures, the smallest number of SNPs in any of the triangulated segments is 1108.[34] Most are much higher in both cM value and SNP numbers. (For more, see the "Segments" section in the "Would You Like Your Data Raw or Cooked?" chapter.)

Four sets of segments triangulate with descendants of two different children of James McManigle and Susan Baecker and at least one representative of Mary Elizabeth Overbeck's family.

33. Segments greater than 7 cM were found to be identical by descent 90% of the time in Brenna M. Henn, et al., "Cryptic Distant Relatives are Common in Both Isolated and Cosmopolitan Genetic Samples," *PLOS One*, published 3 April 2012 (https://journals.plos.org/plosone/article?id=10.1371/journal.pone.0034267), see figure 6.

34. The triangulated segment on chromosome 16 includes Walburn's smallest segment with only 1088 SNPs. See Figure 10.6.

Figure 10.3 shows Moore triangulates on this segment with two Robert McManigle descendants (Barnes and Warmbrodt), but Pierce matches only one of them (Barnes).

Triangulated Matches on Chromosome 2
Solomon McManigle Descendants and Robert McManigle Descendants

Chromosome 2	Start Point	End Point	cM	105 110 115 120 125 130 135
Moore (Solomon > Esther)	106.0	132.9	26.3	
Barnes (Robert)	106.0	132.9	26.3	
Pierce (Solomon > Esther)	107.4	121.7	11.9	
Warmbrodt (Robert)	121.0	132.8	13.7	

Figure 10.3. Triangulation on chromosome 2

Since Sherry and Warmbrodt are first cousins, they count as only one leg of triangulation. Therefore, figure 10.4 shows a triangulation of Hobbs with one Christina McManigle descendant (Barr) and one Robert McManigle descendant (Sherry / Warmbrodt).

Triangulated Matches on Chromosome 5
Solomon McManigle Descendant, Robert McManigle Descendants, and Christina McManigle Descendant

Chromosome 5A	Start Point	End Point	cM	5 10 15 20 25 30
Barr (Christina)	9.1	28.7	24.3	
Sherry (Robert)	9.5	22.4	17.8	
Hobbs (Solomon)	9.1	28.7	24.3	
Warmbrodt (Robert)	9.6	28.7	22.9	

Figure 10.4. Triangulation on chromosome 5

On Chromosome 9 Hobbs and Polk, proposed descendants of Solomon McManigle through his daughter Esther, count as only one leg of triangulation. Figure 10.5 shows both triangulate with a Solomon descendant through daughter Nancy (Walburn) and a William McManigle descendant (Cross). Lee, part of the same leg as Hobbs and Polk, partially matches, although technically does not triangulate.

Triangulated Matches on Chromosome 9

Solomon McManigle Descendants through Esther, Solomon McManigle Descendant through Nancy, and William McManigle Descendant

Chromosome 9	Start Point	End Point	cM	0 2 4 6 8 10 12 14 16 18 20
Walburn (Solomon > Nancy)	0	18.7	40.2	
Hobbs, Polk (Solomon > Esther)	0	18.7	40.2	
Lee (Solomon > Esther)	9.9	18.7	15.4	
Cross (William)	.1	9.4	22.6	

Figure 10.5. Triangulation on chromosome 9

On chromosome 16 Hobbs, Polk, and Lee count as one leg of a triangulation. Figure 10.6 shows all three triangulate with a Solomon descendant through daughter Nancy (Walburn) and a William McManigle descendant (Cross).

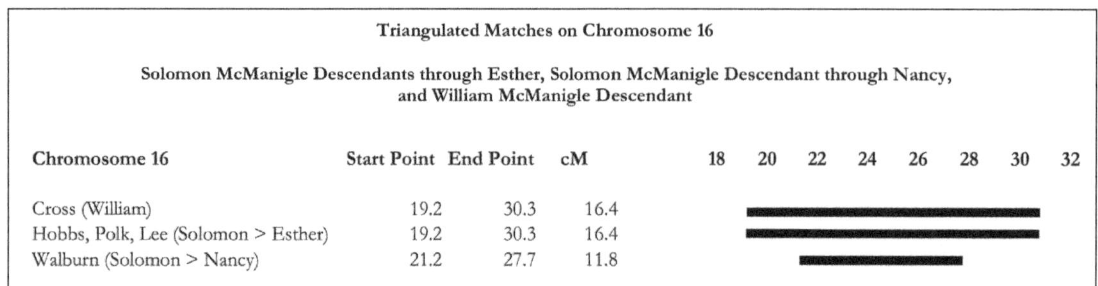

Triangulated Matches on Chromosome 16

Solomon McManigle Descendants through Esther, Solomon McManigle Descendant through Nancy, and William McManigle Descendant

Chromosome 16	Start Point	End Point	cM
Cross (William)	19.2	30.3	16.4
Hobbs, Polk, Lee (Solomon > Esther)	19.2	30.3	16.4
Walburn (Solomon > Nancy)	21.2	27.7	11.8

Figure 10.6. Triangulation on chromosome 16

Two sets of segments on chromosome 11 triangulate Mary Elizabeth Overbeck descendants (Hobbs and Lee) with two different descendant lines of Solomon McManigle and Mary Jane Mason. Figure 10.7 shows chromosome 11[A] where both triangulate with two Solomon descendants through daughter Nancy (Fedorko and Hunter).

Triangulated Matches on Chromosome 11 [A]

Solomon McManigle Descendants through Esther and Solomon McManigle Descendants through Nancy

Chromosome 11A	Start Point	End Point	cM
Fedorko (Solomon > Nancy)	6.9	25.1	29.2+
Hobbs, Lee (Solomon > Esther)	6.9	19.8	19.3
Hunter (Solomon > Nancy)	7.4	25.1	29.2

6 8 10 12 14 16 18 20 22 24 26

Figure 10.7. Triangulation on chromosome 11 [A]

Figure 10.8 shows Chromosome 11[B] where Hobbs triangulates with three Solomon descendants through daughter Nancy (Fedorko, Walburn, and Hunter). Polk and Pierce match only one Nancy descendant (Fedorko).

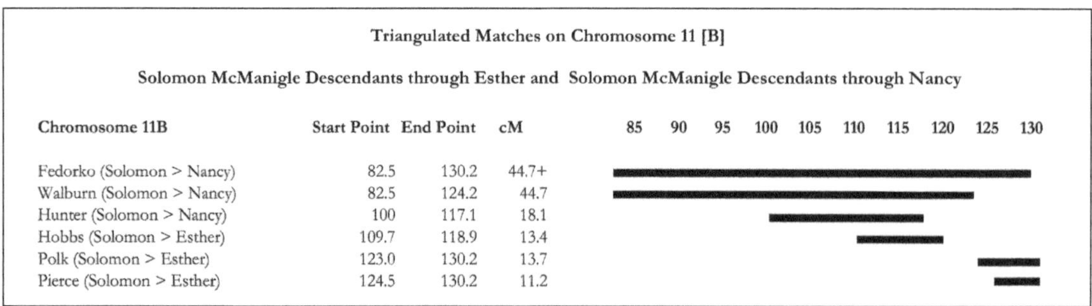

Triangulated Matches on Chromosome 11 [B]

Solomon McManigle Descendants through Esther and Solomon McManigle Descendants through Nancy

Chromosome 11B	Start Point	End Point	cM
Fedorko (Solomon > Nancy)	82.5	130.2	44.7+
Walburn (Solomon > Nancy)	82.5	124.2	44.7
Hunter (Solomon > Nancy)	100	117.1	18.1
Hobbs (Solomon > Esther)	109.7	118.9	13.4
Polk (Solomon > Esther)	123.0	130.2	13.7
Pierce (Solomon > Esther)	124.5	130.2	11.2

Figure 10.8. Triangulation on chromosome 11 [B]

The following shared segments do not strictly triangulate, but it is evident that a longer segment was inherited by Mary Elizabeth Overbeck descendants (Moore, Hobbs, Lee, and Polk) that is shared with other McManigle descendants.

Figure 10.9 shows Chromosome 8[A] where Hobbs, Polk, and Lee match a Christina descendant (Barr) at the first part of the segment, and a Solomon descendant (Moore / Fedorko) shares the second half of the segment. There is insufficient overlap between Barr and Fedorko to form a triangulated group.

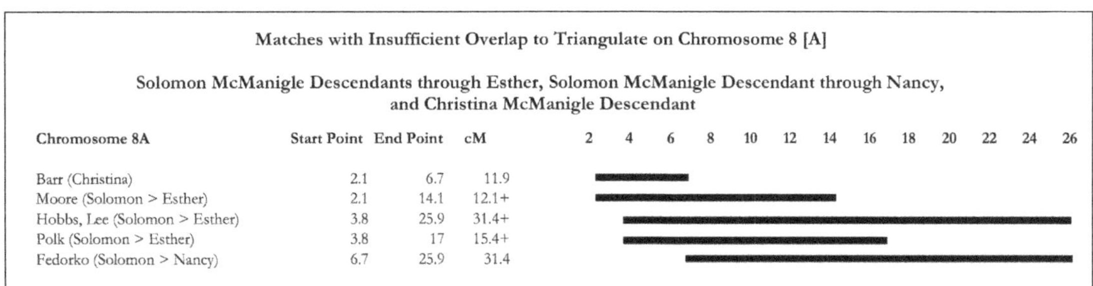

Matches with Insufficient Overlap to Triangulate on Chromosome 8 [A]

Solomon McManigle Descendants through Esther, Solomon McManigle Descendant through Nancy, and Christina McManigle Descendant

Chromosome 8A	Start Point	End Point	cM
Barr (Christina)	2.1	6.7	11.9
Moore (Solomon > Esther)	2.1	14.1	12.1+
Hobbs, Lee (Solomon > Esther)	3.8	25.9	31.4+
Polk (Solomon > Esther)	3.8	17	15.4+
Fedorko (Solomon > Nancy)	6.7	25.9	31.4

Figure 10.9. Triangulation on chromosome 8 [A]

Figure 10.10 shows Chromosome 21 where Polk shares a segment with one Solomon descendant in the first half (Fedorko) and a Solomon descendant (Hunter) in the second half.

Matches with Insufficient Overlap on Chromosome 21

Solomon McManigle Descendants through Esther and Solomon McManigle Descendants through Nancy

Chromosome 21	Start Point	End Point	cM
Polk (Solomon > Esther)	21.1	38.2	27.3
Fedorko (Solomon > Nancy)	21.1	28.8	12.8
Hunter (Solomon > Nancy)	29.6	38.2	14.5

20 22 24 26 28 30 32 34 36 38 40

Figure 10.10. Triangulation on chromosome 21

Shared Mason Ancestry

The relationship of Mary Elizabeth Overbeck descendants to the McManigle family was hypothesized to be through Solomon McManigle for the following three reasons:

- Greater incidence and amounts of shared DNA with Solomon and Mary Jane (Mason) McManigle descendants than with other McManigle descendants.
- DNA matches with the descendants of Nancy Ellen Mason, the sister of Mary Jane Mason.[35]

35. Patricia Lee Hobbs, "Hypothesized Descents from Jacob Mason," *Quotidian Genealogy* (https://quotidiangenealogy.com/files/masonlineages.pdf).

- 1877 adult baptismal record showing that Maria Anna Overbeck, the daughter of Solomon and Maria Johanna (Mason) McManigle, with a birth date one day different than Solomon and Mary Jane (Mason) McManigle's daughter Esther Ann McManigle, was married to Ferdinand Overbeck.

The descent specifically from Solomon and Mary Jane (Mason) McManigle can be further solidified by establishing DNA matching with descendants from the Mason side along with the McManigle matches already identified. Mary Jane (Mason) McManigle's 1912 death certificate states that she was born in 1840 to William Mason and Sara Ann Parson.[36] William Mason is enumerated as a 36-year-old man living in Pinecreek Township, Jefferson County, in 1850.[37] Only two Mason households are found in Jefferson County in the 1840 census: one headed by William and another headed by Jacob Mason, both found in Pine Creek Township enumerated next to each other. William is in the age 20-29 category, and Jacob is aged 60-69.[38] This strongly suggests that William might be Jacob's son.

Other possible Mason relatives can be identified. Jacob Mason is not found in the 1850 Jefferson County census, but a woman who could be his widow is. Betsy Mason, age 77, was born in Pennsylvania and enumerated in the Thomas Hall household in Eldred Township, Jefferson County. Also listed in the household is David Mason, age 34. This might be an example of a common scenario that a widowed mother would live with relatives after the death of her husband and Thomas's wife could be her daughter. Thomas Hall's apparent wife is Polly, age 46.[39] The maiden name of Thomas Hall's wife Polly as Mason is supported by the death certificate of the child enumerated in 1850 as five-year-old Mary Jane in which her parents' names are given as Thomas Hall and Polly Mason.[40] The death certificate of Sarah (Mason) Bowers, a daughter of William Mason, states that her mother's name was Sarah Ann Hall, not Parson as stated in the death certificates of other children: Mary Jane, James, Kate (Caroline), Rosanna, and Liberty.[41] Although not accurately reported, the error

36. "Pennsylvania, Death Certificates, 1906-1964," *Ancestry* (https://search.ancestry.com/search/db.aspx?dbid=51645: accessed 18 December 2016) > 1916 > Certificate Number Range 099261-102540, image 3365, File No. 102361, Mary Jane McManigle.

37. 1850 U.S. census, Jefferson County, Pennsylvania, population schedule, Pine Creek Township, p. 297 (penned), p. 149 (stamped), dwelling No. 758, Family No. 972, William Mason household; citing NARA microfilm publication M432, roll 786.

38. 1840 U.S. census, Jefferson County, Pennsylvania, Pine Creek Township, p. 170 (stamped), William Mason and Jacob Mason; NARA microfilm publication M704, roll 464.

39. 1850 U.S. census, Jefferson County, Pennsylvania, population schedule, Eldred Township, p. 33 (penned), p. 17 (stamped), dwelling no. 59, family no. 64, Thomas Hall household; citing NARA microfilm publication M432, roll 786.

40. "Pennsylvania, Death Certificates, 1906-1964," *Ancestry* (https://search.ancestry.com/search/db.aspx?dbid=5164 : accessed 24 February 2018) > 1926 > Certificate Number Range 106001-109000, image 1435, certificate no. 107263, Mary Jane Young, 10 October 1926.

41. "Pennsylvania, Death Certificates, 1906-1964," *Ancestry* (https://search.ancestry.com/search/db.aspx?dbid=5164 : accessed 24 February 2018) > for Sarah Bowers 1925 > Certificate No. Range 060001-63000, image 574, certificate no. 60525; for James Mason 1916 > Certificate Number Range 076881-080120, image 3414, File No. 80045; for Kate Lindenmuth: 1924 > Certificate Number Range 051001-054000, image 36, File

indicates that a relationship with the Hall family was known. These indicate that Polly (Mason) Hall and David Mason may be siblings of William Mason.

The death certificates of all the children agree that their father, William Mason, was born in Pennsylvania, but Liberty's is more specific stating that his father was born in Jefferson County, Pennsylvania. Therefore, Mason men found in Jefferson County near the time of William's birth are prime candidates for his father.

The censuses show William Mason's ages:

1840—age 20–30[42] (born between 1810 and 1820)
1850—age 36[43] (born ca. 1814)
1860—age 50[44] (born ca. 1810)
1880—age 70[45] (born ca. 1810)

William Mason first appears in the Pine Creek Township, Jefferson County, tax assessments for 1832 indicating that the 1810 birth year implied by the censuses is most likely correct.[46] A man's first appearance on the tax rolls generally signals that he has reached the age of 21.[47]

Jacob Mason was assessed in Pine Creek Township on the first tax list in 1807 and was present continuously thereafter until 1842 (except he was absent from 1835 to 1839). The

No. 51035; for Rosannah Lindenmuth: 1928 > Certificate Number Range 068501-071500, image 1825, File No. 70116; and for Liberty Mason: 1906 > Certificate Number Range 102341-105850, image 1224, File No. 103384.

42. 1840 U.S. census, Jefferson County, Pennsylvania, Pine Creek Township, p. 170 (stamped), William Mason; NARA microfilm publication M704, roll 464.

43. 1850 U.S. census, Jefferson County, Pennsylvania, population schedule, Pine Creek Township, p. 297 (penned), p. 149 (stamped) , Dwelling No. 758, Family No. 972, William Mason household; citing NARA microfilm publication M432, roll 786.

44. 1860 U.S. census, Jefferson County, Pennsylvania, population schedule, Warsaw Township, p. 23 (penned), p. 397 (stamped), Dwelling & Family nos. 169, Wm Mason household : digital image, *Ancestry* (http://interactive.ancestry.com/7667/4289728_00403 : accessed 18 December 2016); citing NARA microfilm publication M653, roll 1118.

45. 1880 U.S. census, Jefferson County, Pennsylvania, population schedule, Warsaw Township, Enumeration District No. 201, p. 20 (penned), dwelling no. 171, family no. 182, William Mason; digital image, *Ancestry*, (https://www.ancestry.com/interactive/6742/4244363-00547 : accessed 6 June 2017); citing NARA microfilm publication T9, roll 375.

46. Jefferson County, Pennsylvania, tax assessment book [no label] beginning in 1807 and continuing through 1840; Jefferson County Historical Society, Brookville.

47. Both of the following show that the age of assessment was 21 in 1811 and 1836 implying it was 21 continuously between those dates. John Purdon, "County Rates and Levies," *Abridgment of the Laws of Pennsylvania, From the Year One Thousand Seven Hundred, to the Second Day of April, One Thousand Eight Hundred and Eleven with References to Reports of Judicial Decisions in the Supreme Court of Pennsylvania* (Philadelphia: Farrand, Hopkins et al, 1811), item 4, Sect. VIII (p. 81) and John Purdon [attributed author], "County and Township Rates and Levies, *Digest of the Laws of Pennsylvania from the year One Thousand Seven Hundred, to the Sixteenth Day of June, One Thousand Eight Hundred and Thirty-six*, 5th edition (Philadelphia: M'Carty & Davis, 1837), item 9, Sect. IV (p. 160); digital image, *Internet Archive* (https://archive.org/details/digestofrevisedc00park: accessed 25 April 2018).

Jefferson County tax lists are complete.[48] Jacob is the only Mason in the tax lists from 1807 to 1821; thereafter, those added were single men:

1821—John Mason, identified as S.M. [single man], is first assessed in Pine Creek Township.
1822—Benjamin Mason, identified as S.M., is first assessed in Pine Creek Township.
1829—George Mason, identified as S.M., is first assessed in Pine Creek Township.
1832—Both David Mason and William Mason first assessed in Pine Creek Township. Neither is marked with the S.M. designation, but David's entries beginning with 1836 do so designate him.
1833—A second Jacob Mason is first listed.[49]

This scenario suggests that Jacob was the father of the newcomers, who were coming of age. Additional evidence confirms the suggestion.

The younger Jacob died in March 1872. His administrator petitioned the court to sell real estate to satisfy debts against the estate. Jacob's heirs were his siblings, "four brothers and three sisters," named as follows: "John Mason, George Mason, William Mason and David Mason, Nancy Long, intermarried with Wm Long, Sallie Smiley widow of Robert Smiley and Susan Cochran, intermarried with David Cochran." The petition specifies the named siblings are those that "survive him."[50] With the exception of Benjamin from the tax lists and Polly, the wife of Thomas Hall from the 1850 census, the names are consistent with this household of Jacob Mason emerging from the tax lists. Polly/Mary had died 5 November 1859.[51]

Since William's son's death certificate gives the information that his father was born in Jefferson County, ca. 1810, and William Mason is enumerated next to Jacob Mason in 1840, consistency supports that William was the son of Jacob.

DNA test results also show not only that these Masons are related, but that Mary Elizabeth Overbeck descendants share DNA segments with descendants of Polly (Mason) Hall, William Mason, and David Mason.[52] This further supports the identification of Mary Elizabeth

48. Jefferson County, Pennsylvania, tax assessment book [no label] beginning in 1807 and continuing through 1840 and tax assessment book 5 (number penciled on binding), 1841 and 1842.

49. Jefferson County, Pennsylvania, tax assessment book [no label] beginning in 1807 and continuing through 1840, tax assessment book 5 (number penciled on binding), beginning with 1841 through 1847, both volumes arranged chronologically for each year and then alphabetically by first letter of surname; Jefferson County Historical Society, Brookville.

50. Jefferson County, Pennsylvania, Orphans' Court Docket 4:431-2, Petition of Wm Goss administrator of Jacob Mason for Sale of Real Estate, May 1873; Register & Recorder's website (http://www.jeffersoncountypa.com/register-recorder), ImageSync, document retrieval system.

51. "Died," *Brookville Jeffersonian* (Pennsylvania), 10 November 1859, p. 2, c. 6; digital image, *Newspapers* (http://newspapers.com/: accessed 25 June 2018).

52. Autosomal DNA analysis, on-request reports, *GEDmatch: Tools for DNA & Genealogy Research*, "One-to-one" comparisons among kits T615126 (Pierce), T514993 (Hobbs), T664909 (Polk), T302206 (Lee), A585788 (Walburn), A080977 (Dolpp), T009046 (Dobson), and T944611 (Burns).

Overbeck's mother being a daughter of Solomon McManigle and Mary Jane Mason. See Figure 10.11 for the hypothesized family structure.

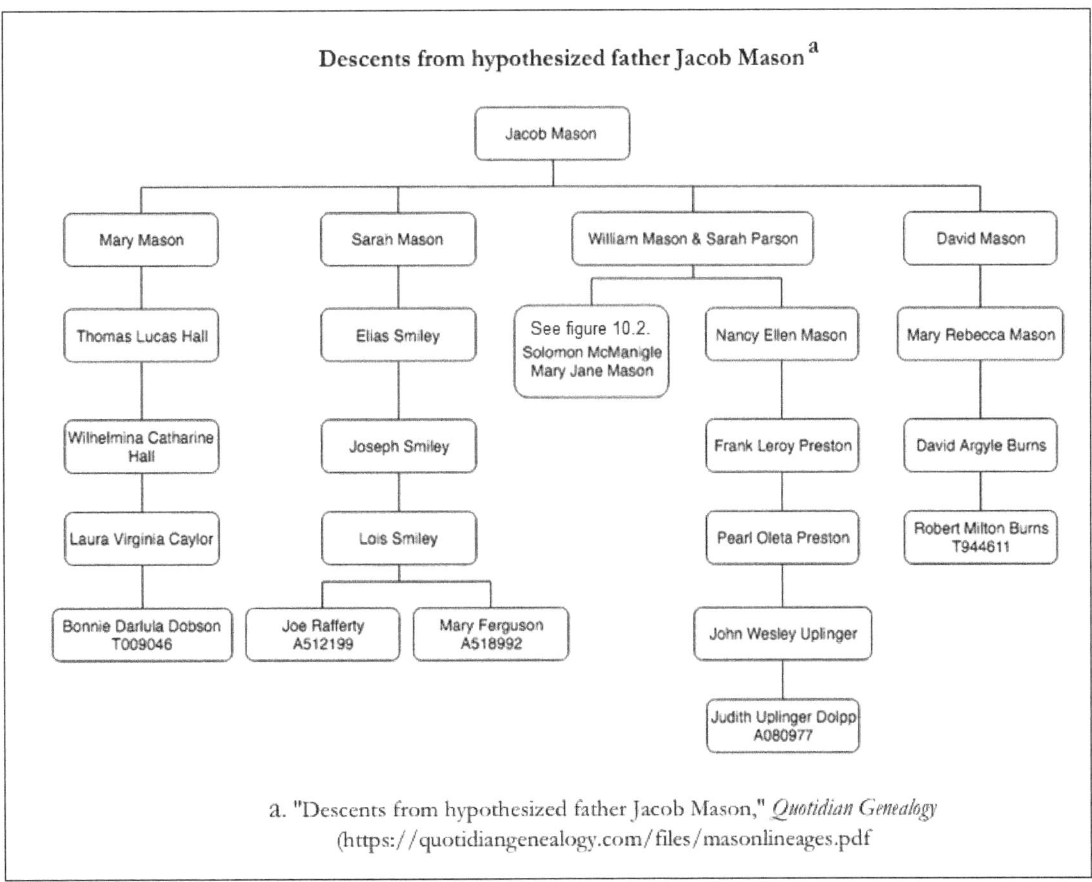

a. "Descents from hypothesized father Jacob Mason," *Quotidian Genealogy* (https://quotidiangenealogy.com/files/masonlineages.pdf

Figure 10.11. Hypothesized selected descendants from Jacob Mason

Figure 10.12 shows one triangulated segment shared by three different Mason descendant lines. A William Mason descendant (Walburn) triangulates with descendants of Polly Mason (Backdorf) and David Mason (Burns). A segment is also shared with another William Mason descendant (Hunter).

Triangulated Matches on Chromosome 5

William Mason Descendant, Polly Mason Descendant, and David Mason Descendant

Chromosome 5B	Start Point	End Point	cM
Hunter (William Mason)	149.9	159	9
Walburn (William Mason)	149.9	169.7	17.1+
Backdorf (Polly Mason)	158.8	169.7	17.1
Burns (David Mason)	158.6	169.1	15.4

150 152 154 156 158 160 162 164 166 168 170

Figure 10.12. Triangulation on chromosome 5

Figure 10.13 shows several segments with two clear triangulations. A David Mason descendant (Burns) triangulates with descendants of William Mason (Hunter) and Sarah Mason (Ferguson). Rafferty and Ferguson are siblings, but they are displayed in separate lines because the segments they inherited from Sarah Mason are different sizes. The sibling descendants of Sarah Mason (Ferguson and Rafferty) triangulate with descendants of two different children of William Mason (Dolpp and the close family group of Hobbs, Polk, and Moore). Hobbs, Polk, and Moore match Dolpp the entire length shown, but the threshold on the one-to-one comparison must be lowered to show that they still match Ferguson and Rafferty after the break indicated in their DNA matching segments. Also, by reducing the matching threshold to 3 cM, it can be seen that both Dolpp and Hobbs's family group match Hunter in the first triangulated group.

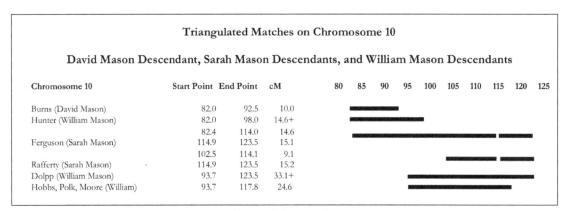

Triangulated Matches on Chromosome 10

David Mason Descendant, Sarah Mason Descendants, and William Mason Descendants

Chromosome 10	Start Point	End Point	cM
Burns (David Mason)	82.0	92.5	10.0
Hunter (William Mason)	82.0	98.0	14.6+
	82.4	114.0	14.6
Ferguson (Sarah Mason)	114.9	123.5	15.1
	102.5	114.1	9.1
Rafferty (Sarah Mason)	114.9	123.5	15.2
Dolpp (William Mason)	93.7	123.5	33.1+
Hobbs, Polk, Moore (William)	93.7	117.8	24.6

80 85 90 95 100 105 110 115 120 125

Figure 10.13. Triangulation on chromosome 10

Despite the lack of triangulation for this segment, figure 10.14 shows that a William Mason descendant (Dolpp), inherited a larger segment that is partially shared by a Polly Mason descendant (Backdorf) and partially shared with a David Mason descendant (Burns).

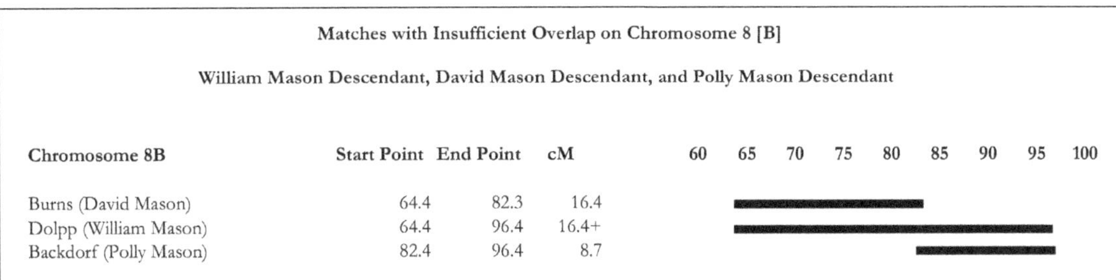

Figure 10.14. Overlap on chromosome 8 [B]

Although Mary Elizabeth Overbeck descendants do not triangulate with the other Mason descendants, they do have matching segments with those same Mason descendants. This also includes matches among William's descendants whose DNA helped establish the identification of the McManigle-Mason couple as the ancestors of the Mary Elizabeth Overbeck descendants. They appear in bold-faced font in table 10.4. Pairs of matching segments for Polly Mason and David Mason are not shown because they share another genealogical line.

Of the seven DNA test takers who descend from McManigles, other than Solomon and wife Mary Jane Mason, none share any DNA with the above Mason descendants.[53]

Table 10.4. Pairs of Shared Segments among Mason descendants

	Chr.	Start Location	End Location	Total cM
Walburn (William Mason) – Dolpp (William Mason)	1	48.8	67.0	21.6
Burns (David Mason) – Hunter (William Mason)	2	78.0	115.4	24.1
Burns (David Mason) – Fedorko (William Mason)	3	103.7	115.5	8.8
Burns (David Mason) – **Pierce** (William Mason)	4	67.3	89.2	20.4
Walburn (William Mason) – Backdorf (Polly Mason)	8	11.7	18.7	12.5
Burns (David Mason) – Hunter (William Mason)	10	82.0	92.5	10.0
Dolpp (William Mason) – **Polk, Lee, Hobbs** (William Mason)	10	93.0	118.0	24.9
Burns (David Mason) – **Hobbs, Lee** (William Mason)	11	0.1	7.0	16.6
Burns (David Mason) – **Pierce** (William Mason)	15	93.4	100.2	21.2
Burns (David Mason) – Fedorko (William Mason)	20	51.3	55.7	11.0

53. Autosomal DNA analysis, on-request reports, *GEDmatch: Tools for DNA & Genealogy Research*, "One-to-one" comparisons among kits A080977 (Dolpp), T009046 (Backdorf), and Burns (T944611) with kits A714224 (Carlson), T522679 (Cross), A798456 (Heasley), A373279 (Warmbrodt), A844758 (Sherry), A255016 (Barnes), and M888331 (Barr).

The Mason DNA shared segments add to the evidence already offered to make a compelling case that the Mary Elizabeth Overbeck progeny descend from the couple Solomon McManigle and Mary Jane Mason.

Traditional Documentary Research

Since previous research in the Jefferson County area focused only on the known surname of Overbeck, research was renewed to determine if any relevant documentary evidence supports the relationship of Mary Elizabeth and her father Ferdinand Overbeck with Solomon and Mary Jane (Mason) McManigle and their daughter Esther Ann McManigle.

- In August 1860, Ferdinand/Fred Overbeck is enumerated in Knox Township, Jefferson County, in his father's home. Two households away, Solomon McManigle with four-year-old daughter Esther also was enumerated.[54]

- In 1870, 15-year-old Ferdinand is enumerated in Rose Township, Jefferson County. Sylvester McManigle lived four households away and Alexander McManigle lived eight households away.[55]

- Not only did Ferdinand and his parents live near other McManigle family members in Rose Township, his grandparents Charles and Amelia Overbeck lived next door to Solomon McManigle and daughter Esther in Brookville in 1870.[56]

- Subsequently, Solomon McManigle's family moved to Rose Township by the time Solomon first appears on the 1875 tax list. The family lived there until 1878 when Solomon McManigle is crossed off the Rose Township list and first appears on the Washington Township assessment.[57] Constantine Overbeck appears on the tax lists in Rose Township for 1869 through 1884.[58]

There is no question that the two families knew each other from—at the latest—1860 when Ferdinand and Esther were very young.

---

54. 1860 U.S. census, Jefferson County, Pennsylvania, population schedule, Knox Township, p. 197 (penned), p. 170 (stamped), Dwelling no. 1411, family no. 1396, Constantine "Ovbeck" and dwelling 1414, family no. 1398, Solomon "McManagal"; citing NARA microfilm publication M653, roll 1118.

55. 1870 U.S. census, Jefferson County, Pennsylvania, population schedule, Rose Township, p. 6 (penned), Dwelling no. 38, family no. 40, Ferdinand Overbeck, and dwelling no. 24, family no. 36, Sylvester McManigle, and p. 5 (penned), Dwelling no. 30, Family no. 32, Alexander McManigle; citing NARA microfilm publication M593, roll 1552.

56. 1870 U.S. census, Jefferson County, Pennsylvania, population schedule, Brookville, p. 38 (penned), Dwelling no. 266, family no. 293, Solomon McManigle, and Dwelling no. 267, family no. 294, Chas. Vanoverbeck; citing NARA microfilm publication M593, roll 1352. That Constantine's father was Charles see, "Pennsylvania, Death Certificates, 1906-1966," Ancestry (https://search.ancestry.com/search/db.aspx?dbid=5164 : accessed 26 February 2018) > 1909 > Certificate Number range 100601-104200 > image 77, Constantine Overbeck, File no. 100672.

57. Jefferson County, Pennsylvania, Rose Twp. 1873–1876 and 1877–1879 and Washington Twp. 1876–1880 [tax assessment lists], arranged by first letter of last name; Jefferson County Historical Society, Brookville.

58. Jefferson County, Pennsylvania, Rose Twp. Volumes for 1868–1872, 1873–1876, 1877–1879, 1877–1879, 1880–1882, and 1883–1885 [tax assessment lists], arranged by first letter of last name.

Solomon McManigle was living in Jefferson County at the time of his death in 1912.[59] His wife Mary Jane also was living in Jefferson County at the time of her death.[60] No probate records or deeds show that any property was transferred to heirs upon their deaths.[61] Only their surviving children are named in their obituaries, and Esther is not among them.[62] Mary Jane's obituary states that she was a member of the M.E. Church in DuBois (in neighboring Clearfield County). The church's surviving records do not go back to 1912.[63]

Solomon's birth family lived in Eldred Township, Jefferson County, in 1850.[64] Sigel—the town identified as Mary Elizabeth (Overbeck) Lee's birthplace—is in Eldred Township. James McManigle, the patriarch of the Jefferson County McManigles, is buried at the Mount Tabor Cemetery in Sigel.[65]

Fred Overbeck and Esther McManigle's Short Life Together

As noted above, "Fred" Overbeck arrived at the state penitentiary on 22 September 1876.[66] He was married and had two children.[67] Esther Ann (McManigle) Overbeck was born in 1858 and married Fred some time before 1 April 1877.[68] Fred Overbeck was discharged from the penitentiary on 21 May 1879.[69] The following year he was living in Brookville with

59. *Ancestry* (https://search.ancestry.com/search/db.aspx?dbid=5164: accessed 22 March 2017) > Pennsylvania, Death Certificates, 1906-1964 > 1912 > 020251-024140> image 2681, Solomon McManigal, File no. 22863.

60. "Pennsylvania, Death Certificates, 1906–1964," *Ancestry* (https://search.ancestry.com/search/db.aspx?dbid=51645 : accessed 18 December 2016) > 1916 > Certificate Number Range 099261-102540, image 3365, Mary Jane McManigle, File No. 102361.

61. Jefferson County, Pennsylvania, electronic index of deeds 1828-1910 and electronic index of wills and orphans' court records on computer; Recorder of Deeds Office. *FamilySearch* (https://www.familysearch.org/search/film/005544507 : accessed 25 April 2018), digital film no. 5544507, Register's and Orphans' Court General Index L-R > image 395, p. 552.

62. "Death of Solomon McManigle," *The Jeffersonian Democrat*, 24 March 1912, p. 1, c. 3 and "Mrs. Solomon McManigle," *Brookville Republican*, 19 October 1916; Jefferson County History Center, newspaper clipping collection arranged in file folders by surname.

63. Email from Lori Chambers, administrative assistant for the First United Methodist Church to Patti Hobbs, 6 July 2017. Identification of specific church from *Insurance Maps of DuBois, Clearfield County, Pennsylvania* (New York City: Sanborn Map Co., 1896), Sheet 7.

64. 1850 U.S. census, Jefferson County, Pennsylvania, population schedule, Eldred Township, p. 31 (stamped), Dwelling No. 45, Family No. 49, James "McManigil"; citing NARA microfilm publication M432, roll 786. See death certificate for father's name of James McManigle.

65. *Find A Grave* (https://findagrave.com : accessed 8 March 2018), "James McManigle," Memorial ID 145114594; records of Mount Tabor Cemetery, Sigel, Jefferson County, Pennsylvania.

66. Admission and Discharge Book, 1872–1900, Fred Overbeck, No. 5276; MM B 120, Western State Penitentiary Population Records, RG-15, Pennsylvania State Archives, Harrisburg; microfilm #407.

67. Descriptive List, 1876–1880, Fred Overbeck, No. 5276; MM B 128, Western State Penitentiary Prison Population Records, RG-15, Pennsylvania State Archives, Harrisburg; microfilm #9.

68. Immaculate Conception Catholic Church (Brookville, Pennsylvania), "1877–1905 Baptisms," arranged chronologically, 1 April 1877, Maria Anna Overbeck; Parish rectory, Brookville

69. Admission and Discharge Book, 1872–1900, Ferdinand Overbeck, No. 5276; MM B 120, Western State

wife Eliza J., age 24, and daughter Ida M., age 8.[70] Esther Ann McManigle, age 22, was living with her parents in Washington Township, Jefferson County, with her marital status given as "single."[71]

From this census, one would never suspect that Esther had been married and given birth to at least one child. There is one clue that things are not quite as they appear. The 1880 Solomon McManigle household includes two sons named John. One is John, age 8, and the other is John J., age 4. John J. does not appear on Solomon McManigle's listing of his living children in 1898 for his Civil War pension.[72] He might have died in the meantime, but more likely he was the child of a daughter in the family. Four different newspapers in April and May 1898 reported that Joel, Joseph, or Frank Overbeck had drowned at Curwensville when he was working rolling logs off the breast of a dam and fell in. One paper reported that he'd been living in Curwensville (Clearfield County) for three years. The local paper in Brookville is likeliest to provide the correct name, although one of the others agrees that his name was Joel Overbeck and that he was the son of "Ferd" Overbeck of Kane, Pa."[73] This could cause some conflict—if Ida of the 1880 census was also his daughter, there would be three children that potentially could be Fred's, not the two he claimed when arriving at the penitentiary. Or it would not cause conflict if she was not Fred's child after all. It's also possible that John J.'s age in the 1880 census is off by a year and he was born after Fred was incarcerated.

The Walburn Connection

No records directly connect the Solomon McManigle family and their daughter Esther to Mary Elizabeth Overbeck. But the two families' movements provide strong evidence of a relationship. Terry Walburn and his wife Judy have researched the McManigle family for many years. It was known in the family that two Walburn brothers married two McManigle sisters: Emanuel Walburn married to Nancy McManigle and A. Samuel Walburn married to Esther McManigle.[74] Without this information, it would have been almost impossible to know what

Penitentiary Population Records, RG-15, Pennsylvania State Archives, Harrisburg; microfilm #407.

70. 1880 U.S. census, Jefferson County, Pennsylvania, population schedule, Brookville, Enumeration District 186, p. 41, dwelling and household nos. 351, Fred Overbeck household; citing NARA microfilm publication T9, roll 1136.

71. 1880 U.S. census, Jefferson County, Pennsylvania, population schedule, Washington Township, Enumeration District No. 200, p. 23 (penned), p. 257 (stamped), dwelling no. 199, family no. 211, "Soloman McNigle."

72. Pension questionnaire, 4 July 1898, Solomon McManigle, pension no. S.C. 740572, RG 15, NA–Washington.

73. He was Joel in *Jeffersonian Democrat* (Brookville, Pennsylvania), 28 April 1898, p. 5, c.1. Joseph in "A Watery Grave," *The Daily Journal* (Philipsburg, Pennsylvania), 28 April 1898, p. 4, c. 2. Joel Overbeck item, *Raftsmans Journal* (Clearfield, Pennsylvania), 4 May 1898, p. 3, c. 3. Frank in "Drowned at Curwensville," *Clearfield Republican*, 29 April 1898, p. 3, c. 1. The last three available in Clearfield County Historical Society microfilmed newspapers, Clearfield.

74. DNA match Terry and Judy Walburn were the recipients of photos and information from a Walburn relative Margaret Virl (Kessler) Sundie. Margaret was the great granddaughter of David Walburn, brother to Emanuel Walburn and A. Samuel Walburn. Her lineage to David is documented through the following: David's death certificate names the same parents (Simon Walborn and Mary Solada) as Samuel Walburn's: Maryland Department of Health, Certificate of Death, no. 42703, David Walburn (1918); Division of Vital

happened to Esther, the daughter of Solomon and Mary Jane McManigle and former wife of Fred/Ferdinand Overbeck.

In 1880, the Solomon McManigle family including daughter Esther lived in Washington Township, Jefferson County.[75] By December of that year, they had moved across the county border into the borough of DuBois, Clearfield County. Solomon is listed in the DuBois tax rolls from 1880 through 1889.[76] In 1891 A. Samuel Walburn filed an affidavit for Solomon McManigle's Civil War pension, stating that A. Samuel had known Solomon since 1881, usually living within a quarter-mile of him.[77] It seems likely that this is when A. Samuel became Esther Ann (McManigle) Overbeck's second husband. In 1884 Solomon bought a lot in DuBois on Isabella Street (now Jared).[78] From 1891 through 1893 Solomon and Samuel Walburn were assessed in Hamlin Township, McKean County, where Mary Elizabeth Overbeck lived when she married Ira Lee. In 1895 Samuel Walburn was assessed for taxes in Penn Township, Clearfield County, as well as from 1896 through 1901 along with Solomon McManigle. Esther Ann Walburn died in Penn Township in 1897.

Records, Baltimore and "Pennsylvania, Death Certificates, 1906-1964," *Ancestry* (https://search.ancestry.com/search/db.aspx?dbid=5164 : accessed 25 April 2018) > 1937 > Certificate Number Range 112851-1160000 > image 3305, File No. 115580, Samuel Walborn. Margaret's father was William H. Kessler as shown in her obituary: "Margaret V. Sundie," *Courier Express*, 17 July 2006, np; scanned copy emailed by Carol Laughlin of the DuBois Area Historical Society. William Kessler's mother Mary Ann Walburn is named in "Pennsylvania, Death Certificates, 1906–1964,"*Ancestry* (https://search.ancestry.com/search/db.aspx?dbid=5164 : accessed 27 February 2018) > 1964 > Certificate Number Range 114101-116950 > image 3164, William H. Kessler, File No. 115772-64. Mary Ann's father is named as David Walburn in "Pennsylvania, Death Certificates, 1906–1964,"*Ancestry* (https://search.ancestry.com/search/db.aspx?dbid=5164 : accessed 27 February 2018) > 1943 > Certificate Number Range 086401-089100 > image 2832, Mary Kessler, File No. 88647.

75. 1880 U.S. census, Jefferson County, Pennsylvania, population schedule, Washington Township, Enumeration District No. 200, p. 23 (penned), p. 257 (stamped), dwelling no. 199, family no. 211, "Soloman McNigle"; citing NARA microfilm publication T9, roll 1136.

76. Clearfield County, Pennsylvania, Tax Assessment Lists for the Borough of DuBois, 1881–1889, each year's list arranged alphabetically by first letter of last name, "Sol McMonigle"; microfilm in County Assessor's office, Clearfield.

77. A. S. Walburn, Neighbor's or other Person's Affidavit, 24 November 1891, Solomon McManigle, pension no. S.C. 740572, RG 15, NA–Washington.

78. Clearfield County, Pennsylvania, Deed Book 72:88, W.N. and Isabella J. Prothers to Solomon McManigle, 7 June 1884; Recorder of Deeds office, Clearfield. The deed identifies the property as Lot 16 of the Stone addition. The modern location was identified with the assistance of GIS manager for Sangamon County, Illinois, Tracy Garrison, subject to additional information to be verified. Lot 16, Stone addition information was verified by the deed of the current owner identified through the Landex system on the Recorder of Deeds office website referencing #0074-003-000-0531 for 315 S. Jared Street found on the *Clearfield County Public Map Viewer* (http://gis.clearfieldco.org/flexviewers/publicmapviewer/ : accessed 27 February 2018).

Table 10.5. Tax Assessments in Jefferson, Clearfield, and McKean Counties		
	Solomon McManigle	**Walborns**
1881–1883	DuBois, Clearfield County*	Henry, Dubois* Samuel, Sandy Township, Clearfield County* Emanuel, Sandy Township*
1883–1884	Franklin, corner Washington, DuBois*[a]	Henry, Jared corner of Weber Av. DuBois*[a] Samuel, DuBois* Emanuel, Sandy Township*
1885	Dubois*	Henry, Dubois* Samuel, DuBois* Emanuel, Sandy Township*
1886	DuBois*	Henry, Dubois* Samuel, Sandy Township* Emanuel, Sandy Township*
1887–1888	DuBois: Morrison bel Valley[b]; Lot 16, Stone Addition[c] DuBois*	Henry, DuBois: Jared corner of Weber Av.[b] DuBois* Samuel, Sandy Township* Emanuel, Sandy Township*
1889	DuBois*	Henry, Dubois* Samuel, Sandy Township tax list (crossed off)* Emanuel, Dubois tax list, crossed off* Sandy Township list*
June 1890	Dubois, Veteran's schedule[d]	
1891	Hamlin Township, McKean County**	Samuel, Hamlin Township** Affidavit on behalf of Solomon.[e]
1892	Hamlin Township**	Samuel, Hamlin Township**
September 1892	**Mary Elizabeth Overbeck of Mt. Jewett (Hamlin Township) married Ira N. Lee of Mt. Jewett.[f]**	
1893	Hamlin Township** DuBois, mortgage and claimant affidavit[g]	Samuel, Hamlin Township**
1895		Samuel, Penn Township, Clearfield County* Emanuel, Penn Township*
1896–1897	Penn Township*	Samuel, Penn Township, Clearfield County Emanuel, Penn township, Clearfield County

Table 10.5. Tax Assessments in Jefferson, Clearfield, and McKean Counties		
	Solomon McManigle	**Walborns**
September 1897	**Esther Walborn, age 39, born in Brookville, died in Penn Township, Clearfield County, 13 September 1897.[h]**	
1898	Penn Township*	Samuel, Penn Township* Emanuel, Penn Township*
1899	Penn Township*	Samuel, Penn Township* Emanuel, Penn township*
1900	Penn Township[i]	Simon, Penn Township[i]
1901	Penn Township*	Emanuel, Penn township* Simon, Penn Township*
1905	Warsaw Township, Jefferson County[j]	

All tax records give various forms of the given names and sometimes initials. They are standardized in this table.

* From tax records in Clearfield County, Pennsylvania, Assessment of Property," records grouped by township, and entries recorded alphabetically by first letter of last name, Penn Township, 1892–1900, entries for McManigle and Walburn, DuBois Boro, 1881–1891, entries for McManigle and Walburn, Sandy Township 1881–1891, entries for Walburn and McManigle; County Assessor's office microfilm, Clearfield.

** From tax records in McKean County, Pennsylvania, Annual Assessment of McKean County Incl Hamlin Township, in volumes containing 1888–1892, 1893–1896, search for McManigle, Walburn, and Lee; County Assessor's office, Smethport.

a. *Boyd's Williamsport City Directory: including DuBois and Lock Haven … 1883-1884* (Pottsville, PA: W. Harry Boyd, 1883), entries for McManigle and Walburn; scans emailed by the DuBois Area Historical Society, DuBois.

b. *Boyd's Williamsport City Directory: including DuBois and Lock Haven … 1887-1888* (Pottsville, PA: W. Harry Boyd, 1887), entries for McManigle and Walburn; scans emailed by the DuBois Area Historical Society, DuBois.

c. Clearfield County, Pennsylvania, Deed Book 72:88, W.N. and Isabella J. Prothers to Solomon McManigle, 7 June 1884.

d. 1890 U.S. census, Clearfield County, Pennsylvania, Special Schedule Surviving Soldiers, Sailors, and Marines, and Widows, Etc. Dubois, Enumeration District 89, p. 5, first line, Solomon McManigal; citing NARA microfilm publication M123, roll 85.

e. A. S. Walburn, Neighbor's or other Person's Affidavit, 24 November 1891, Solomon McManigle, pension no. S.C. 740572, RG 15, NA–Washington.

f. State of New York, Department of Health, Return of a Marriage, Lee-Overbeck, 8 September 1892; Bureau of Vital Statistics, Albany.

g. Solomon and Mary Jane are "of DuBois" in Clearfield County, Pennsylvania, Mortgage Book T:15, Solomon and Mary J. McManigle to Mutual Building and Loan, 16 February 1893; Recorder of Deeds office, Clearfield. Also Solomon McManigle, Claimant Affidavit, Civil War pension no. S.C. 740572, RG 15, NA–Washington shows his residence as DuBois.

h. Clearfield County, Pennsylvania, Death Record 1 & 2:277, line 7, Ester A. Walburn, 13 September 1897; FHL microfilm 1463016.

i. Simon Walburn, Samuel's father is six households away from the Solomon McMcManigle household: 1900 U.S. census, Clearfield County, Pennsylvania, population schedule, Penn Township, Enumeration District 77, Sheet 8 B, Dwelling no. 133, family no. 136, Solomon McManigle; Dwelling no. 139, family no. 143, Simon Walburn; citing NARA microfilm publication T623, roll 1396.

j. Jefferson County, Pennsylvania, [tax assessments] Warsaw Twp. 1905, arranged alphabetically by first letter of surname, S McManigal; Jefferson County Historical Society, Brookville.

Not only did the two families go to the same general locations, they usually lived near each other.

In Dubois, Henry Walburn's property was on Jared at the corner of Weber.[79] The description of the lot he purchased makes it clear that it lies on the northeast corner of that intersection.[80] Solomon's was at the modern address of 315 Jared.

Figure 10.15: Taken from T.M. Fowler, *Du Bois, Clearfield County, Pennsylvania* (Morrisville, PA: Fowler & Moyer, 1895). Map selection large enough to capture street names for identification.

In McKean County, a more specific location was identifiable because the tax assessments include the original warrant number of the property on which the taxpayers resided.

Table 10.6. Locations for Tax Assessments in Hamlin Township, McKean County, Pennsylvania			
	1891[a]	**1892[a]**	**1893[b]**
I. N. Lee			Warrant no. 2703
S. E. McManigle	Warrant no. 2690	Warrant no. 2703	
A. S. Walburn	Warrant no. 2690	Warrant no. 2690	

a. McKean County, Pennsylvania, Annual Assessment of McKean County 1888-1892 Incl Hamlin Township, arranged by year and then alphabetically, entries for S. E. McManigle and A.S. Walburn; Tax Assessor's office, Smethport.

b. McKean County, Pennsylvania, Annual Assessment of McKean County incl. Hamlin Township, McKean County 1893, arranged alphabetically, entries for I. N. Lee.

These two warrants are adjoining each other in Hamlin Township indicating that Solomon's move from 1891 to 1892 was not far, and both places were close enough to Ira N. Lee for him to have met Mary Elizabeth Overbeck.

79. *Boyd's Williamsport City Directory: including DuBois and Lock Haven … 1883–1884* (Pottsville, PA: W. Harry Boyd, 1883), entry for Henry Walburn; scans emailed by the DuBois Area Historical Society, DuBois.

80. Clearfield County, Pennsylvania, Deed book 22:240–1, Jared B. and Annie L. Evans to Henry Walburn; Recorder of Deeds office, Clearfield.

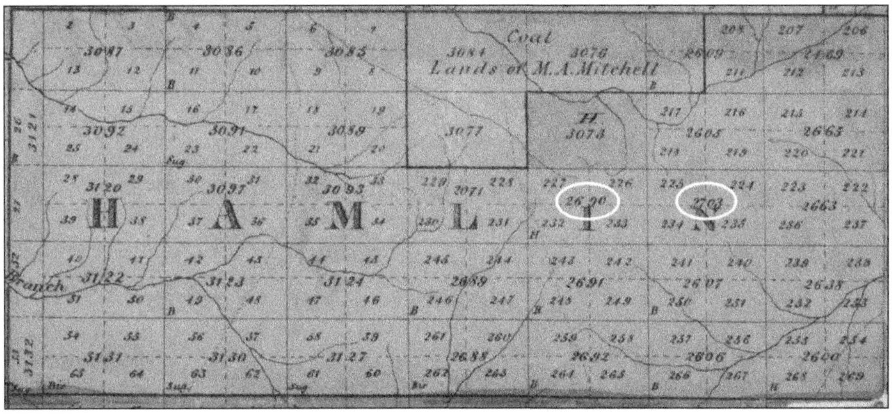

Figure 10.16: W. K. King, *Map of M'Kean County, Pennsylvania* (Philadelphia: A Kollner, 1857); digitized map, *Library of Congress* (https://www.loc.gov/item/2012590204/ : accessed 8 March 2018).

From 1881 until Esther's death in 1897, the lives of Samuel Walborn and Solomon McManigle ran in parallel paths. They likely met in DuBois, Clearfield County. Samuel's brother Henry lived only blocks away from where the McManigles lived. The Walburns and McManigles moved together by the end of 1890 to Hamlin Township, McKean County. That move set up the converging paths that led to the marriage in September 1892 of Esther's daughter Mary Elizabeth Overbeck "of Mt. Jewett," born in Jefferson County, to Ira Lee who had come from Clarion County. Ira first appeared on the McKean County tax lists for Hamlin Township for 1893, compiled after the oath of the assessor on 23 November 1892, which meant he could have arrived in Hamlin Township any time after the previous year's oath in November 1891. He also was "of Mt. Jewett" when he married Mary Elizabeth Overbeck in September.

After leaving Hamlin Township, Samuel Walburn moved to Penn Township, Clearfield County, near Grampian. Solomon first went back to DuBois for a couple of years before following to Penn Township. There he remained through the death of his daughter Esther in 1897 until 1901 when he moved back to Jefferson County.

Nothing in Esther's death record tells of her marriages or that she was the daughter of Solomon. But when Samuel remarried in 1903, he stated that his former wife died 13 September 1897.[81] The age of Esther Walborn (39 at her death on 13 September 1897) matches the age of Esther McManigle born February 1858. Esther was buried in the Friends Cemetery in Grampion, but no probate, cemetery records, or obituaries exist to reveal more details about this woman who lived in the shadows of her father and husband.[82] Also hidden in the

81. "Pennsylvania, County Marriages, 1885–1950," *FamilySearch* (https://www.familysearch.org/search/collection/1589502: accessed 27 February 2018) > access through search or catalog, digital film no. 004460699 > image 2533, Marriage License Docket 19:535, Sam'l Walborn and Josephine Witherow, 1903.

82. "Pennsylvania Probate Records, 1683–1994," *FamilySearch* (https://www.familysearch.org/search/collection/1999196: accessed 28 February 2018) > Clearfield > Estate Index 1823-1919 vol RS-WXYZ > image 374, Register's & Orphan's Court Index WXYZ: 31-1, no Walburn entries. Two massive collections of obituaries are indexed in binders at the William B. Alexander Research Center of the Clearfield County

shadow of her grandfather, Mary Elizabeth Overbeck made her presence known only when she married Ira Newell Lee in 1892.

Summary

- No direct evidence exists that points to the mother of Mary Elizabeth (Overbeck) Lee, born in 1874 in Jefferson County, who married Ira Lee as a resident of Mt Jewett, Hamlin Township, McKean County, Pennsylvania, in 1892.
- DNA evidence points to descent from James McManigle and Susan Baecker through several matching DNA segments with descendants of their children.
- DNA evidence also shows genetic relationships to the Mason family of Jefferson County, relatives of Mary Jane Mason, the wife of James and Susan's son Solomon McManigle.
- Only one of Solomon and Mary Jane (Mason) McManigle's children was of an age to be Mary Elizabeth Overbeck's mother: Esther Ann McManigle, born February 1858.
- Esther/Maria Anna Overbeck, wife of Ferdinand/Fred and daughter of "Salomon" McManigle and Maria Johanna Mason, born February 1858, was baptized in the Immaculate Conception Catholic Church in Brookville in 1877.
- Esther McManigle was living with her birth family in 1880, but the youngest child John J. carries the same name as one of the children of Solomon and Mary Jane, but is not listed as a child on Solomon McManigle's affidavit listing his children. This child is likely Joel who died in 1898 in a log-rolling incident and was identified as the son of "Ferd" Overbeck.
- The McManigle family moved to DuBois by 1881 where they crossed paths with Samuel Walburn whose brother also lived in DuBois. Walburn family information states that the two brothers Emanuel and Samuel Walburn married sisters: Esther and Nancy McManigle.
- In 1891 A. S. Walburn filed an affidavit on behalf of Solomon McManigle for his Civil War pension stating that Walburn had known McManigle since 1881 and had lived near him ever since.
- Samuel Walburn and Solomon McManigle and their families followed parallel migration paths from DuBois, Clearfield County; to Hamlin Township, McKean County; to Penn Township, Clearfield County, over the course of almost twenty years.

Historical Society in Clearfield. The first is titled "Obituaries Index" and the second is title "Supplemental Obituaries Index Vol. 1 to 12." There were no entries for Esther Walborn. According to the death record, Esther is buried at the Friend's Cemetery near Grampian, Penn Township, Clearfield County. The woman who holds the cemetery records searched her files but found no information about the burial of Esther Walburn: phone call and cemetery visit on 7 and 8 July 2017.

- Esther Walborn died on 13 September 1897 in Penn Township, Clearfield County, age 39, placing her birth in 1858.
- Samuel Walborn remarried in 1903 giving the death date of his previous wife as 13 September 1897.
- Both Samuel Walburn and Solomon McManigle and their families lived in Hamlin Township, McKean County, in 1892 when Mary Elizabeth Overbeck met and subsequently married Ira N. Lee.

Strong DNA evidence connecting Mary Elizabeth Overbeck descendants with the McManigle and Mason families of Jefferson County, Pennsylvania, provides overwhelming support for their descent from Solomon McManigle and his wife Mary Jane Mason. Compelling indirect evidence that Esther Ann McManigle was the mother of Mary Elizabeth Overbeck is provided by the baptismal record showing that their daughter Esther Ann was married to Ferdinand Overbeck and the subsequent parallel migration trails of Solomon and Esther's second husband, Samuel Walborn, in three different locations including Mary Elizabeth Overbeck's residence at the time of her marriage.

Seeking Further Evidence

For further corroboration, descendants from Esther and Samuel Walburn should be DNA tested. However, they appear to have had only one child: James Walburn, born about 1895.[83] No evidence of his life or death after 1910 has been found. No likely candidates are apparent from a search in the 1920 census, World War I Draft Registrations, or Clearfield County tax records.[84] No listing for James is found in the obituary index of the Clearfield County Area Historical Society.[85] No probate record was left by A. Samuel Walborn.[86] When Samuel Walborn died 25 December 1937, his obituaries named no family members.[87]

83. 1910 U.S. census, Clearfield County, population schedule, Clearfield Ward 2, Enumeration District 63, sheet 33 A, household 298, dwelling 698, family no. 708, James Walborn; digital image, *Ancestry* (http://interactive.ancestry.com/7884/4449849_00965 : accessed 27 September 2016).

84. "1920 United States Federal Census," database and images, Ancestry (https://search.ancestry.com/search/db.aspx?dbid=6061), search using search variations James Walburn/Walborn with and without year of birth, James born around 1895 limited to Clearfield County. "United States World War I Draft Registration Cards, 1917-1918," *FamilySearch* (https://www.familysearch.org/search/collection/1968530 : searched 8 September 2017)

Used "James Walburn" born from 1894 through 1896 in Pennsylvania. I also searched with "Clearfield County" as place of residence with no last name entered. Clearfield County, Pennsylvania, Tax Assessors Lists, Clearfield Ward 2, 1916–1918, microfilm at Tax Assessors office, Clearfield.

85. Obituaries Index, and Supplemental Obituaries Index, William B. Alexander Research Center of the Clearfield County Area Historical Society in Clearfield.

86. Clearfield County, Pennsylvania, Register's & Orphan's Court Index W-X-Y-Z, searched p. 11 (for A. Samuel), p. 81 (for Samuel); Orphan's Court office, Clearfield. Waypoints on *FamilySearch* inaccurately convey that the probate records go only to 1919. This index extends to at least 1961.

87. "Samuel Walborn," *Clearfield Progress*, 27 December 1937, p. 8, c. 3 and "Samuel Walborn," *Clearfield Times*, 31 December 1937, p. 8, c. 5; State Library of Pennsylvania microfilmed newspapers.

Accumulating more DNA evidence might further enhance these conclusions. Of the four children of Ira and Mary Elizabeth (Overbeck) Lee who lived to adulthood, only descendant Beverly Pierce agreed to be tested. Patricia Hobbs, Carol Polk, and Debbie Lee are siblings; each may share differing segments of DNA with cousins, but these siblings and Beverly Pierce represent only two lines descended from Mary Elizabeth Overbeck.

Because of DNA testing, the mother of Mary Elizabeth Overbeck was identified even though her first and last name were unknown. Through connections made with cousins sharing DNA, even her face is now known by her descendants.[88]

Figure 10.17. Photograph of Esther Ann (McManigle) Overbeck Walburn from the collection of Terry and Judy Walburn

88. Photograph in the collection of Terry and Judy Walburn, North East, Pennsylvania; photo scanned by Patricia Lee Hobbs at the Walburn home, July 2017.

Writing about, Documenting, and Publishing DNA Test Results

Thomas W. Jones, PhD, CG, CGL, FASG, FUGA, FNGS

"I can't write without a reader. It's precisely like a kiss—you can't do it alone."
John Cheever (*Christian Science Monitor*, 24 October 1979)

Authors write not just because they want to write. They write for readers. To achieve their goal of being read, authors envision their readership. They keep readers in mind throughout the writing process and address that audience.

The readership of those writing about DNA includes genealogists with research and writing experience. Some have expertise with genetic research, others with documentary research, and some with both. The readership also likely includes family historians with little experience. In most cases, the readership also includes unknown future researchers and unborn descendants of families who are the subjects of that research and writing. Authors must make their explanations and documentation clear to that broad audience.

DNA testing, still a relatively new resource for family researchers, adds to a genealogical readership's diversity and complicates the writing process. Issues include unfamiliar words and acronyms, reasoning processes that do and do not resemble reasoning from documentary research, and relatively high densities of numbers, which readers can find baffling or repellant. Furthermore, DNA testing involves living people to a greater degree, and in significantly larger numbers, than most documentary genealogical research. Living people's privacy rights complicate writing about DNA more than writing solely about people in the past.

Genealogists' writing should meet general standards for writing about research findings and the genealogy field's generally accepted standards.[1] Enabling other researchers to replicate and verify—or overturn—research findings, reasoning, and conclusions, genealogical writers lay trails of data presentation, documentation, and explanation. At the same time, their writing engages readers with varying levels of expertise.

Despite issues of specialized technical content and diverse readerships, researchers can successfully present DNA test results. They can document and explain their work, and they can publish it in suitable venues. For that to occur, they first must write.

Writing about DNA Test Results

Genealogists write about DNA test results when DNA has helped them solve a research problem, typically a challenging problem. The essay explains how assembled research findings support a conclusion about an unobservable relationship, identity, or both.[2] "Without bias or preconception," the essay argues for supporting a conclusion and defends its likely accuracy.[3] Such proofs can take many forms, formal and informal, including articles, blog postings, segments of family histories, and reports to file or for paying or pro bono clients.

Writers initially might have only a general idea of the shape their final product will take. Following a process that gradually transforms rough writing into a polished essay, they apply principles for effective writing and addressing the rights of living people.

The process: start rough, rewrite, polish

"Writer's block" is a familiar phenomenon. Starting can be the writer's hardest task. Beginning is easiest when the writer focuses on first steps, not the finished product.

Writers can begin by jotting down words related to their topic. For some authors, describing why they are writing a piece—their objective—is helpful. Others list major points. Some find an outline to be an effective way to begin. Whatever the starting point, written words, even if a jumble, will lead to more.

Once writing begins, words and sentences flow. They help clarify the author's thinking. Eventually the writer groups sentences into tentative paragraphs and then organizes those paragraphs into tentative sections and subsections. The process can reveal areas of overlap and repetition, incomplete or absent data and documentation, cloudy explanations, inconsis-

All URLs were accessed 23 December 2018 unless otherwise indicated.

1. Style recommendations in this chapter follow *The Chicago Manual of Style*, 17th ed. (Chicago: University of Chicago Press, 2017), which the genealogy field has adopted. For that adoption, see Board for Certification of Genealogists, *Genealogy Standards*, 2nd ed. (Nashville, Tenn.: Ancestry.com, 2018), 8, standard 6, "Format."

2. Thomas W. Jones, "Proof Arguments and Case Studies," in Elizabeth Shown Mills, ed., *Professional Genealogy: Preparation, Practice, and Standards* (Baltimore, Md.: Genealogical Publishing, 2018).

3. Board for Certification of Genealogists (BCG), *Genealogy Standards*, 2nd ed., 36, standard 63, "Honesty."

tencies, confusing data, and disjointed transitions. As the writer repairs each problem—one at a time—the essay begins to take shape.

As a work's shape emerges, authors rearrange and rewrite. They will discard some material and add better ideas. They might remove data from a narrative and place it in a table, or vice versa. A guide for nonfiction writers says, "Rewriting is the essence of writing well."[4] Effective authors revise again and again until their written product flows logically in an orderly sequence from beginning to end.

After a work's content, documentation, sequence, and shape stabilize, writers begin polishing. They ensure that paragraphs begin with topic sentences and flow from one to the next. They check for textual references to all tables and figures. They reword for clarity and efficient communication. They check citations, grammar, spelling, and wording. They proofread, again and again and again. Eventually, they complete their polishing and make their work available for others to read. Then writers incorporate changes based on readers' feedback.

Content

Genealogical reports presenting DNA test results include essential content:

- Focus on genealogical research questions
- Background information pertinent to each question and its answer
- Discussion of assembled research findings, including DNA test results, sequenced to lead from a starting point to each research question's answer
- Statement of the answer

When genealogists have augmented documentary research with DNA test results, or when they have used DNA testing to help answer a genealogical question that paper-based research alone could not answer, their written reports include further components. Those components include explaining why the research question had been unanswered and discussing why they opted to use DNA evidence to help answer the research question.

Genealogical writers do not describe unproductive research, including inconclusive DNA test results, except when negative findings help support a conclusion. They also avoid digressions that do not advance their case from a question to a defensible answer.

At logical points in the essay, the author presents the test takers whose DNA helped answer the research question. They identify those people to the extent that permissions from those people allow. Identifications typically include documenting the test takers' descents from people whose relationships pertain to the research question, its answer, or both.

4. William Zinsser, *On Writing Well: The Classic Guide to Writing Nonfiction* (New York: Collins, 2006), 83.

At logical points in the essay, genealogists present the DNA test results that support, directly or indirectly, the research question's answer. They identify the testing company, the chip used, or both. They also identify any third-party algorithms and tools used. Enabling readers to assess their conclusion's validity, genealogists report data precisely:

- For example, atDNA results can include data about chromosomes, segments, SNPs, total amounts of shared DNA, and admixture estimates.
- Similarly, mtDNA results can include regions tested, RSRS or CRS comparison values, marker values shared by test takers, and haplogroup designations.
- Y-DNA results can include the number and names of STR markers tested, the resulting values, marker values shared by test takers, SNP identifications, and haplogroup designations.

Other data also might be relevant, for example, atDNA "pileup" regions, mutation rates of Y-DNA markers, and results of calculations for admixture or endogamy. Only when writers have a valid reason for altering statistics reported to them do they transform DNA data—for example, from centimorgans to percentages or vice versa. In those cases, genealogists mention the transformation and explain its rationale.

In their reasoning and writing, genealogists integrate documentary and DNA data. They explain how both kinds of data relate to a solution and how they reinforce each other. They also provide their rationales for resolving conflicts within DNA and documentary data, between them, or both.

Clear writing

Genealogists write for diverse readerships about complex DNA test results and hypothesized family relationships. For readers to understand the writing, it must be clear.

Explanations are direct and precise. Sentences focus on one thought each. The writer minimizes long sentences, supporting clauses, and passive-voice verbs. Statements are straightforward, and usually their wording is positive. For example, *the match rate is lower than expected* (stated positively) is easier for readers to process than *the match rate is not as high as expected* (stated negatively).

Writers should explain terms and concepts that many readers would not understand. At the same time, however, they avoid "over explaining"—providing more depth or detail than readers need for following the meaning and reasoning.

Throughout the essay, the genealogist maintains focus on the genealogical problem and its solution. Digressions and extraneous information will confuse readers and bury important points. To avoid those problems, authors ruthlessly delete extraneous information. They prune away all words that are not necessary to communicate their meaning.

Writers addressing genealogical conclusions keep themselves in the background. Readers are interested in people whose relationship is in question and the people whose identities and DNA help establish relationships. Therefore, subjects of most sentences are those people. Secondarily, sentence subjects include genealogical and genetic data. Authors avoid referring to themselves, even indirectly.

Authors may, of course, express opinions when results are inconclusive. Rather than referring to themselves, however, writers use words of tentativeness—like *perhaps, could be*, or *might*—to show readers that they are expressing an opinion, not drawing a conclusion from data.

Clear writing is most comprehensible when it appears within an overall structure that helps readers stay with the writer from beginning to end.

Structure

Shapeless writing loses readers. Conversely, structure helps readers understand what they are reading and why they are reading it. It also helps them follow the author's reasoning and understand their conclusions. For this reason, genealogy standard 68 states, "Genealogists organize written presentations and discussions of assembled research results into structured parts and logical sequences."[5] Structure includes the essay's overall shape from beginning to end; divisions, connectedness, and flow within the essay; and the visual separation of textual narrative and supporting documentation.

More than two millennia ago, the scientist and philosopher Aristotle wrote that "the proper structure of the Plot [is a] ... whole ..., which has a beginning, a middle, and an end."[6] That same structure applies to genetic-genealogy writing today, whether the essay is an article, blog, chapter, or report:

- *Beginning.* The writing opens with a discussion of the genealogical question that DNA testing helped answer. Showing that the research started on a firm foundation, this section identifies and documents the people who are central to the question—for example, someone with unknown parentage. The section also puts the problem in relevant contexts, like historical setting, record-keeping issues, and prior failed research efforts. The beginning stops short of presenting the new findings that helped solve the problem. Many essays do not mention DNA testing in the beginning section; others touch on it.

- *Middle.* The writing's middle section presents the research findings, including relevant DNA test results and documentary evidence. The author unfolds the findings not in the sequence they were found or learned, and not in the sequence of the researcher's reasoning. Instead, with knowledge of the problem's solution, the author builds this section on the foundation laid in the writing's first section. Then the

5. BCG, *Genealogy Standards*, 2nd ed., 38, standard 68, "Structure."
6. *The Poetics of Aristotle*, part 7, trans. A. H. Butcher, 3rd ed. rev. (New York: McMillan, 1902), 31.

author proceeds, layer by layer, to the solution. Perhaps after trying and discarding several attempts, the writer finds a way to sequence the data and explanations for readers to easily follow. The unfolding is linear, with no cross-references to information that had been or will be discussed. The result will be an assemblage of evidence combining DNA and documentary evidence to support the author's conclusion.[7] This will be the essay's longest section. DNA test results might play a minor role, or they might be central to the unfolding proof argument. Tying documentary findings together, genealogists often present DNA data late in the middle section. The section closes with a statement of the problem's solution.

- *End.* The essay's last section summarizes the genealogical problem and its solution. It places the conclusion in any relevant familial or research context. If the research was methodologically interesting—as genetic-genealogy cases often are—the end section also summarizes the methodology.

Each of the three sections contains one or more paragraphs. The authors of arguably the best guide to writing in the English language advise authors to "Make the paragraph the unit of composition."[8] A paragraph is a group of sentences, usually three or more, pertaining to a topic. Paragraphs of one or two sentences can be effective, but when overused they make writing choppy and hard to follow.

Usually the first sentence of each paragraph states the paragraph's topic, giving the paragraph focus. That topic sentence also gives consumers a reason to read the paragraph. Authors help paragraphs flow from one to the next by ending paragraphs with words foreshadowing the next paragraph and by beginning paragraphs with words referring to the previous paragraph.

Paragraphs break prose into readable chunks. On a printed page they break at least once, usually several times. In electronic media—online articles, blogs, and reports, for example—they will break frequently.

Besides a beginning-middle-end structure and paragraphing, the structure of genealogical writing includes visual separation of narrative text and the author's documentation. Genealogists reading a narrative like seeing the documentation simultaneously, but without its intruding into the narrative. Reference-note numbers—typically superscripted without punctuation and placed usually at the ends of statements—connect each reference-note with words that a citation supports. In print media, citations usually appear in reference notes at the bottom of the page—footnotes. Reference notes are less accessible when placed at the

7. Thomas W. Jones, "Reasoning from Evidence," in Elizabeth Shown Mills, ed., *Professional Genealogy: Preparation, Practice, and Standards* (Baltimore, Md.: Genealogical Publishing, 2018), 281–87.

8. William Strunk Jr. and E. B. White, *The Elements of Style*, 4th ed. (Needham Heights, Mass.: Longman, 1972), 15–16.

end of an article, book, or report—endnotes. In electronic media, reference notes can appear in popups when the reader clicks on, mouses over, or touches a reference-note number.

Enhancements: tables and figures

Readers struggle to understand two kinds of data in essays incorporating DNA test results. When writers discuss DNA in prose, readers do not readily visualize similarities, differences, and patterns in the number-dense data. They also cannot easily visualize familial relationships among DNA test takers and the people whose relationships genealogists have hypothesized or established, when described in narrative form. Tables and figures, however, graphically arrange those complex data elements, helping readers understand genetic and genealogical data and how they support the writer's argument.

Tables for numerical data

Tables are the ideal format for presenting numerical data, like most DNA test results. Resembling spreadsheets, tables are two-dimensional portrayals of information. Vertical columns and horizontal rows show categories, layers, or subdivisions of a variable. Column headings and row stubs succinctly describe the respective column's or row's content. The headings can have more than one level, and the row stubs can be layered. Cells—each intersection of a column and a row—contain the data.[9] The table's title generically describes the cells' contents. See figure 11.1 for the parts of a complex table.

The insert-table feature of word-processing software is ideal for creating tables because they stabilize the table's arrangement. In contrast, when tabs and spaces align a table's data, the arrangement can change when the author moves the table to another platform or format or shares it with an editor, a publisher, or another researcher.

Uncluttering tables helps readers comprehend them. For example, tables contain lines only where lines minimize confusion. Also, the lines are light in visual "weight"—a thickness of .25 points is typical. If a table seems number dense, authors can divide the data among two or more tables; for example, one for atDNA-segment locations and another for segment characteristics. See figures 11.2 and 11.3 for examples of tables showing comparisons of atDNA test results. For an example showing Y-DNA results, see figure 11.4.

9. For guidelines for table preparation, see *Chicago Manual of Style*, 17th ed., 148–70, sections 3.47–88, "Tables." Also, "Manuscript Preparation – Tables," *The University of Chicago Press Journals* (https://www.journals.uchicago.edu/cont/prep-table).

Table Title (Describes Cell Contents)

	COLUMN HEADING, LEVEL 1		COLUMN HEADING, LEVEL 1	
	Column heading, level 2	*Column heading, level 2*	*Column heading, level 2*	*Column heading, level 2*
Row stub, layer 1				
Row stub, layer 2	Cell (data)	Cell (data)	Cell (data)	Cell (data)
Row stub, layer 2	Cell (data)	Cell (data)	Cell (data)	Cell (data)
Row stub, layer 1				
Row stub, layer 2	Cell (data)	Cell (data)	Cell (data)	Cell (data)
Row stub, layer 2	Cell (data)	Cell (data)	Cell (data)	Cell (data)

Table documentation goes here.

Figure 11.1. Structure of a complex table

Selected Autosomal DNA Matches to Joanne Gram

TESTEE'S NAME	TOTAL cMs IN SHARED SEGMENTS LARGER THAN 7 cMs	NUMBER OF SHARED SEGMENTS LARGER THAN 7 cMs	LARGEST SHARED SEGMENT
Roberto Gasca	162.7	11	25.3 cMs
Hugo Jaramillo	131.2	3	60.0 cMs
Ana "Anita" Gasca	108.4	6	32.1 cMs
Maria Teresa Ruiz	65.2	5	15.8 cMs
Alejandro Macías	51.7	3	23.1 cMs
Jorge Zúñiga	35.9	1	35.9 cMs
Evaristo Ruiz	29.1	2	15.3 cMs

For identification of matching kits, see "'One-to-many' matches," *GEDmatch: Tools for DNA and Genealogy Research* (gedmatch.com), for kit A767282 (Joanne Gram). For the tabulated data, see "'One-to-one' compare," *GEDmatch*, kit A767282 compared individually to kits T363242 (Roberto Gasca), M538273 (Hugo Jaramillo), A187233 (Ana "Anita" Gasca), M694610 (Maria Teresa Ruiz), A382232 (Alejandro Macías), A600364 (Jorge Zúñiga), and T472651 (Evaristo Ruiz). The acronym cMs refers to centiMorgans, a measure of genetic linkage based on recombinant frequency. See "CentiMorgan," *International Society of Genetic Genealogy Wiki* (isogg.org /wiki/Centimorgan).

From Karen Stanbary, "Rafael Arriaga, A Mexican Father in Michigan: Autosomal DNA Helps Identify Paternity," *National Genealogical Society Quarterly* 104 (June 2016): 87. Published here with the author's permission.

Figure 11.2. Table comparing atDNA results in unpaired rows

Pairs of Greenfield Autosomal-DNA Test Results

TRACED GREENFIELD ANCESTOR	DNA DONOR	LONGEST COMMON SEGMENT >7, IN CENTIMORGANS	SERIAL SNP COUNT	ESTIMATED GENERATIONS TO COMMON ANCESTORS
Nathaniel Thomas	Cheryl Mulder Elizabeth Shawen, through Lovilla	7.6 (chromosome 1)	717	7.4
Nathaniel Thomas	Descendant X Gerald Greenfield, through Caleb	14.9 (chromosome 2)	1543	5.0
Nathaniel Thomas	Thomas Jones Martha Marx, through Lovilla	12.9 (chromosome 3)	3469	5.1
Nathaniel Thomas	Thomas Jones Frances Hansen, through Lovilla	12.9 (chromosome 3)	3437	4.5
Nathaniel Thomas	Nancy Judd Sharon Bennett, through Luther	14.9 (chromosome 17)	672	5.0
Nathaniel Thomas	Nancy Judd Frances Hansen, through Lovilla	14.3 (chromosome 18)	1500	5.0
Nathaniel Thomas	Thomas Jones Frances Hansen, through Lovilla	14.4 (chromosome 18)	3441	[shown in row 4]
Nathaniel Thomas	Descendant X Frances Hansen, through Lovilla	7.2 (chromosome 18)	1616	7.5

Sources: "GEDmatch.Com Autosomal Comparison," on-request listings, *GEDmatch: Tools for DNA and Genealogy Research* (v2.gedmatch.com), kits A190412 (Shawen), A839038 (Hansen), M123945 (Marx), F202780 (Jones), F299963 (Mulder), F329609 (Greenfield), F329613 (Bennett), M115137 (Judd), and M201030 (Descendant X). *Ancestry.com* tested Hansen, Marx, and Shawen; *23AndMe* tested Descendant X and Judd; and *Family Tree DNA* tested Bennett, Mulder, Greenfield, and Jones.

From Thomas W. Jones, "Too Few Sources to Solve a Family Mystery? Some Greenfields from Central and Western New York," *National Genealogical Society Quarterly* 103 (June 2015): 97. Published here with the author's permission.

Figure 11.3. Table comparing atDNA results in paired rows

Y-Chromosome Haplotype of
William Lee Who Died in 1717 in Richmond County, Virginia

LOCATION (DYS)	VALUE (ALLELES)	LOCATION (DYS)	VALUE (ALLELES)	LOCATION (DYS)	VALUE (ALLELES)
19a	14	442	12	462	11
19b	–	444	12	463	25
385a	11	445	12	464a	15
385b	13	446	13	464b	15
388	12	447	25	464c	17
389-1	13	448	19	464d	17
389-2	29	449	30	464e	–
390	24	452	30	464f	–
391	10	454	11	GGAAT1	10
392	13	455	11	YCA II a	19
393	13	456	17	YCA II b	23
426	12	458	19	Y GATA A10	12
437	15	459a	9	635	23
438	12	459b	10	Y GATA H4.1	22
439	12	460	11		
441	13	461	11		

Source: "Search for Genetic Matches," *Ysearch* (http://www.ysearch.org), for users 9VR6B (descendant of Charles Lee) and ERFK2 (descendant of elder William Lee). Ancestry (formerly Relative Genetics) performed both tests. Values for the following markers have been modified to conform to Ysearch nomenclature: DYS441, DYS442, YGATA A10, Y GATA H4.1. See "Markers that may need to be converted," *Ysearch* (http://www.ysearch.org/conversion_page.asp).

From Judy Kellar Fox, "Documents and DNA Identify a Little-Known Lee Family in Virginia," *National Genealogical Society Quarterly* 99 (June 2011): 91. Published here with the author's permission.

Figure 11.4. Table showing Y-DNA results

Charts for genealogical relationships

Narrative descriptions of familial relationships confuse readers, especially when the researcher uses DNA testing to distinguish more than one hypothesized relationship among test takers. The remedy is to chart the relationships. When the chart's format is conventional and uncluttered, readers readily understand the relationships. For an example, see figure 11.5.

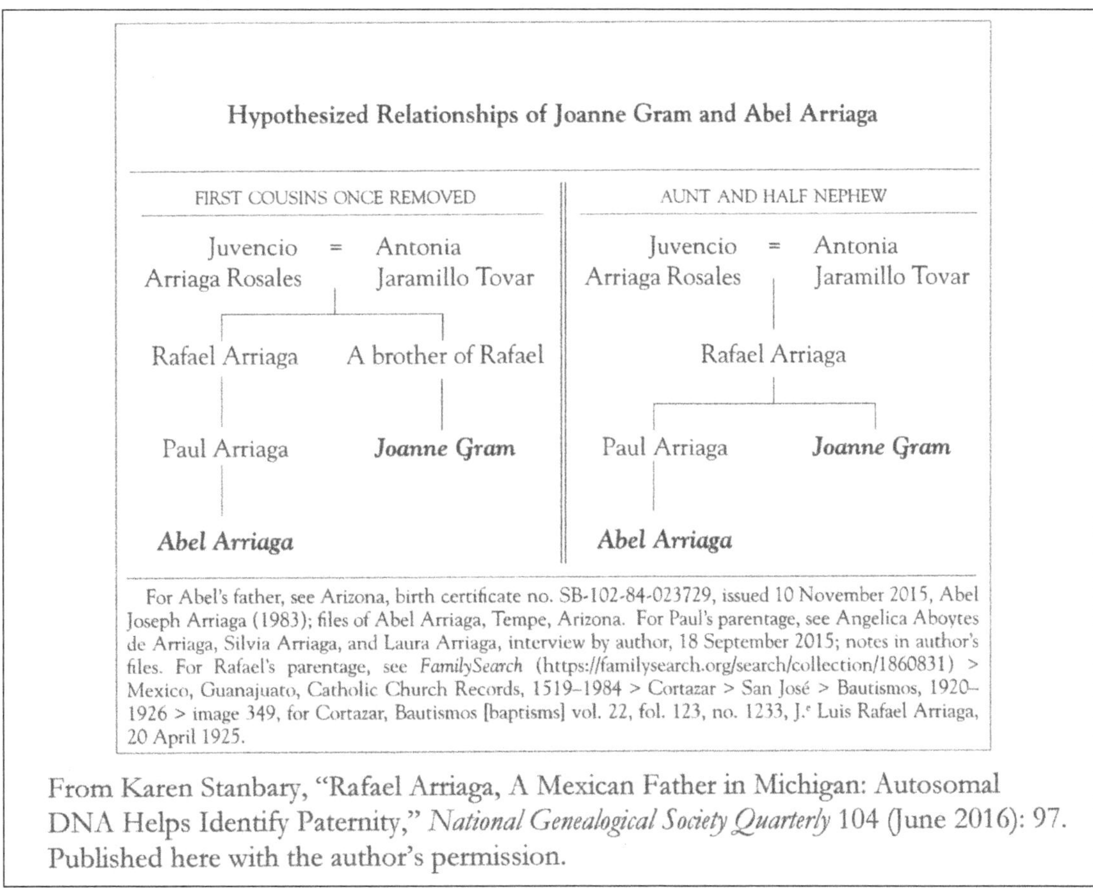

From Karen Stanbary, "Rafael Arriaga, A Mexican Father in Michigan: Autosomal DNA Helps Identify Paternity," *National Genealogical Society Quarterly* 104 (June 2016): 97. Published here with the author's permission.

Figure 11.5. Uncluttered chart illustrating hypothesized familial relationships

Genealogical charts are works of art and precision. People in the same generation are aligned horizontally, and vertical space between generations remains constant. Vertical lines connect generations, and horizontal lines overarch siblings. Horizontal spacing avoids crowding sibling groups and separates them from each other. An equal sign or the abbreviation *m.* (for *married to*) connects spouses or mates. The horizontal placement of an intergenerational vertical line shows whether children descend from one parent or the other—if both, the vertical line aligns with the connector between the parents.[10] Genealogical charts show only people and data essential to illustrating the writer's points; extraneous data is omitted. Highlighting DNA test-takers with bold, color, italics, or underlining adds clarity. See figure 11.6 for a chart documenting the test-takers' descents. For a landscape-oriented figure with test-takers' names bolded, see figure 11.7.

10. *Chicago Manual of Style*, 17th ed., 146–47, sect. 3.46, "Genealogical and pedigree charts," and figure 3.10, "A genealogical chart."

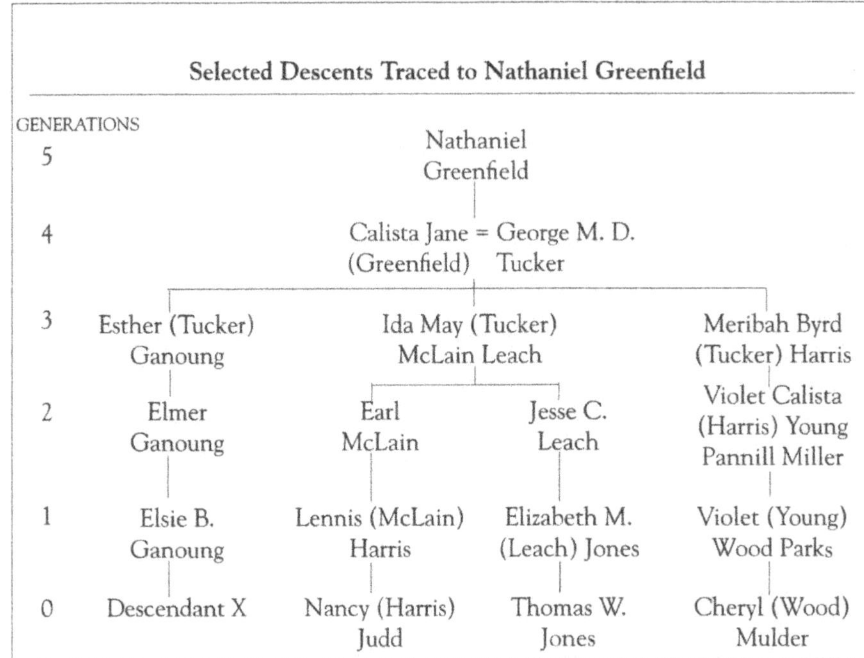

Sources: For Calista's parentage and husband, see the text.

For Esther, Ida, and Meribah's parents, see George M. D. Tucker, questionnaire 3—402, 1 August 1898; in George M. D. Tucker (Pvt., Cos. C and F, 3rd Mich. Cav., Civil War), pension no. S.C. 874,447, Case Files of Approved Pension Files 1861–1934 . . . , Civil War and Later Pension Files; Department of Veterans Affairs, Record Group 15; National Archives, Washington, D.C.

For Elmer's parents, see Kent Co., Mich., Returns of Marriages in the County of Kent for the Quarter Ending March 30 A.D. 1911, p. 11, Ganoung-Nelson, 25 February 1911; digital image, *FamilySearch* (https://www.familysearch.org) > Michigan, Marriages, 1868–1925 > 004209154 > image 394. For Elsie's, see 1920 U.S. census, Cook Co., Ill., population schedule, Chicago, Ward 24, enumeration district (ED) 1354, sheet 2B, dwelling 31, family 52, Elmer "Ganong" household; National Archives and Records Administration (NARA) microfilm T625, roll 335.

For Earl's parents, see Van Buren Co., Mich., Return of Births in the County of Van Buren for the Year Ending 31 December 1876, p. 256, no. 890, Earl McLain; digital image, *FamilySearch* > Michigan, Births, 1867–1902 > 004206431 > image 424. For Lennis's parents, see 1920 U.S. census, Sandusky Co., Ohio, pop. sch., Green Creek Twp., Clyde Village, ED 84, sheet 6A, dwell. 161, fam. 167, Earl McLain household; NARA microfilm T625, roll 1428.

For Jesse's parents, see Allegan Co., Mich., Return of Births in the County of Allegan for the Year Ending 31 December A.D. 1878, p. 7, no. 40, "Jessie D." Leach; digital image, *FamilySearch* > Michigan, Births, 1867–1902 > 004206453 > image 22. For Elizabeth's, see 1920 U.S. census, Sandusky Co., Ohio, pop. sch., Green Creek Twp., Clyde Village, ED 84, sheet 11B, dwell. 325, fam. 342, Jesse C. Leach household.

For Violet Harris's parents, see St. Clair Co., Mich., Return of Marriages in the County of St. Clair for the Quarter Ending June 30 A.D. 1922, record 17202, Pannill-Young, 14 June 1922; digital image. *FamilySearch* > Michigan, Marriages, 1868–1925 > 004210110 > image 352. For Violet Young's mother, see 1920 U.S. census, St. Clair Co., Mich., pop. sch., Port Huron, ward 4, precinct 8, ED 117, sheet 18B, dwell. 408, fam. 572, John Harris household; NARA microfilm T625, roll 795.

From Thomas W. Jones, "Too Few Sources to Solve a Family Mystery? Some Greenfields from Central and Western New York," *National Genealogical Society Quarterly* 103 (June 2015): 91. Published here with the author's permission.

Figure 11.6. Portrait-oriented chart documenting test-taker descents

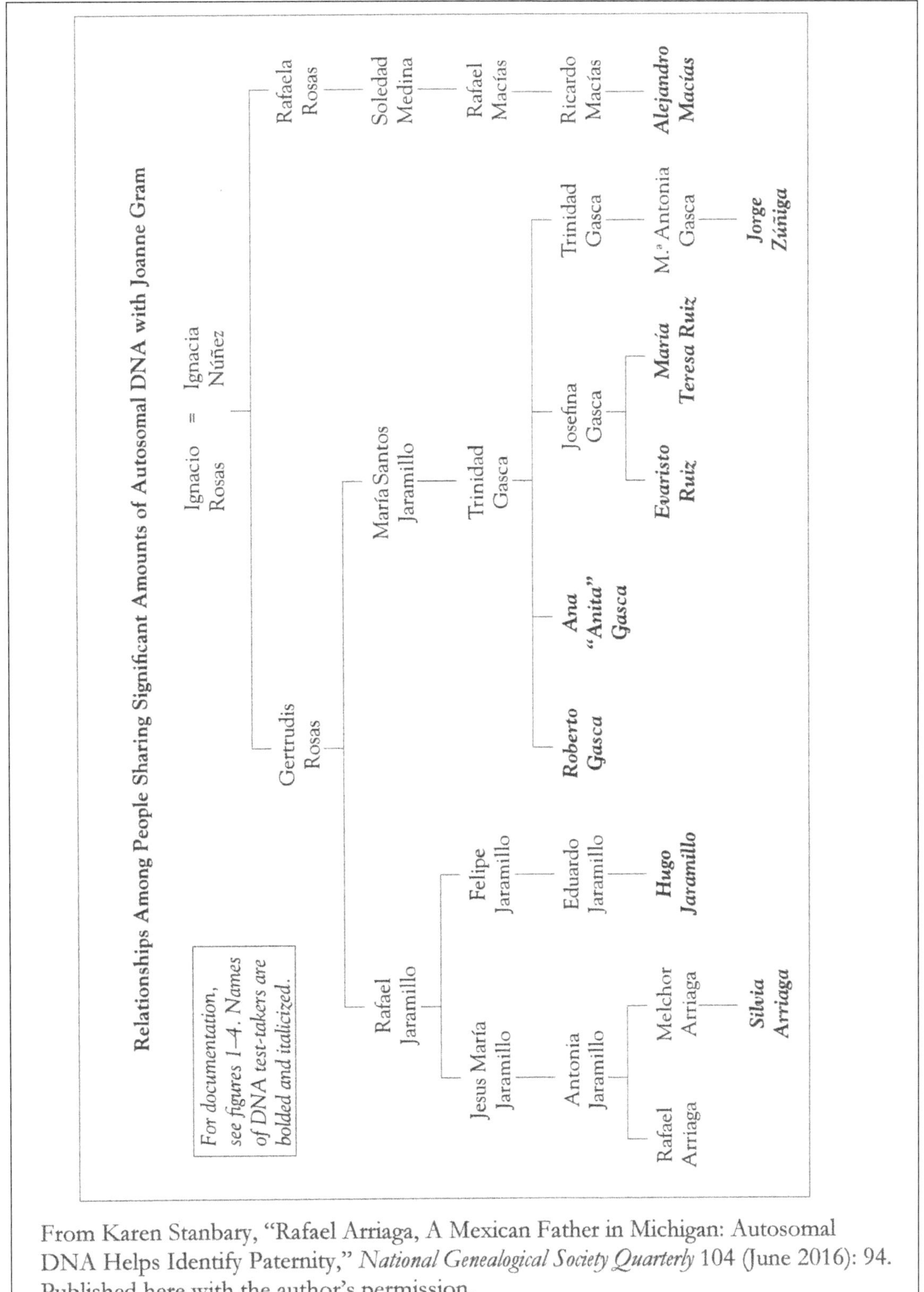

From Karen Stanbary, "Rafael Arriaga, A Mexican Father in Michigan: Autosomal DNA Helps Identify Paternity," *National Genealogical Society Quarterly* 104 (June 2016): 94. Published here with the author's permission.

Figure 11.7. Landscape-oriented chart documenting test-taker descents

Because genealogical charts are artwork, often for publication, they should be stable and editable. Proprietary software and genealogy database programs produce stable genealogical charts, but they are not easily editable and sometimes they do not meet the needs of readers of genealogists' writing. Those charts also would not meet the style and font requirements of most genealogy journals. They can be, however, easier to use during the research process and for less formal publication.[11]

Creating a genealogical chart for a formal context requires careful planning, but clear, logical, and attractive arrangements also result from trial-and-error. Researchers can create charts of professional quality in word-processing programs by using the insert-table feature and by darkening or inserting lines where needed. They can also use drawing programs, like Microsoft PowerPoint and Adobe Illustrator, or typesetting programs, like Abode InDesign and Microsoft Publisher. Converting to an editable PDF stabilizes the drawing for sharing with others. Converting to an image format also facilitates sharing but complicates editing—adding a new test-taker's data, for example, or changing fonts to match a publication's specifications.

Shared considerations for charts and tables

Genealogists document each relationship they assert. In genealogical charts, citations show the basis for the vertical lines connecting people of successive generations. Citations also document the connection between spouses or mates. Documentation similarly supports data in tables.

Writers usually create tables and figures as "stand-alone" elements with self-contained documentation, because they do not know exactly where an enhancement will appear in the written product until the work is ready for publication or sharing. At an early stage, the author will not know whether a citation in an enhancement will be a first (full) or subsequent (shortened) citation in the essay's overall context. Thus, the first or only citation in a table or figure has a full form, and a subsequent citation to that source in the same table or figure will have a shortened form. Also, complete documentation facilitates exporting a figure or table from one context to another, for example from a blog to a chapter or report, or from an article to a textbook.

Writers integrate tables and figures into their narrative. In formal publications, they refer to the enhancement or its content, with a callout in a logical place, like *See table 1*, or *figure 2 shows that. . . .* If documentation in an enhancement supports a writer's statement, the callout replaces a reference note, thus minimizing repetition of citations between enhancements and the narrative. Tables and figures can appear in electronic media as pop-ups at the callout. In print, the enhancement appears in the next available space after the callout, ideally on the same page or a facing page.

11. Examples include draw.io, *Flowchart Maker & Online Diagram Software* (https://www.draw.io/); *LucidChart* (https://www.lucidchart.com/); Microsoft Excel, Powerpoint, and Word SmartArt and drawing tools; Progeny Genealogy, Inc., *Charting Companion* (https://progenygenealogy.com/products/family-tree-charts.aspx); and *TreeDraw* (http://treedraw.spansoft.org/).

To avoid confusion with the narrative's reference-note numbers, writers conventionally use lowercase reference-note letters in enhancements, starting with letter *a* in each table or figure. Like reference-note numbers in the narrative, reference-note letters are superscripted in the enhancement.[12] The superscripted letters typically follow the information they document.

Printed tables and figures can be smaller than a page, with prose narrative flowing alongside or beneath them. They also might fill a page or two facing pages. Mid-page placement sometimes needlessly intrudes into flowing narrative. Placement at the bottom of the page can visually confuse the enhancement's documentation with the narrative's footnotes. Portrait orientation is the default, because readers in print and electronic media will not need to turn the page sideways to read it. Where data or an image is wider than the between-margins horizontal page width, landscape orientation is acceptable. Formatting, including typefaces and use of lines, should be consistent throughout a work's tables and figures.

In formal publications, like journal articles and book chapters, writers number tables (rows and columns) separately from figures (charts, images, maps, and the like). The first table in an article, blog, or chapter is table 1, and the first figure is figure 1. The consecutive numbering thereafter can result in a figure with a low number following a table with a higher number or vice versa.

Sharing information about living people

Genealogy and genetic genealogy standards and codes of ethics require permissions to publish information, including genetic data, about living people:

> I will keep confidential any personal or genealogical information given to me, unless I receive written consent to the contrary.[13]
>
> Genealogists respect all limitations on reviewing and sharing DNA test results imposed at the request of the tester [test taker]. For example, genealogists do not share or otherwise reveal DNA test results (beyond the tools offered by the testing company) or other personal information (name, address, or email) without the written or oral consent of the tester.[14]
>
> Genealogists share living test-takers' data only with written consent to share that data.[15]

Genealogists, therefore, request permissions from living people before disclosing their identifying information.[16] Besides DNA test takers, those living people include owners of family

12. *Chicago Manual of Style*, 17th ed., 755–56, sect. 14.24, "[Note] Numbers in text versus numbers in notes."
13. BCG, "The Genealogists Code," in *Genealogy Standards*, 2nd ed., 50.
14. *The Genetic Genealogy Standards* (http://www.geneticgenealogystandards.com/).
15. *Genealogy Standards*, 2nd ed., 32, standard 57, "Respect for Privacy Rights."
16. See "Ethical Underpinnings of Genetic Genealogy" chapter for more on permissions. Also, Debbie Parker Wayne, "DNA Analysis Consent Forms," *Deb's Delvings*, 28 September 2017 (http://debsdelvings.blogspot.com/2017/09/dna-analysis-consent-forms.html). Also, Blaine T. Bettinger, "Informed Consent Agreement

Bibles, photographs, cemetery lots, and other privately held artifacts, information, and records. Each of them has the right to give or withhold permission for someone else to use what they own or share information about it.

Authors have the option of anonymizing DNA test takers and private holders of documentary information. Extensive anonymization, however, risks creating unverifiable data, which could appear false. It also can prevent replication, a universal requirement for reputable research conclusions.

Obtaining permissions to share personal information in writing, therefore, is highly desirable. This is especially true for genealogists, whose work relies on relatively large numbers of living people—DNA test takers and, sometimes, their living relatives. If a few test takers deny permission to share their personal data, the written work's credibility will not suffer. If most of them deny permission, however, the written work likely would not be credible or professionally publishable. Widespread refusal, however, is rare.

When requesting permissions, authors exert no pressure. They tell the information-holder exactly what details they would like to share, the form and venues where the data will appear, and who the audience will be. They also explain any risks of sharing the information in writing. To the extent possible, genealogists also show test takers how their published data will appear. Regardless of whether permission is given or withheld, courtesy suggests offering the person a copy of the published work.

If a living person withholds permission, the author can use genetic data without personally identifying information. The unshared information includes names, locations, close relatives, personally identifiable genetic characteristics, and DNA kit numbers. When someone withholds permission to share identifying information, the genealogist should redact not only the person's identifying information but also parental information that could identify the person.

Discussion of anonymized DNA test results is least awkward when the writer gives an obvious pseudonym to each test taker who declined to allow sharing of personal information. False names that look like real names can confuse readers, especially when other test takers permitted sharing their names. Coded names based on test results or position in a family tree are hard to remember and follow, especially if the codes are similar or complex. Distinct and simple pseudonyms like *Test-taker 1* and *Descendant A* are less likely to confuse readers. In a Y-DNA study, the surname belonging to a multigenerational group does not identify a specific person. Especially when comparing surnames and Y-chromosome test results, the genealogist can refer to, and chart, permission-declining men with the group surname and a number, like *Jones1* and *Jones2*.

---

and Beneficiary Agreement," *The Genetic Genealogist*, 15 February 2018 (https://thegeneticgenealogist.com/2018/02/15/informed-consent-agreement-and-beneficiary-agreement/).

Documenting DNA Test Results

The *Chicago Manual of Style*, adopted by the genealogy field as its style guide, emphasizes the importance of documenting:

> Whether quoting, paraphrasing, or using others' ideas to advance their own arguments, authors should give explicit credit to the source of those words or ideas. This credit often takes the form of a formal citation.[17]
>
> Ethics, copyright laws, and courtesy to readers require authors to identify the sources of direct quotations or paraphrases and of any facts or opinions not generally known.... Source citations must always provide sufficient information either to lead readers directly to the sources consulted or ... to positively identify them.[18]

Genealogy Standards makes similar points:

> Genealogists use citations to identify the sources of all substantive information and images they gather, use, or plan to gather or use, except sources of "common knowledge" beyond dispute, such as the years of major historical events.[19]
>
> Documentation is fundamental to planning and executing genealogical research, collecting and recording data, and compiling research results.... The Genealogical Proof Standard requires complete and accurate citations to the source or sources of each information item supporting a claim that a conclusion is proved.[20]

Any writing about research is incomplete without documentation. Genealogists, therefore, use citations to document all sources they use, including those providing DNA data and comparisons, supporting the relevant relationship of DNA test takers and their ancestors, and showing the basis for their quotations, paraphrases, and ideas. Citations to those sources appear in research-based writing.

Basics

Genealogists use authoritative sources for their genetic and documentary data and interpretations. For event-based data, they cite, to the extent possible, *(1)* information from eyewitnesses and *(2)* records made soon after the events. For genetic data, they cite reputable testing companies and third-party tools. They also cite reputable genealogists, geneticists, genetic genealogists, and published works—online or in books and journals. Without such sourcing, research reports and other writing would be incomplete.

---

17. *The Chicago Manual of Style*, 17th ed., 708, sect. 13.3, "Giving credit and seeking permission."
18. Ibid., 743, sect. 14.1, "The purpose of source citations."
19. BCG, *Genealogy Standards*, 2nd ed., 5, standard 1, "Scope."
20. Ibid., 6, standard 3, "Purposes."

Genealogists, like researchers in any field, thoroughly and precisely document their work. To provide that documentation, they use citations in reference notes—preferably footnotes.[21] Along with the author's data and narrative explanations, reference-note citations show, primarily, that the author's reasoning and conclusions have a sound basis and that their research was thorough. Secondarily, reference-note citations enable researchers to consult sources and information that an author used.

For documentation to serve its purposes, it must be clear, complete, and usable. Citations that follow standard formats have those three characteristics.

Creating reference-note citations to genetic-genealogy sources

Reference-note citations typically answer three to five questions. As authors examine or use sources, they ask the questions[22]

> ***Who*** *created the source?* The answer typically is the source's author or authors (perhaps a private person, like an author or creator of a third-party tool) or an organization (like a DNA testing company or the *International Society of Genetic Genealogy*).
>
> ***What*** *is the source?* The answer usually is the source's formal title, capitalized like a headline and, if the source is a publication, italicized. The answer also can include the title of a part of a source that an author used—for example, an article within a journal, a chapter within a book, or a web page within a website. Titles of source parts typically are cited within quotation marks, capitalized as they appear in the source, and not italicized. The answer to *What* also includes any descriptive words that would help readers understand the source's qualities.
>
> ***When*** *was the source created or viewed?* This answer is the source's copyright, publication, or "last updated" date, if it has one. If the source is online, the answer is the author's most recent download or viewing date.
>
> ***Where is*** *the source?* If the source is a book, the answer is the publisher's city, state, and name. If a journal article, the answer is the issue's year and month or season of publication. If a web page, the answer is the least cumbersome and most specific stable URL.
>
> ***Where in*** *the source?* This answer focuses on the location within a source of specific information documenting an author's statement or data. If the source is numbered, like a page or digital image, the answer is the page or image number. If the image is on a website with multiple same-number images, the answer includes a path or way-

21. Genealogists follow "humanities-style citation principles" modeled by *Evidence Explained*, 3rd ed. rev., and *The Chicago Manual of Style*, 17th ed. See BCG, *Genealogy Standards*, 2nd ed., 8, standard 6, "Format."

22. BCG, *Genealogy Standards*, 2nd ed., 7–8, standard 5, "Citation Elements." Also, Thomas W. Jones, *Mastering Genealogical Documentation* (Arlington, Va.: National Genealogical Society, 2017).

points through nested levels to the specific numbered image. The answer also can be a URL for a specific page or image. It also can be a description of search terms, a location within a website, or the running of a DNA comparison or report. The author may include in the answer a description or summary of the information of interest.

Sample citations

Although genealogists use unpublished documentary records for part of their research, virtually all sources specific to genetic genealogy are publications—books, journals, and websites. The longstanding format for reference-note citations of books provides a pattern for citing most online sources, and therefore most genetic-genealogy sources:

> Blaine T. Bettinger and Debbie Parker Wayne, "Genetic Genealogy, Standards, And Ethics," *Genetic Genealogy in Practice,* NGS Special Publication No. 120 (Arlington, Va.: National Genealogical Society, 2016), 11–18.

That citation begins with the answer to *Who*—the coauthors' names. If either of them had a role in the chapter or book's preparation other than author—for example, compiler, editor, or translator—that descriptive information—abbreviated as comp., ed., or trans.—would follow the personal name. If both had the same role, the designation would follow both their names.

The citation continues with the answer to *What*—in this case the chapter title in quotation marks and the book title in italics. The chapter title and the descriptive words after the book title are optional—authors can add such words to make the citation more informative or specific. Other common possibilities for descriptive words, as applicable, include abbreviations like 2nd ed. (second edition), rev. (revised), and vol. 2 (volume two). Commas separate the chapter title, book title, and descriptive information, but no comma precedes the opening parenthesis.

The citation's publication details appear on the back of most books' title pages. Answering *Whereis* and *When,* they appear in the citation in parentheses. A colon conventionally separates the city and state from the publisher's name. A comma sets off the copyright year.

A comma after a citation's closing parenthesis sets off the answer to *Wherein.* For books, that answer usually is a page number or numbers. A description of information on the page is unconventional for a published book, but it is allowable when it helps readers understand the source, locate the information, or both.

As the preceding example shows, reference-note citations are capitalized and punctuated like sentences. They begin with capital letters and end with periods. Commas separate phrases, and parentheses set off supplementary information. Only proper nouns—book and chapter titles and names of people, places and publishers—are capitalized.

The pattern for citing a journal article resembles the pattern for citing a book:

> Karen Stanbary, "Rafael Arriaga, a Mexican Father in Michigan: Autosomal DNA Helps Identify Paternity," *National Genealogical Society Quarterly* 104 (June 2016): 85–98.

The content and placement of the answers to the *Who* and *What* citation questions resemble those in citations to books. Descriptive information after the journal title includes the volume number, not set off with punctuation in *Chicago* style. Citations to journals generally omit the parenthetical publisher name—the answer to *Whereis*—because journal titles typically include the publisher's name. The parenthetical information—including the date, month, or season pointing to a specific issue—answer the *When* citation question. The colon after the parenthetical matter is a longstanding convention for indicating volume and page numbers without using the words or abbreviations for *volume* and *page*. (For example, 12:34 is conventional shorthand for volume twelve, page thirty-four.) Those numbers answer the *Wherein* citation question.

Citations to online material closely reflect the conventional format for citing books:

> International Society of Genetic Genealogy (ISOGG), "Autosomal DNA Statistics," *ISOGG Wiki* (http://www.isogg.org/wiki/Autosomal_DNA_statistics : viewed 23 December 2018).

That example, like citations to books, answers the questions in the sequence *Who, What, Whereis,* and *When*. It omits a separate answer to *Wherein* because the URL, also answering *Whereis*, links to the specific web page. If an author wanted to point to specific data on that page, a description of that content or its place on the page could follow the closing parenthesis and a comma, like a page number in a book citation.

That same format applies to citing reports of DNA test results and comparisons:

1. "'One-to-many' matches," on-request reports, *GEDmatch: Tools for DNA & Genealogy Research* (https://www.gedmatch.com/ : 23 December 2018), atDNA match list for kit T829710.

2. "'One-to-one' Compare," on-request reports, *GEDmatch: Tools for DNA & Genealogy Research* (https://www.gedmatch.com/ : 23 December 2018), comparisons among kits T829710, T275027, M201030, and M115137.

3. "GEDmatch Chromosome Segment Matching," on-request report (https://www.gedmatch.com/chrom_match_4c.php?id=80802&kit_num=A451308 : compared 23 December 2018), matching atDNA segments for kits T829710, T538455, and T316184.

4. Jones Y-DNA Surname Project, "Jones Y-DNA–Y-DNA Colorized Chart," *Family TreeDNA* (https://www.familytreedna.com/public/jonessurname?iframe=ycolorized : viewed 23 December 2018), comparison of kits 202780 and 323269.

All four citations follow the basic book-citation format. Citations 1–3 omit a separate answer to *Who* (GEDmatch) because the information appears elsewhere in the citation. Citations 1 and 2 use the URL for the website because the URL for the comparison would link to an error message. The URLs for citations 3 and 4, however, when pasted into logged-in browsers, link directly to the referenced information.[23]

Citation forms

The reference-note citation examples above are full (or long-form) reference-note citations. Answering, as applicable, the five citation questions, they are formatted to document specific data, facts, and findings.

Genealogists also use a shortened (or short-form) reference-note citation format. It serves the same documentation purposes as full reference-note citations. When citing a source previously cited in full form in the same article, chapter, essay, figure, or table, genealogists use shortened citations.[24] In print media shortened citations conserve space, thus reducing paper and mailing costs. They are optional in electronic media.

Shortened reference-note citations typically answer, as applicable, only the *Who, What,* and *Wherein* citation questions. They shorten the answer to *Who* by using, for example, an author's surname only or a county without its state. They shorten the answer to *What* by using only the first few words of the title. The answer to *Wherein* can vary from the first, full citation. Shortened citations contain enough specifics to remind readers of the prior full citation:

> Bettinger and Wayne, "Genetic Genealogy, Standards, and Ethics," 15.
>
> Stanbary, "Rafael Arriaga," 87.
>
> ISOGG, "Autosomal DNA Statistics," *ISOGG Wiki.*
>
> "'One-to-many' matches," kit T829710.
>
> "'One-to-one' Compare," comparisons among kits T829710, T275027, and M201030.
>
> "GEDmatch Chromosome Segment Matching," kits T829710, T538455, and T316184.
>
> Jones Y-DNA Surname Project, "Jones Y-DNA – Y-DNA Colorized Chart," kits 202780 and 323269.

---

23. For more examples, see Elizabeth Shown Mills, *Quicksheet: Citing Genetic Sources for History Research,* laminated folder (Baltimore, Md.: Genealogical Publishing, 2015).

24. *Chicago Manual of Style*, 17th ed., 757–59, "Shortened Citations," sects. 14.29–14.33. Also, Mills, *Evidence Explained*, 3rd ed., 64–66, "Short Citations, Creating," sects. 2.43–2.45.

The third citation format that genealogists use is for source lists (also called reference lists or bibliographies).[25] The purpose of such lists is to tell readers the sources the author consulted or to provide students and others a list of resources related to a specific topic. Because those lists do not document specific facts, they usually omit the answer to the *Wherein* citation question. For the same reason, they often need not refer to specific articles, blog postings, chapters, pages, or other divisions within a source. Because source lists typically are alphabetized, a first or sole author's name appears last-name first.

Unlike full and shortened reference-note citations, which are formatted and capitalized like sentences, source-list citations conventionally are formatted and capitalized like paragraphs. The citation conventionally is indented with a hanging indent, and each element within the citation is formatted like a sentence, beginning with a capital letter and ending with a period:

> Bettinger, Blaine T. and Debbie Parker Wayne. "Genetic Genealogy, Standards, and Ethics." *Genetic Genealogy in Practice,* NGS Special Publication no. 120. Arlington, Va.: National Genealogical Society, 2016.
>
> Stanbary, Karen. "Rafael Arriaga, a Mexican Father in Michigan: Autosomal DNA Helps Identify Paternity." *National Genealogical Society Quarterly* 104 (June 2016).
>
> International Society of Genetic Genealogy (ISOGG). "Autosomal DNA Statistics." *ISOGG Wiki.* http://www.isogg.org/wiki/Autosomal_DNA_statistics : 2018.
>
> *GEDmatch: Tools for DNA & Genealogy Research.* https://www.gedmatch.com/ : 2018.
>
> Jones Y-DNA Surname Project. "Jones Y-DNA – Y-DNA Colorized Chart." *FamilyTreeDNA.* https://www.familytreesdna.com/public/jonessurname?iframe=ycolorized : 2018.

Authors can create shortened reference-note citations and reference-list citations by starting with full reference-note citations and subtracting the answers that do not apply. For source-list citations, they also rearrange the answer to the *Who* citation question for proper alphabetizing.

25. *Chicago Manual of Style*, 17th ed., 776–84, "Bibliographies," sects. 14.61–14.71. Also, Mills, *Evidence Explained*, 3rd ed., 43, sect. 2.4, "Citations, Types of."

Restricted-access citation issues

Outside the context of notes for a researcher's own use, citing sources that others cannot access has little value. Writing based solely on inaccessible sources is indistinguishable from fiction. The author who relies solely on inaccessible sources tells readers, in effect, "trust me"—an approach that raises doubts. Genealogists, therefore, avoid citing sources and DNA test results that others cannot consult—for example, a personal DNA match list on 23andMe, Ancestry, or FamilyTreeDNA; or comparisons and analyses of personal test results accessible only via one password.

By transferring as much of their restricted test results as they can, authors can bypass restricted-access issues. They upload results to venues where readers can access the data with their own passwords or with no password. Examples include *GEDmatch* and FamilyTreeDNA's surname projects.

Both of those sites require no subscription or fee for access, but fees are not the issue. Citing a source for which a reader must pay a fee is not ideal, but it is respectable. Researchers in reputable research fields routinely cite sources behind pay walls when they have no better alternative. Out of consideration to their readers, however, authors do try to minimize citing such sources.

In some cases, transferring all the relevant data to an accessible site is impossible. For example, genealogists regularly cite privately-held genealogical records with access restrictions. Similarly, a DNA tool unique to private data or outside a test-taker's privacy preference could prevent a genealogist from transferring data to an accessible venue. Transferring away from an inaccessible tool also could limit the data's meaningfulness. When transferring all relevant data is impossible, the author's goal is to make at least enough publicly available to ensure that the written work has credibility.

Besides transfer to publicly accessible venues, genealogists can present anonymized data. With permissions, they also can show anonymized screenshots of the results of analyses run on inaccessible data.

If readers, with or without paying a fee, can access a researcher's data, then citing the accessible venue is meaningful. Citing sources that readers can access gives credibility to written work. When readers can see the data, and that it supports the researcher's conclusions, they find the writer's conclusions believable. The most dubious readers, when the data is accessible, can replicate the author's reasoning and see that the conclusions are sound.

PUBLISHING AND SHARING DNA TEST RESULTS

Writers have two options for sharing their work: private sharing and public sharing.

Private sharing has risks, especially when the work is not in its final form. Even when written research results are labeled as "preliminary" or "tentative," readers can give the results more credibility than they merit. When a writer asks a recipient to keep shared information private, the recipient might not honor that request. Erroneous, incomplete, or unsupported conclusions can become widespread.

Writers, of course, can share finished products privately. In that circumstance, they run less risk of unintentionally distributing errors. The most common example of privately sharing a polished finished product is a research report written for a client, either for pay or pro bono.

Unlike sharing privately, sharing publicly is publishing. Writers who opt to publish make their research results available to the public. They or a publisher typically sell the work, but they also can distribute it for free. Either way, writers have two broad distribution options:

- *Self-publishing.* Authors distribute their written material themselves. When writers self-publish their research online, they typically manage access and publicity. If the self-published work is printed, for sale, or both, writers pay printing, shipping, publicity, and any sales and financial accounting costs. Examples of self-published works include blogs, monographs, newsletters, and books, including narrative family histories and skeletal genealogies.

- *Working with a publisher.* The writer enters into a contract with a party that oversees the work's printing, provides publicity, and handles sales, distribution, and any financial accounting. Examples of professionally published research works include books and, more often, journals. Book publishers typically pay the writer a royalty—a contract-specified percentage. Journal publishers usually do not pay genealogical authors. Instead they provide copies of the publication or a similar quid pro quo. Many academic journals charge authors to publish their research results.

Of the various publication options that researchers have, the most common—and the most respectable—is an article in a peer-reviewed research journal:

> Scholarly journal articles are original, *primary* publications. This means that they have not been previously published, that they contribute to the archive of scientific knowledge, and that they have been reviewed by a panel of peers.... To ensure the quality of each contribution—that the work is original, valid, and significant—scholars in the subspecialties of a field carefully review submitted manuscripts.[26]

26. *Publication Manual of the American Psychological Association*, electronic ed. (Washington, D.C.: American Psychological Association, 2017), sect. 8.01, "Peer Review." Italics are in the original.

Perhaps even more than a written work's documentation, peer review—vetting—establishes a written work's credibility. The evaluators are hand-picked researchers with expertise related to an author's research. After assessing a submission's research, reasoning, documentation, and writing, they make recommendations to publish or not publish the paper. If field reviewers believe the work has merit, they make recommendations for improving it.

Book authors or their publishers often seek the expert evaluation and validation that peer reviewers provide. Research journals, however, require it. Typically, those critiques are "masked" (when the author's identity is hidden from the reviewers) or "double blind" (when neither the author nor the reviewers know the others' identities). That anonymity prevents bias based on an author's reputation, whether abysmal, stellar, or in between. It also helps reviewers avoid repercussions from negative critiques.

Researchers who self-publish without seeking expert critiques take risk. Their work could contain omissions or errors that a field reviewer could helpfully point out. Even if the work were perfect, an uncritiqued article or blog posting would not receive the respect and credibility of an article published in a peer-reviewed research journal.

The genealogy field has several peer-reviewed research journals. They use editorial boards, hand-picked reviewers, or both to critique, accept, and reject submissions. The most highly respected of those journals include *The Genealogist, The National Genealogical Society Quarterly, The New England Historical and Genealogical Register, The New York Genealogical and Biographical Record,* and *The American Genealogist.*[27] Several of those journals have published articles reporting results of genetic genealogy research. For a listing, see the "Recommended Reading" section. Vetted journals with a greater genetic than genealogical focus include the *Journal of Genetic Genealogy* and the *Surname DNA Journal.*[28]

Conclusion

When genealogists, geneticists, and genetic genealogists use DNA test results to help establish genealogical conclusions, they are genealogical researchers. When they write about that research, they become scholarly writers. When their written work helps present-day and future researchers and members of the families that they have studied, they have met their research and writing goals.

---

27. For similar assessments, see David L. Greene, "Scholarly Genealogical Journals in America—Part One: Five National Journals," *The American Genealogist* 61 (July/October 1985): 116–20. Also, Eugene Aubrey Stratton, "The Scholarly Journals," *Applied Genealogy* (Salt Lake City: Utah, 1988), 29–39. Also, "Genealogical Writer's Market; Part I: The Five U.S. National Journals." *Association of Professional Genealogists Quarterly* 8 (September 1993): 61–63. Also, "Genealogical Writer's Market; Part II: The Five U.S. Regional Journals." *Association of Professional Genealogists Quarterly* 8 (December 1993): 93–95.

28. *JoGG: Journal of Genetic Genealogy* (http://jogg.info/). Also, *Surname DNA Journal* (http://www.surnamedna.com/).

Authors of scholarly genealogy works can choose to publish their writing themselves, or they can work with a book or journal publisher. Whatever their choice, they should aim for venues that will give credibility, longevity, and professionalism to their research findings and writing.

Using DNA test results to advance family histories bridges two disciplines, each with its own standards, terminology, and conventions. The genetic-genealogist's audience includes practitioners in either of those disciplines, practitioners in both, students of methodology, and laypeople interested in research results. For scholarly writing about genealogy to satisfy those audiences for generations to come, it must be exceptionally clear and precise. It must be structured and thoroughly and clearly documented. To be ethical and professional, it also must respect the rights of living people.

Successful genealogists consider and address their diverse readership. Like kissing, writing cannot be done alone. It requires readers.

Ethical Underpinnings of Genetic Genealogy

Judy G. Russell, JD, CG, CGL

An Overview of Genealogical Ethics

Ethical issues and privacy considerations are nothing new in genealogy. For as long as people have been researching the hidden details of the past, ethical researchers have recognized the potential for discoveries and disclosures that can impact family members. The birth of a couple's full-term first child some four months after the wedding. The incarceration of an ancestor for bootlegging or arrest for prostitution. The reality of enslavement or destitution. At a minimum, the fact that great-grandfather was not one of three brothers who came to America together and then, respectively, went north, south, and west. Or that great-great-grandmother actually was not a Cherokee princess. For these reasons, genealogical ethics codes have long imposed constraints to protect the living—balanced against the need for truth and accuracy in reporting results.

Ethical constraints for genealogists

Key among these constraints are those suggested by the National Genealogical Society in its "Guidelines for Sharing Information with Others" which provide, in relevant part, that ethical genealogical researchers

- respect the restrictions on sharing information that arise from the rights of another ... as a living private person; ...
- inform people who provide information about their families how it may be used, observing any conditions they impose and respecting any reservations they may express regarding the use of particular items;
- require evidence of consent before assuming that living people are agreeable to further sharing or publication of information about themselves;

- convey personal identifying information about living people—such as age, home address, genetic information, occupation, or activities—only in ways that those concerned have expressly agreed to;
- recognize that legal rights of privacy may limit the extent to which information from publicly available sources may be further used, disseminated, or published; ... and
- are sensitive to the hurt that information discovered or conclusions reached in the course of genealogical research may bring to other persons and consider that in deciding whether to share or publish such information and conclusions.[1]

Some ethical decisions are easy. Not posting information online about living people without consent is a simple ethical matter. So is returning every piece of paper in an archival file to the same place where we found it, no matter how much we might wish to take one record home with us.

But other decisions are far more difficult. Do we tell our cousins that their father's first marriage was not annulled until years after he married their mother? Do we disclose to those other cousins that their beloved grandfather was not a war hero as they were told, but rather in the stockade for the duration?

Clearly, making ethical decisions as a genealogist is often not a simple black-and-white matter. As genealogists, balancing our interest in the truth against our concern for the hurt that disclosures may cause brings us all too often into shades of gray.

Nowhere is that more true than when we try to make ethical decisions when DNA is included in the genealogical mix. When a DNA test discloses that the man who raised you, sent you to college, walked you down the aisle at your wedding, and dotes on his grandchildren is not in fact your biological parent, do you have an ethical obligation to tell him so? Does the answer change if he is eighty-five years old and frail as opposed to sixty years old and healthy? Does it change if he is now in the process of estate planning? What are your ethical obligations as the family genealogist and DNA expert when your grandmother's will specifies that an inheritance is to go to her grandchildren "share and share alike" and you—and you alone—know there is one more grandchild out there? When an adoptee contacts you because you are a DNA match and it is clear to you—but not to the adoptee—that the unknown parent is your first cousin, what is your obligation to the adoptee? To your cousin?

Making ethical decisions only gets more complicated when we try to get input from others. Those whose advice we may seek, in internet discussion groups, on special-interest mail lists, or on social media platforms such as *Facebook* or *Twitter*, are often not experts in law or ethics and, often, not even experts in genetic genealogy. The advice we receive may be driven by

All URLs were accessed on 30 September 2018 unless otherwise indicated.

1. "Guidelines for Sharing Information with Others," *National Genealogical Society* (https://www.ngsgenealogy.org/wp-content/uploads/NGS-Guidelines/Guidelines_SharingInfo2016-FINAL-30Sep2018.pdf).

a personal or group agenda, and it may have its origins in ethical values very different from those of the genealogical community as a whole.

All of these factors come into play when we try to determine what the ethical course is in confronting issues as test takers ourselves, as genealogists who ask others to test to advance our research interests, and as researchers who interact with third-parties based on evidence we glean from DNA testing.

Ethical Issues Affecting Test Takers

Individuals who take DNA tests face a variety of ethical issues and concerns, from ensuring their own informed consent for the tests to the sharing and use of their results with and by others. Confronting these issues in advance, planning for the unexpected result, and ensuring that those who test are as prepared as they can be for what the results may show will mitigate the consequences of these difficult decisions.

Informed consent

By definition, informed consent is "an agreement to do something or to allow something to happen, made with complete knowledge of all relevant facts, such as the risks involved or any available alternatives."[2]

In the DNA context, informed consent is at the center of genealogical ethics. A specific paragraph in Genetic Genealogy Standards reflects this ethical mandate. It provides, in relevant part "genealogists only obtain DNA for testing after receiving consent, written or oral, from the tester. In the case of a deceased individual, consent can be obtained from a legal representative. In the case of a minor, consent can be given by a parent or legal guardian of the minor. However, genealogists do not obtain DNA from someone who refuses to undergo testing."[3]

DNA testing in the absence of informed consent poses numerous ethical and even legal concerns. Every testing company takes steps to ensure that those who test, and those who submit samples on behalf of others, do so only where there is informed consent. In every case, the terms of service require consent. AncestryDNA states that each adult who tests "must create their own account (and) may also be required to explicitly consent to the processing of sensitive personal information when they activate their DNA kit."[4] At 23andMe, an account holder may only submit a personal saliva sample or a sample from someone for

2. Wex, *Legal Information Institute*, Cornell Law School (http://www.law.cornell.edu/wex), "informed consent."

3. Paragraph 2, 'Standards for Obtaining, Using, and Sharing Genetic Genealogy Test Results: Testing With Consent' in *Genetic Genealogy Standards* (http://www.geneticgenealogystandards.com/). Note that a footnote to this paragraph recognizes that exceptions may be made for situations where DNA testing is specifically mandated by law or court order.

4. Paragraph 1, Eligibility to Use the Services in "Ancestry Terms and Conditions," effective 5 June 2018, *Ancestry* (https://www.ancestry.com/cs/legal/termsandconditions).

whom the account holder has legal authority to agree.[5] A user at MyHeritage must represent that "any DNA sample you provide and any information that you transfer or upload that associates an individual with his/her DNA Results are either your DNA or the DNA of a person for whom you are a legal guardian or have obtained legal authorization to provide their DNA to us."[6]

At a minimum, informed consent requires a level of understanding of the nature of the tests, the limits of the information they can provide, and an understanding of the terms and conditions imposed by the testing company.

In all cases, informed consent also requires an understanding of the legal implications of DNA testing.

Finally, there are additional special issues that must be taken into consideration when the person being tested is incapable of providing informed consent.

In general

For truly informed consent, the test taker must understand what DNA tests can and cannot do. The test taker must also read, understand, and accept the terms of service of the testing company chosen.

These elements are set out explicitly in Genetic Genealogy Standards. Indeed, the Standards begin with a paragraph titled "Company Offerings" that provides: "Genealogists review and understand the different DNA testing products and tools offered by the available testing companies, and prior to testing determine which company or companies are capable of achieving the genealogist's goal(s)."[7] In a later section on "Standards for the Interpretation of Genetic Genealogy Test Results," there are additional standards for different types of tests,[8] the limits of what those tests can show,[9] the reality that "there is frequently more than one possible interpretation of DNA test results,"[10] and the simple fact that "no single piece of evidence, including evidence gathered from DNA testing, alone constitutes genealogical

5. Paragraph 3, Prerequisites in "Terms of Service," *23andMe* (https://www.23andme.com/about/tos/).

6. 'The Service: DNA Services' section, "Terms and Conditions," *MyHeritage* (https://www.myheritage.com/FP/Company/popup.php?p=terms_conditions).

7. Paragraph 1, 'Standards for Obtaining, Using, and Sharing Genetic Genealogy Test Results: Company Offerings,' *Genetic Genealogy Standards*.

8. Paragraph 13, 'Standards for the Interpretation of Genetic Genealogy Test Results: Different Types of Tests,' *Genetic Genealogy Standards*.

9. Paragraphs 15–18, 'Standards for the Interpretation of Genetic Genealogy Test Results: Different Types of Tests: Limitations of Y-DNA Testing,' 'Limitations of mtDNA Testing,' 'Limitations of Autosomal DNA Testing,' and 'Limitations of Ethnicity Analysis,' *Genetic Genealogy Standards*.

10. Paragraph 19, 'Standards for the Interpretation of Genetic Genealogy Test Results: Different Types of Tests: Interpretation of DNA Test Results,' *Genetic Genealogy Standards*.

proof."[11] These issues are more fully discussed in the "Drowning in DNA? The Genealogical Proof Standard Tosses a Lifeline" and "Correlating Documentary and DNA Evidence to Identify an Unknown Ancestor" chapters.

The Genetic Genealogy Standards also expressly impose an ethical requirement that "Genealogists review and understand the terms and conditions to which the tester consents when purchasing a DNA test."[12] This review and understanding is critical to informed consent, because company terms of service may greatly impact the information available about a DNA test or test taker. Terms of service are on a take-it-or-leave-it basis. Our choices are to accept the terms or not test with that company. A test cannot be processed without the account holder agreeing to accept the terms.

Some terms of service provisions are essentially uniform across the entire industry. For example, as noted above, every company requires consent of the person tested, or that person's legal representative, in order to process a test kit. The companies require that a test taker provides the company with permission to process the test sample and provide the results to the test taker.

Other terms of service are specific to one company. At 23andMe disputes with the company must be resolved by binding arbitration, rather than court action, and are subject to California law.[13] At FamilyTreeDNA the terms of service limit jurisdiction and venue over disputes to the courts of Texas.[14]

The terms of service control how much information and what kind of information is collected about each tested person and how that information can and cannot be shared by the company. In particular, the terms of service will specify whether any information can be disclosed by the company to third parties—and, if so, what information and in what form. AncestryDNA notes that it "may disclose user information in an aggregated form as part of the Services or our marketing, or in scientific publications published by us or our research partners. For example, we might note the percentage of immigrants in a State that are from a particular geographic region or country. Such publications will never include Personal Information."[15] At 23andMe the terms specify that it may share with third parties a test taker's "Aggregate Information, which is information that has been stripped of your name and con-

11. Paragraph 20, 'Standards for the Interpretation of Genetic Genealogy Test Results: Different Types of Tests: DNA as Part of Genealogical Proof,' *Genetic Genealogy Standards*.

12. Paragraph 5, 'Standards for Obtaining, Using, and Sharing Genetic Genealogy Test Results: Terms of Service,' *Genetic Genealogy Standards*.

13. Paragraph 28b, 'Miscellaneous: Applicable law and arbitration' in "Terms of Service," *23andMe* (https://www.23andme.com/about/tos/).

14. Paragraph 22, 'Applicable Law,' "Terms of Service," *FamilyTreeDNA* (https://www.familytreedna.com/legal/terms-of-service).

15. Paragraph 6, 'When We Share Your Information: A note about aggregated data' in "Ancestry Privacy Statement," *AncestryDNA* (https://www.ancestry.com/cs/legal/privacystatement).

tact information and combined with information of others so that you cannot reasonably be identified as an individual."[16] FamilyTreeDNA states that it does not share identifying DNA data with third parties without the express consent of the individual tested.[17]

Additional permissions may be requested, on a voluntary basis, for uses of our information beyond those needed for processing the tests. At 23andMe,[18] AncestryDNA,[19] and MyHeritageDNA[20] there are research projects for which they actively seek additional permissions. In each case, the permissions allow the company to use individual-level information, rather than aggregated data, for research and product development. The permissions are purely voluntary and may be withdrawn at any time. However, data already disclosed to a third party after permission was given cannot be recalled.

A person considering DNA testing, or asking another to test, should never simply click through terms of service without carefully reading the terms.

Legal implications of testing

Because of the nature of DNA testing, its legal implications must be understood thoroughly in order for informed consent to be given. These include both the potential misuse of DNA results to discriminate against test takers and the potential that genealogical test results may be used by or disclosed to police or law enforcement agencies.

Protection against the misuse of DNA-related data to discriminate against test takers is very much a matter of localized law: the legal protections governing direct-to-consumer DNA testing vary from jurisdiction to jurisdiction and country to country. In the United States, for example, the primary statute protecting the individual privacy of test takers is the Genetic Information Nondiscrimination Act (GINA).[21] In Canada, a similar genetic nondiscrimination act took effect in May 2017.[22] Both of these statutes protect individuals from having to disclose the results of genetic testing, in a limited fashion. Both offer protections in the workforce and, to some degree, with respect to insurance.

In the United States, GINA's provisions may be augmented by additional protections on a state level.[23] Privacy rules adopted in connection with the Health Insurance Portability and

16. Paragraph 4c, 'Information we Share with third parties: Aggregate Information' in "Full Privacy Statement," *23andMe* (https://www.23andme.com/about/privacy/#full-privacy-statement).

17. "FamilyTreeDNA Privacy Statement," *FamilyTreeDNA* (https://www.familytreedna.com/legal/privacy-statement).

18. "Research Consent Document," *23andMe* (https://www.23andme.com/about/consent/).

19. "AncestryDNA Informed Consent," effective 14 December 2017, *Ancestry* (https://www.ancestry.com/dna/en/legal/informedConsent/v3).

20. "DNA Informed Consent Agreement," updated 6 May 2018, *MyHeritage* (https://www.myheritage.com/dna-informed-consent-agreement).

21. "Genetic Information Nondiscrimination Act of 2008," 122 *Stat.* 881 (21 May 2008).

22. "Genetic Non-Discrimination Act," *S.C.* 2017, c. 3 (4 May 2017).

23. Alaska Statutes, §18.13.010, "Genetic testing." Also, §§16.12.1201- to -1208, "Genetic Information," Delaware Code. See also §§629.101-629.201, "Genetic Information," Nevada Revised Statutes.

Accountability Act of 1996 (HIPAA) also provide some protection to the privacy of test takers. The rule was broadened to define "genetic information" to include not merely our own tests but tests taken by family members as well.[24]

These protections, however, do not reach every person who might test under every circumstance. For example, only employers with fifteen or more employees are subject to the provisions of GINA in the United States.[25] Smaller companies are not covered by the act. The act also does not cover life, disability, or long-term health insurance.[26]

It is important to note, moreover, that anti-discrimination laws governing use of genetic data may very well change over time and there is no way to know or predict what those changes may be.

Another consideration is the potential that DNA-related data may be disclosed to the police or other law enforcement agencies. This issue had been simmering in the background since the advent of genealogical DNA testing. It leapt into the headlines in April 2018 with the revelation that DNA data from a genealogical database had been used by police in California to identify a suspect in what was called the Golden State Killer case. This is a decades-old serial murder-and-rape case.[27] The database was eventually identified as *GEDmatch*,[28] and the details of how and by whom the match was identified only disclosed much later.[29]

Some testing companies responded by clarifying their terms of service to prohibit the use of their DNA testing and matching services for law enforcement or forensic purposes.[30]

24. "Modifications to the HIPAA Privacy, Security, Enforcement, and Breach Notification Rules Under the Health Information Technology for Economic and Clinical Health Act and the Genetic Information Nondiscrimination Act; Other Modifications to the HIPAA Rules," 78 *Fed. Reg.* 5565 (25 January 2013), U.S. Government Printing Office (https://www.gpo.gov/fdsys/pkg/FR-2013-01-25/pdf/2013-01073.pdf).

25. "Facts About the Genetic Information Nondiscrimination Act," U.S. Equal Employment Opportunity Commission (https://www.eeoc.gov/eeoc/publications/fs-gina.cfm).

26. Sarah Zhang, "The Loopholes in the Law Prohibiting Genetic Discrimination," *The Atlantic*, 13 March 2017 (https://www.theatlantic.com/health/archive/2017/03/genetic-discrimination-law-gina/519216/). See also "What is genetic discrimination?," *Genetics Home Reference*, U.S. National Library of Medicine (https://ghr.nlm.nih.gov/primer/testing/discrimination).

27. Sam Stanton, Benjy Egel and Ryan Lillis, "Update: East Area Rapist suspect captured after DNA match, authorities say," *Sacramento (Calif.) Bee*, online edition, 25 April 2018 (https://www.sacbee.com/news/local/crime/article209779364.html).

28. Taylor Hatmaker, "DNA analysis site that led to the Golden State Killer issues a privacy warning to users," *Techcrunch.com*, 28 April 2018 (https://techcrunch.com/2018/04/27/golden-state-killer-gedmatch/).

29. Matthias Gafni, "Exclusive: The woman behind the scenes who helped capture the Golden State Killer," *Mercury (Calif.) News*, online edition, 24 Aug 2018 (https://www.mercurynews.com/2018/08/24/exclusive-the-woman-behind-the-scenes-who-helped-capture-the-golden-state-killer/).

30. See 'Use of 23andMe Personal Genetic Service for Law Enforcement Casework and Forensics' in "23andMe Guide for Law Enforcement," *23andMe* (https://www.23andme.com/law-enforcement-guide/). See also 'FamilyTreeDNA User Conduct - Unlawful and Prohibited Use of the Services,' "Terms of Service," *FamilyTreeDNA* (https://www.familytreedna.com/legal/terms-of-service). See also 'The Service: DNA Services,' "Terms and Conditions," *MyHeritage* (https://www.myheritage.com/FP/Company/popup.php?p=terms_conditions).

Despite those prohibitions, every testing company provides, in its terms of service, that it may disclose DNA results and identifying information if it is required to do so by law or court order:

- 23andMe states that "Personal Information may be subject to processing pursuant to laws, regulations, judicial or other government subpoenas, warrants, or orders" and it "may be required to disclose Personal Information in coordination with regulatory authorities in response to lawful requests by public authorities, including to meet national security or law enforcement requirements…"[31]

- AncestryDNA notes that it may disclose personal information if "reasonably necessary to … [c]omply with valid legal process (e.g., subpoenas, warrants)…"[32]

- FamilyTreeDNA states that it "may share your Personal Information if we believe it is reasonably necessary to … comply with a valid legal process (e.g., subpoenas, warrants)…"[33]

- MyHeritageDNA notes that it will make disclosures of personal information "if required by law, regulatory authorities, [or] legal process…"[34]

The number of cases in which data disclosure has actually occurred as the result of legal process or court order issued to a testing company is exceedingly small and many relate to credit card and other fraud.[35] By contrast, the law enforcement use of the third-party database *GEDmatch*—the database used in the Golden State Killer case—expanded dramatically after the site responded by changing its terms of service to expressly allow such use. The policy was revised on 20 May 2018 to allow uploading of data from "DNA obtained and authorized by law enforcement to either: (1) identify a perpetrator of a violent crime against another individual; or (2) identify remains of a deceased individual"; violent crime is defined as "homicide or sexual assault."[36]

---

31. Paragraph 4e, 'As required by law' in "Full Privacy Statement," *23andMe* (https://www.23andme.com/about/privacy/#full-privacy-statement).

32. Paragraph 7, 'When Do We Share Your Information and Who are the Recipients?' in "Your Privacy," *AncestryDNA* (https://www.ancestry.com/cs/legal/privacystatement).

33. Paragraph 5D, 'How FamilyTreeDNA shares your information: For Legal or Regulatory Process' in "FamilyTreeDNA Privacy Statement," *FamilyTreeDNA* (https://www.familytreedna.com/legal/privacy-statement).

34. 'Will MyHeritage disclose any of my personal information to third parties?' in "MyHeritage Privacy Policy," *MyHeritage* (https://www.myheritage.com/FP/Company/popup.php?p=privacy_policy).

35. See generally "Transparency Report," updated 15 May 2018, *23andMe* (https://www.23andme.com/transparency-report/). See also "Ancestry 2017 Transparency Report," *Ancestry* (https://www.ancestry.com/cs/transparency).

36. 'Raw DNA Data Provided to GEDmatch' in "Terms of Service and Privacy Policy," revised 20 May 2018, *GEDmatch* (https://www.gedmatch.com/tos.htm).

The methodology used to identify a suspect using *GEDmatch* is the same methodology genealogists use to identify their own matches or to identify biological parents and other family members in cases of adoption, donor conception, and more. It begins with an upload of raw DNA test data. For a genealogist, the data comes from a testing company. For the police, raw data equivalent to the testing companies' data can be created from crime scene DNA samples and then uploaded to the website. In both cases, a list of matches is generated from all other kits in the database. A genealogist then builds out the family trees of the matches to determine where the source of the uploaded suspect-DNA fits into the family trees of the matches. That analysis assists police in identifying potential suspects of the right age and gender who may have been in the area at the time of the crime. The police then obtain a sample from the suspect to compare to the crime scene sample.

By the fall of 2018, more than a dozen arrests had been announced by law enforcement agencies throughout the United States based on suspects identified through that methodology and the *GEDmatch* database. One Virginia company, Parabon Nanolabs, has established a service called Snapshot under the direction of genetic genealogists expressly for law enforcement agencies to use genealogical DNA testing for criminal investigations.[37] By the end of 2018, the genealogical testing company FamilyTreeDNA had followed *GEDmatch*'s lead and changed its terms of use to permit law enforcement use of testing and matching services for a wider variety of crimes.[38]

Ethical issues underlie this use of the *GEDmatch* and FamilyTreeDNA databases. Those issues include the lack of informed consent for this use of the databases, the potential for expansion of law enforcement use beyond the specified crimes, and the potential for misidentification by untrained or unskilled researchers. First, many test takers uploaded their data to the databases before law enforcement uses were allowed by the terms of service and had no notice that their results might be used now and in the future to identify crime suspects among their relatives, even their own descendants in generations to come. More than a few uploaders were deceased by the time the terms of use provided for law enforcement use and thus cannot be contacted for consent. Second, although the terms of use state that the databases may be used only in cases of violent crimes, FamilyTreeDNA had adopted the definition of a federal statute that includes a wide variety of crimes in which no actual injury occurred[39] and *GEDmatch* candidly admits that it is "unable to guarantee that users will not find other ... new genealogical and non-genealogical uses."[40] Finally, untrained or unskilled researchers may misidentify individuals as potential suspects who will then be subjected to unwarranted law enforcement scrutiny.

---

37. "Parabon Snapshot, Advanced DNA Analysis," *Parabon Nanolabs* (https://snapshot.parabon-nanolabs.com/).

38. Paragraph 6B(xii), 'Your Use of the Services: Requirements for Using the Services' in "Terms of Service," updated December 2018, *FamilyTreeDNA* (https://www.familytreedna.com/legal/terms-of-service). During early 2019 revision of this section was under discussion in online public forums.

39. See 18 United States Code § 924(e)(2)(B).

40. "Terms of Service and Privacy Policy," revised 20 May 2018, *GEDmatch* (https://www.gedmatch.com/tos.htm).

Going forward, then, informed consent requires understanding and acceptance of the type of data access that the companies may provide to police and other law enforcement agencies under court order or other legal mandate. In cases like that of FamilyTreeDNA and *GEDmatch*, and other companies that may adopt this practice, it also requires understanding of and express acceptance of the use of uploaded data by law enforcement for investigative purposes. This is an area where changes will likely occur rapidly as companies respond to customer demands.

Special issues

There are special issues that must be taken into consideration in cases where the person tested is incapable of giving individualized informed consent. These include children, incompetent adults, and postmortem cases.

In general, a parent or legal guardian may consent to DNA testing for his or her minor child. Actual legal authority is required. A grandparent generally does not have legal authority to consent to the testing of a grandchild, unless the grandparent has legal custody of the child. A noncustodial parent should consult with the custodial parent before testing a minor child.

Some of the testing companies require a representation that the minor has agreed to the testing. AncestryDNA limits DNA testing to persons thirteen years and older, and requires the parent or legal representative to expressly state that the test has been discussed with the minor "and the minor has agreed to the collection and processing of their saliva."[41]

In the case of an incompetent adult, legal authority to consent to testing generally rests with the person's legal representative, which may include a spouse, an adult child, or someone with either or both power of attorney and medical power of attorney. The same is true in postmortem cases: legal authority to consent rests with the deceased's legal representative, which, again, may include a spouse, an adult child or children, or with the executor or administrator of the deceased's estate.

From an ethical perspective, however, both as to incompetent adults and in postmortem situations, there may well be a distinction between what is legally permitted and what is ethically right. In these cases, consideration must be given to the expressed objections or reservations of the individual during his or her lifetime. While the legally-recognized privacy rights of the individual may be compromised by incompetence or terminated by death, the ethical decision-making should take into account the expressed desires of the individual and only in the most compelling fact-sensitive situations will the ethical decision be to override the wishes of the individual.

41. Paragraph 1, 'Eligibility to Use the Services' in "Ancestry Terms and Conditions," effective 5 June 2018, *Ancestry* (https://www.ancestry.com/cs/legal/termsandconditions).

Sharing results

One of the most serious, significant, and potentially life-changing ethical lapses in the area of genealogy is the careless sharing of test result data without having received specific consent of the person whose data is being shared. Test takers receive access to much information that relates to other people. They get the names of their matches, information about haplogroups and ethnicity of those matches, and in some cases, even specifics about health-related and other markers. Sharing this information with anyone other than the match himself or herself can have enormous implications and is a serious ethical lapse.

The simple reality of modern mass communications is that, once information is shared, there is no control over how it may be shared further. Once information becomes available online, it may very well be available online forever. Trying to sanitize DNA results by stripping off personally-identifying information or even using an alias is no guarantee that the information will remain private. Stated simply: "complete anonymity of DNA tests results can never be guaranteed."[42] There have already been documented studies showing how anonymized data can be re-identified with a specific test taker by specialized researchers.[43]

At the same time, DNA data carries with it the potential for life-changing disclosures. DNA test results often disclose unexpected information. A test taker may discover previously unknown relatives or previously unknown ethnicity. DNA testing may establish that a purported biological parent or grandparent has no biological relation to a child. It may disclose that an individual carries the potential or even the certainty for developing a dread disease.[44] As more fully explored in the "Uncovering Family Secrets: The Human Side of DNA Testing" chapter, the risk of unexpected results is high enough that the Genetic Genealogy Standards explicitly caution:

> Genealogists understand that DNA test results, like traditional genealogical records, can reveal unexpected information about the tester and his or her immediate family, ancestors, and/or descendants. For example, both DNA test results and traditional genealogical records can reveal misattributed parentage, adoption, health information, previously unknown family members, and errors in well-researched family trees, among other unexpected outcomes.[45]

That caution is underscored by the testing companies that offer specific warnings to test takers. For example, 23andMe cautions that "You may learn information about yourself that

42. See Paragraph 6, 'Standards for Obtaining, Using, and Sharing Genetic Genealogy Test Results: Privacy,' *Genetic Genealogy Standards.*

43. Erika Check Hayden, "Privacy protections: The genome hacker," *Nature* 497: 7448, 8 May 2013 (https://www.nature.com/news/privacy-protections-the-genome-hacker-1.12940).

44. Leah LaPerle Larkin, "Cystic Fibrosis: A Case Study in Genetic Privacy," *The DNA Geek*, 20 March 2017 (http://thednageek.com/cystic-fibrosis-a-case-study-in-genetic-privacy/).

45. Paragraph 12, 'Standards for the Interpretation of Genetic Genealogy Test Results: Unexpected Results,' *Genetic Genealogy Standards.*

you do not anticipate. This information may evoke strong emotions and has the potential to alter your life and worldview. You may discover things about yourself that trouble you and that you may not have the ability to control or change (e.g., your father is not genetically your father, surprising facts related to your ancestry, or that someone with your genotype may have a higher than average chance of developing a specific condition or disease). These outcomes could have social, legal, or economic implications."[46] AncestryDNA warns: "You may discover unanticipated facts about yourself or your family when using our Services that you may not have the ability to change (e.g. you may discover an unknown genetic sibling or parent, surprising facts about your ethnicity, or unexpected information in public records)."[47]

Information disclosure, and the risk of disclosing unexpected results, is not limited to the person taking the test. Genetic genealogy testing and its matching systems cannot work without disclosing some of our information to those persons we match and without disclosing their information to us. That will inevitably also pose a risk of disclosing information about their family members—and our own. Availability of third-party sites and tools to analyze DNA test results adds to the risk of disclosure.

For genealogists, therefore, the price of genealogically-useful information is the risk involved in allowing matches access to some data for matching purposes. The ethical mandate that attaches to that access is that information about another person should never be disclosed to any third person without the test subject's express consent.

Disclosing information about matches

The ethical imperative to do no harm to living persons is at the heart of all genealogical ethics. Protecting the privacy interests of living people is reflected in every genealogical ethics code. From the information-sharing guidelines of the National Genealogical Society[48] through the ethics codes of certifying and accrediting organizations,[49] a wide variety of genealogical organizations have addressed this issue and speak with one voice: we protect the living.

This mandate is also reflected in the Genetic Genealogy Standards, which dictate that "Genealogists respect all limitations on reviewing and sharing DNA test results imposed at the

46. Paragraph 5, 'Risks and Considerations Regarding 23andMe Services' in "Terms of Service," *23andMe* (https://www.23andme.com/about/tos/).

47. 'Important Things for You to Understand When You Use Our Services' in "Ancestry Terms and Conditions," effective 5 June 2018, *Ancestry* (https://www.ancestry.com/cs/legal/termsandconditions).

48. "Guidelines for Sharing Information with Others," National Genealogical Society (https://www.ngsgenealogy.org/wp-content/uploads/NGS-Guidelines/Guidelines_SharingInfo2016-FINAL-30Sep2018.pdf).

49. "Genealogist's Code of Ethics," Board for Certification of Genealogists (https://bcgcertification.org/ethics/code/). See also "Code of Ethics," International Commission of the Accreditation of Professional Genealogists (https://www.icapgen.org/professional-ethics/code-of-ethics/).

request of the tester. For example, genealogists do not share or otherwise reveal DNA test results (beyond the tools offered by the testing company) or other personal information (name, address, or email) without the written or oral consent of the tester."[50]

This mandate is often at odds with our enthusiasm and desire to share our genealogical findings with others. We find information on a testing company website and want to share it with our family members. We decide to upload our data to a third-party site for analysis and sharing and gain even more information about others that we want to share. We grab a screen capture of our results compared to our matches to send by email or post online. Or we use screen captures to try to teach others about the joys of using DNA in genealogical research.

In every one of these cases, sharing the information of persons other than ourselves is an ethical violation if we do not have consent. Despite the value of DNA to our research, the ethical bottom line remains: information with respect to our matches should not be disclosed to any third person without the consent of the match.

Company sharing systems

The sharing systems provided by testing companies make information about our matches available to us. The extent of the information made available is not always clear at the time the test is taken. For that reason, each person tested who consents to receive matching information must understand the potential that significant amounts of data may be disclosed to matches. Our matches may receive our full names and email addresses along with some portion of our DNA information. They may be able to see our family trees. Only by opting out of matching completely can a person be tested without the risk of having any information revealed to others. That, of course, eliminates genealogical uses of the DNA data.

Yet while every testing company makes information available in its matching system, it does so only to and about those we match. Putting ourselves into a matching system, then, serves only as consent to disclose information to—and receive information from—our matches. The fact that we have tested with company A is not a grant of general consent to disclosure of our information to every other person who has tested with company A. Nor is it a grant of consent by our matches for their information to be disclosed to any third party, not even our family members. Nor is it a grant of consent for our matches to disclose our information to any third party, including their family members.

Whenever information from a company sharing system is to be disclosed to any person other than a match, express consent should be requested and received, either orally or in writing. The request should be specific and state precisely what use will be made of the information. Any limitations or restrictions placed on sharing of information by the match must be scrupulously honored.

50. Paragraph 8, 'Standards for Obtaining, Using, and Sharing Genetic Genealogy Test Results: Sharing Results,' *Genetic Genealogy Standards*.

If we wish to make use of a screen capture of data from a company sharing system, all identifying information with respect to a match should be eliminated. At a minimum, that means using an imaging program to blur the photographs, screen names, real names, addresses, and any other identifying data of our matches.

Third-party access and tools

A wide variety of third-party data sharing and analysis sites make understanding and using DNA evidence in genealogy easier. But they also pose ethical considerations beyond those involved in testing company websites. Particularly because most third-party tools and third-party DNA data sites do not have the same degree of privacy controls built into their systems as the testing companies do, sharing of information to or from those sites requires careful analysis of the ethical underpinnings.

Just as one example, the popular DNA data sharing and analysis site *GEDmatch* allows those who have tested with one company to upload their data and find matches who may have tested with other companies. Registration is open to anyone, and does not require that the registered user have DNA data uploaded to the site. Once registered, any user of the site can see who the matches are for any *GEDmatch* kit where the number is known and can see the kit numbers for all of those matches. Any of those kit numbers can then be used to run any of the utilities on that site. That includes, just as one example, the "Are Your Parents Related?" tool that can easily disclose whether the test taker is the product of incest.

Thus, uploading data to a site like *GEDmatch* poses a risk of disclosing information some would prefer to keep to themselves, and sharing one's own *GEDmatch* kit number poses a risk as well. Sharing anyone else's *GEDmatch* kit number without the express consent of the kit owner violates the privacy rights of that person.

Moreover, the fact that we may manage or have access to another's DNA test results on a company's website does not by itself constitute authority to upload that person's data to any other website. Access to another's test data at AncestryDNA, for example, by itself does not give us consent to transfer that data to another testing company or to upload it for use in one of the third-party tools.

Finally, an individual's use of a third-party site or tool does not generally constitute a general grant of consent to the use of the DNA information outside of that site or tool or its disclosure to others. There are exceptions—some sites clearly state that any use of the site is entirely public. The Personal Genome Project for example makes it clear that data uploaded to that site for scientific research is public and will be available in a publicly-accessible format. It adds "neither anonymity nor confidentiality of participant identities or their data are promised."[51]

51. *The Personal Genome Project* (http://www.personalgenomes.org/us).

Ethical Issues Affecting Test Requesters

There are additional ethical considerations that come into play whenever we ask another person to take a test on our behalf or to allow us to use any portion of their genetic data. At a minimum, we must ensure that the person testing is giving informed consent to the test and to our use of the DNA test data. We must scrupulously honor any limits or restrictions another person places on agreement to test. Particularly in an adoption or donor conception situation, we must explain truthfully our reasons for asking for the test, even if the truth about those reasons might jeopardize the chances of securing agreement to the test. Finally, we must be aware of the ethical issues underpinning the ownership and control of DNA data when disputes arise between test-taker and test-manager and with respect to ownership and control after a test taker's death.

Asking another to test: ensuring informed consent

Clearly the obligation to ensure that informed consent is given for testing rests principally and primarily with the person tested. However, when the test is requested and arranged by another person, that other person—generally a genealogist—has his or her own ethical obligation to ensure that the test taker is actually informed and consenting to testing.

This obligation begins with telling the truth. We must be candid about what the test could show, including the risk of unexpected results. We must honestly answer any questions the person has and not mislead the person about the potential for privacy issues simply because we are afraid that the person will say no. We simply cannot secure informed consent if we misrepresent what we are asking the other person to do or what the risks are.

The best way of securing and ensuring informed consent on the part of a person we want to test for our benefit is to provide an informed consent form for that person to sign indicating voluntary agreement to test and allowing access to the results. Authorization for our access to the test results should be express and clear and set off in a separate paragraph that the person can initial. That provision may include how long access will continue—whether for a limited time or for our lifetime. It may also address the method by which access permission may be revoked.

Among the issues that should be addressed in an informed consent form are

- That the test taker has read and understands the terms of service of the testing service chosen for the DNA test
- The degree of access the test taker wants to the results
- Whether the test taker wishes to use a pseudonym or real name
- Whether the test taker wants to know if there are unexpected results from DNA testing

- Whether the test taker allows the test data to be transferred to any other testing company and if so which company

- Whether the test taker allows the test data to be used at any third-party tool or database and if so which tool or database

- Whether any additional use can be made of the test results and data, such as sharing with family members, inclusion in genealogical research for certification or accreditation, publication in books or journal articles, or lecturing

The consent form should include

- A place for the individual to specifically initial a paragraph indicating that he or she has read the Genetic Genealogy Standards

- A paragraph indicating understanding that DNA testing can have unexpected results ranging from previously unknown relatives to misattributed parentage

A sample informed consent form developed by genealogist Blaine T. Bettinger[52] is available for free use under a Creative Commons license.[53] The license allows the form to be adapted for individual use. One change that should be considered is the inclusion of an express reference to the testing company's terms and conditions. As noted above, those terms of service are on a take-it-or-leave-it basis; the only choices are to accept the terms or not test with that company. If a person we are asking to test for our benefit does not wish to accept the terms, that test cannot go forward. There is no ethical option to by-pass the decision when a person does not agree to the terms.

If the person elects not to test or to limit the use of the test results, we must scrupulously honor the decision. We can try to persuade the individual to give consent, or broader consent, but cannot allow our enthusiasm for DNA evidence to draw us across the line into bullying conduct.[54] Even if this person is the last possible candidate for a particular Y-DNA line, or the only living descendant of a maternal-line ancestor who could have the mitochondrial DNA we need to break down a brick wall, we have to accept that "no" means "no."

---

52. Blaine T. Bettinger, "Informed Consent Agreement and Beneficiary Agreement," *The Genetic Genealogist*, 15 February 2018 (https://thegeneticgenealogist.com/2018/02/15/informed-consent-agreement-and-beneficiary-agreement/).

53. For an explanation of Creative Commons licenses, see "About The Licenses," *Creative Commons* (https://creativecommons.org/licenses/).

54. Roberta Estes, "No (DNA) Bullying," *DNAeXplained — Genetic Genealogy*, 15 May 2013 (https://dna-explained.com/2013/05/15/no-dna-bullying/).

Special issues

Additional issues come into play where the person we wish to test is not or may not be personally competent to agree to testing, or when the test is sought in a case of unknown parentage, such as an adoption or donor conception case.

Competence

Where an individual we wish to test is clearly incapable of giving informed consent due to incompetence, our ethical obligation is clear: we simply cannot take a test sample and have it tested any more than we could take a DNA sample from someone else's child. As a matter of law, ethics, and the contractual representations required by the testing companies, we cannot proceed with the testing without securing the authority of the person serving as legal representative of the incompetent person.

It is much more difficult where the question of the person's competence or incompetence is not clear. An elderly relative could well have what we would consider to be good days and bad days, agreeing to our requests on good days and objecting to them on bad days. Assessing competence in such a circumstance clearly falls into the "shades of gray" ethical category.

There are general standards in use in the medical and legal fields for assessing decisional capacity. In general, four criteria may be used:

1. Did the person actually express a choice or a preference?
2. Did the person show understanding of what that choice meant by using words, phrases, ideas, and information appropriate to a decision?
3. Did the person use logical processes to consider benefits and risks?
4. Did the person show an appreciation of those risks and benefits including the potential emotional impact and consequences?[55]

In the final analysis, when we arrange for a test in such circumstances, we have an ethical obligation to act in the best interests of the person being tested. If we have doubts, we should not proceed without additional legal authority even if that means we are unable to get the test we want.

Adoption and donor cases

Our desire to have an individual test may have its most consequential implications for ethical decision-making in cases where DNA testing is sought in order to identify the unknown parent of an adoptee or the donor in a donor conception case. In many cases, it may appear that

55. "Policy and Procedures for Assessing Capacity To Consent for Research: Standards for Capacity," 10 July 2002, UC Davis Alzheimer's Disease Center (https://www.ucdmc.ucdavis.edu/clinicaltrials /StudyTools/ Documents/ResearchCapacityPolicy7=02.pdf).

our chances of securing agreement to a DNA test will be severely impacted if the person we have approached is informed that the reason we are asking for the test is to try to solve a case of unknown parentage. It is not at all clear, from a legal perspective, whether our failure to disclose the specific reason behind a test request would negate or invalidate informed consent, as long as the risks of testing in general were disclosed.

From an ethical perspective, however, the rules are clear: we have an ethical obligation to "inform people who provide information about their families how it may be used."[56] That is inconsistent with hiding our purpose in asking someone to test. In an unknown parentage or donor conception case, failing to disclose or downplaying the risk that DNA testing would identify the person tested as a close relative or even the parent does not inform that test taker how the information may be used.

It may be true that telling the truth behind the testing request may mean the test is not performed. That may be the ethical price that has to be paid.

Ownership of test samples and data

Ethical considerations also come into play in the ownership and control of the test sample and the data that results from DNA testing. When a DNA test is paid for or when it is requested by someone other than the person tested, conflicts may arise between the person tested and the person on whose behalf the test was done. There are also issues that come into play when the tested person dies without establishing a beneficiary for the DNA sample and results.

Even when a DNA test is paid for by someone other than the person tested, the test sample itself and the data resulting from the test belong to the individual tested. Ethically, that person may not be denied access to the results. The Genetic Genealogy Standards are specific on this point: "Genealogists believe that testers have an inalienable right to their own DNA test results and raw data, even if someone other than the tester purchased the DNA test."[57]

Issues can arise between the person who tested and the person who paid for the test, with the result that the test taker wishes to terminate the access of the payor to the results. Only one company ever addressed this issue and has since removed that provision from its terms of service.[58] Thus, those who pay for tests to be taken by others need to address this with the test takers—in advance and in writing.

---

56. "Guidelines for Sharing Information with Others," *National Genealogical Society* (https://www.ngsgenealogy.org/wp-content/uploads/NGS-Guidelines/Guidelines_SharingInfo2016-FINAL-30Sep2018.pdf).

57. Paragraph 3, 'Standards for Obtaining, Using, and Sharing Genetic Genealogy Test Results: Raw Data,' *Genetic Genealogy Standards.*

58. FamilyTreeDNA once had a provision in its terms of service requiring a test taker to reimburse the payor before terminating access. See "Gene by Gene Privacy Policy: Privacy and Confidentiality Specific to Family Tree DNA-3rd Party Payments and Account Access," May 2018, *FamilyTreeDNA*, via Wayback Machine (https://web.archive.org/web/20180503150303/https://www.familytreedna.com/privacy-policy). That provision is not in the current terms.

When a test is merely requested by a third person, but paid for by the person tested, ownership of the sample and control of data access remains entirely with the individual tested. Any access granted to any third person can be revoked at any time for any reason.

Two additional considerations come into play when the person tested dies without designating a beneficiary to own and control the DNA sample and results. Both the right to consent to additional tests and the right to access existing results are at stake in this circumstance.

The problem is that—without a beneficiary designation—ownership and control of a test sample and related DNA data after the death of the individual tested is not legally clear. The sample and information may be regarded as personal property to be passed under the residuary clause of the will of the deceased or by way of the intestate laws of the jurisdiction where the deceased resided. In either case, that may sever the legal authority of the account manager to order additional tests or even access the existing results. At a minimum, it may greatly complicate the ownership and control of the DNA data by fragmenting decision-making authority among many people.

The Genetic Genealogy Standards recommend that all genealogists "designate a beneficiary to manage test results and/or stored DNA in the event of their death or incapacitation."[59] As of the time of writing this chapter, FamilyTreeDNA was the only company that had a system for designating a beneficiary directly on its website. However, its beneficiary form can be adapted for use with tests from other companies. Sample forms prepared by genealogists such as Debbie Parker Wayne[60] and Blaine T. Bettinger[61] and available under Creative Commons licenses[62] may be used for this purpose.

Any form used should include, at a minimum, a clear identification of the person or persons authorized to act with respect to the DNA testing sample, results, and data when the test taker is incapacitated or deceased. It should include the authority to perform all of the acts with respect to the DNA test that the person tested could have taken on his or her own. It may include specific provisions for accessing the data, analyzing and downloading test results, uploading results to third-party websites or tools, and using any of the DNA related information for publication or lecturing either with or without an alias used for the test takers identity. The form must be signed by the test taker and, for the strongest chance of having the document recognized in court should that become necessary, the signature should be witnessed and may even be notarized.

59. Paragraph 11, 'Standards for Obtaining, Using, and Sharing Genetic Genealogy Test Results: Designating a Beneficiary,' *Genetic Genealogy Standards*.

60. Debbie Parker Wayne, "Permission Form and Designation of Beneficiary," 2017 (http://debbiewayne.com/pubs/PERM_beneficiary_form_blank.pdf).

61. Blaine T. Bettinger, "Informed Consent Agreement and Beneficiary Agreement," *The Genetic Genealogist*, 15 February 2018 (https://thegeneticgenealogist.com/2018/02/15/informed-consent-agreement-and-beneficiary-agreement/).

62. For an explanation of Creative Commons licenses, see "About The Licenses," *Creative Commons* (https://creativecommons.org/licenses/).

Where a beneficiary has been named, that beneficiary generally will have the legal and ethical authority to consent to additional testing and to access or grant third-party access to the testing data.

In the absence of a beneficiary designation, however, additional testing of the sample may be conducted ethically only with the informed consent of the legal representative of the deceased. Access to the test data may similarly be controlled by the legal representative, and not the individual who previously managed the account.

One open legal question with respect to the designation of a beneficiary is the ability to name a contingent beneficiary in case the original person named dies or is incapacitated. From a legal and ethical perspective, there is no reason why the beneficiary designation could not be in favor of an entity as opposed to an individual, such as "the Project administrator of the Baker DNA project." Then as long as the Baker DNA project continued in existence and had an administrator, that administrator could continue in the role of beneficiary of that account. Such a designation, however, has not been tested in any court. Alternatively, the beneficiary form itself could grant to the beneficiary the express right to name a successor beneficiary. That right, as well, has not been tested.

Despite these unresolved issues, from both a legal and an ethical perspective, having an estate plan for our own tests and for any tests performed on others at our expense or request that provides for a specific beneficiary for DNA-related matters is strongly advised.

Ethical Issues Affecting Third Parties

In many instances, genealogical research utilizing DNA results cannot succeed without input from persons who have not tested. Contact with persons who have not tested raises ethical concerns, particularly for the privacy of those who have tested. Moreover, DNA testing can disclose information not just about the persons who test but also about their family members, including health-related information. What are our ethical obligations to protect, or to disclose, such information? These are among the most difficult and perplexing ethical questions that we face as genealogists—and there are no clear right or wrong answers.

Adoption and donor cases

Contact with those who have not tested in adoption and donor conception cases may be essential to solving these unknown parentage cases. Yet they pose ethical dilemmas that are not easily resolved. When and under what conditions do the rights of the child surrendered for adoption override the confidentiality laws of the jurisdiction in which the adoption took place? When and under what conditions may we reach out to suspected siblings, aunts and uncles, cousins, and grandparents when the likely parent has refused any contact? In a donor conception case, when and under what circumstances should the anonymity promised to the donor be breached?

There are no easy answers to these questions. DNA testing can almost certainly answer the question of biological parentage. It is an accepted maxim within the genealogy community that every individual is entitled to know his or her biological identity. It is equally an accepted maxim within the community that no one has the right to force contact or a relationship on another person. Where the line is to be drawn between knowing one's identity and demanding a relationship is a question better left to the individual circumstances of each case and the individual decision-making of the persons most directly impacted.

In general, in cases of unknown parentage, the decision as to whether, how, and when to approach a possible family member should be left to the adopted individual. If the services of an intermediary are required, a person trained in counseling, particularly in the adoption context, is the preferred choice to make the first approach.

Health disclosures

Contact with third parties in order to provide health information based on genealogical DNA testing will almost never be the ethical choice. There are few if any cases in which genealogical DNA tests can result in evidence of a health condition so clear and so compelling that disclosure to a third party might be considered ethically appropriate.

With a very few exceptions, health-related DNA results are not crystal-clear. The majority of heritable conditions do not result from a single gene but rather from a combination of the presence and absence of a number of genes. In most cases, these genetic indicators are not all included in genealogical DNA tests, leaving the interpretation of the significance of the genetic indicators that are included in doubt.[63] Many genes related to health conditions only indicate probabilities of a condition developing and not certainties. Other factors, such as environment and diet, also affect whether a genetic predisposition will result in manifestation of a condition. With few exceptions, genealogists lack the necessary training to properly evaluate the medical consequences of DNA test results.

Even in a case where the evidence of health conditions is clear and compelling, with few exceptions, genealogists also lack the necessary training in genetic counseling to undertake the difficult and emotionally-charged task of disclosing the health data in a way that might be beneficial to the third party.

Our lack of medical and counseling expertise makes us singularly inappropriate purveyors of health information.

63. Leah LaPerle Larkin, "Cystic Fibrosis: A Case Study in Genetic Privacy," *The DNA Geek*, 20 March 2017 (http://thednageek.com/cystic-fibrosis-a-case-study-in-genetic-privacy/).

Conclusion

Ethical decision-making in genealogy is not a black-and-white matter but rather one where most questions fall somewhere into various shades of gray. Nonetheless, knowing and applying the ethical standards of the genealogical community in general and the genetic genealogy community in particular will help us reach the best conclusion that we can based on the information we have at the time. Our decision-making will not be perfect but advance planning for unexpected results and keeping in mind the ethical imperative that we do no harm to living persons will keep us on an ethical path.

Uncovering Family Secrets: The Human Side of DNA Testing

Michael D. Lacopo, DVM

Introduction

In May 1923, a 29-year-old married Italian immigrant with two children witnessed the marriage of a friend to his teenage bride. Four months later the immigrant impregnated his friend's wife. Her husband abandoned her. She and the immigrant's wife gave birth to children six months apart.

In the spring of 1946, a 29-year-old married woman in an unhappy marriage had a one-night stand with a 19-year-old on a weekend pass from the U.S. Navy. She divorced her first husband, and rapidly married the younger man who she thought was the father. She was wrong.

Near Christmas 1913, the 33-year-old wife of a Wisconsin railroad conductor often away from home took a lover and conceived a child. Her husband doted on the girl he accepted as his daughter until his death in a railroad accident in 1921.

An Ohio farm girl just shy of her sixteenth birthday married her much older boyfriend in 1956 when she found out she was pregnant. The child was not his.

Anyone who has delved into their family history or listened to their great-aunts gossip after a few too many eggnogs during Christmas dinner knows that all families whisper about some version of the previous anecdotes. No matter how pristine and upright a family presents to the outside world, skeletons lurk in the closets.

One thing makes all the preceding examples special. They are all from my extended family, and they were all uncovered unexpectedly via DNA testing. The Italian immigrant was a known philanderer, but even he may not have been aware that the child born in 1924 was his. The Wisconsin girl died in 2012 as the beloved matriarch of four children and a slew of grandchildren and great-grandchildren. All had repeatedly heard the stories of her beloved father who died when she was a young girl. She never suspected he was not her father. Sure, there were whispers and doubts about the paternity of the Ohio girl, but it was all just vicious gossip.

In less than a decade, the use of commercial DNA testing has pushed thousands of skeletons out of closets. The frequency with which misattributed parentage discoveries are reported on genetic genealogy *Facebook* groups, in human-interest news stories, and among genealogists has increased so rapidly as to become virtually passé. The vagaries of human sexuality are nothing new, but now it is documentable and discoverable. It is an exciting new world in genealogy.

But not necessarily for those whose lives are altered by these findings.

Background

Uncovering family secrets is not new. Genealogists have unearthed court proceedings for bastardy and fornication in musty old courthouses for centuries. Diaries and letters have hinted at extramarital affairs. Simple math has revealed children born less than nine months after their parents' marriages. Furthermore, obvious and previously known breaks in a biological lineage represented by adoption, sperm donation, unwed mothers, documented foundlings, and known infidelity have long been doggedly pursued by genealogists in search for answers. Birth-parent and adoptee research and support groups have existed for decades. Paper trails left in the wake of a child's birth have been sought and petitioned for release to reconnect the biological chain broken years earlier.

Genealogists work with records and documentation. An extramarital fling rarely leaves a trace, especially when the participants have all died. Many secrets have gone to the graves with those who brought children into the world by means not prescribed by the societal norm of marriage and procreation with one's spouse. For genealogists, no documents mean no answers. Furthermore, documents that actually yield no indication nor hint of misattributed parentage are accepted at face value, and no questions are raised.

DNA testing has changed the playing field.

DNA Testing

Y-DNA testing of men gained momentum with the founding of FamilyTreeDNA in 2000. Surname groups followed almost immediately. Unexpected mismatches also followed. Y-DNA results that did not yield the expected surname matches met with, at best, nebulous responses. Perhaps the number of markers tested were too few to give reliable surname matches. Perhaps no other male had tested with the surname of interest. Or perhaps this indicated a break in the paternal biological line. If Y-DNA testing implies misattributed parentage, it does not reveal when it may have occurred. Only testing multiple males of the same family surname descended from a variety of branches at different generations can identify the misattributed paternal ancestor. Pursuing questionable Y-DNA test results may not give the test taker a sense of immediacy, because the break in the paternal line could have occurred centuries before. Yet, before the rise of autosomal DNA testing, Y-DNA testing revealed modern-day secrets.

> Most researchers had traced the Hinds line back to an Adam Hinds born in Virginia around 1765. That was everybody's brick wall. My line to Adam was my paternal line. A distant Hinds cousin I had corresponded with had done a Y-DNA test. Being the compulsive spender I am, I bought my first Y-DNA test from FamilyTreeDNA: a 12-marker test. When I got my results, there were no Hinds matches. I knew my 12-marker test wasn't the most reliable, so I upgraded to 37 markers, then 67 markers. Still no Hinds matches. I DID have a perfect match with a guy with the surname of Irving. Furthermore, seven of my top ten matches had the surname Irving. I figured there was an NPE in my line. Now I was going to find it.
>
> I mentioned [my] genealogy quandary to my mother. I told her about the Y-DNA test and no Hinds matches, but the perfect match to Irving. When I mentioned the latter surname, she blurted out "Randall?" When I asked who that was, she mentioned it was a guy the year behind my father in high school. After months of my mother adamantly insisting that I was a Hinds, she broke down and told me "it was possible" I was an Irving.[1]

Autosomal DNA testing has been the game changer in the field of genealogy as it pertains to the revelation of family secrets. When 23andMe, AncestryDNA, and FamilyTreeDNA dropped their autosomal DNA testing prices in 2012, exploring one's genetic heritage became inexpensive and readily available to the public. MyHeritage launched its testing service in 2016. Prices continue to drop. The exponential rise in database numbers continues unabated. By the autumn of 2018, the results of over twenty million test takers could be

All URLs were accessed 26 January 2019 unless otherwise indicated.

1. All stories shared with the author have had names and identifying information changed to protect their identity.

viewed in the databases of these four sites.[2] The sheer size of the databases makes it likely that a match within fourth cousins will be identified. It also makes it likely that surprises will be discovered.

The Genetic Genealogy Standards

A committee of eighteen preeminent genealogists presented the Genetic Genealogy Standards in Salt Lake City, Utah, on 10 January 2015.[3] Twenty-one topics intended "to provide standards and best practices for the genealogical community to follow when purchasing, recommending, sharing, or writing about the results of DNA testing for ancestry" were provided to help the researcher navigate the use of this new and exciting genealogical tool. Item 12 involves "Unexpected Results." It states:

> Genealogists understand that DNA test results, like traditional genealogical records, can reveal unexpected information about the tester and his or her immediate family, ancestors, and/or descendants. For example, both DNA test results and traditional genealogical records can reveal misattributed parentage, adoption, health information, previously unknown family members, and errors in well-researched family trees, among other unexpected outcomes.

While this bullet point addresses the obvious, it goes no further in recommending what the test taker does with those unexpected results; nor does it address the responsibility of the genealogist when those they test are confronted with such surprises.

Identity and Self

Identity as a psychological phenomenon gained special attention with the work of psychologist and psychoanalyst Erik Erikson (1902–1994). He described the formation of the ego identity, or "the self." In the 1970s and 1980s, Henri Tajfel and John Turner expounded upon the theories of self. They formed their social identity theory by looking outward to the social groups in which we claim legitimate membership. The study of self is a complex discipline within the realm of psychology warranting the formation in 2002 of the International Society for Self and Identity.[4]

The families that raise us forge our sense of *personal identity* from an early age. Those we mimic and emulate during our formative years help form our personal identity. Our *social identity* is a product of the groups in which our inclusion defines us: race, religion, occupation, and ethnicity. "One's ethnic identity is defined as that part of the totality of one's

2. Leah Larkin, "Database Sizes—September 2018 Update," *The DNA Geek*, 3 September 2018 (http://thednageek.com/database-sizes-september-2018-update/).

3. *The Genetic Genealogy Standards* (http://www.geneticgenealogystandards.com/).

4. International Society for Self and Identity (ISSI): An Interdisciplinary Association for Social and Behavioral Scientists (http://www.issiweb.org/).

self-construal made up of those dimensions that express the continuity between one's construal of past ancestry and one's future aspirations in relation to ethnicity."[5] Being citizens of a country almost exclusively populated by the descendants of immigrants, this last group is one with which most Americans identify. The primary marketing strategy employed by many DNA-testing companies aimed at American audiences is to tap into this ethnic pride. For many, it is a significant identifier.

So what happens when Patrick O'Donnelly takes a DNA test and finds out his father was Jewish? The story that defines his very entry into the world changes. The individual self defined by the relationship with his parents shifts. His cultural and ethnic identities are immediately erased. DNA surprises can create an immediate identity crisis of magnificent proportions. This trauma is most acute with the discovery of misattributed parentage of the test taker or within recent generations of the test taker. However, it can also be manifested by the discovery of a previously unknown half-sibling. While the test taker's identity may remain unchanged, the possible discovery of parental infidelity can change the test taker's perception of the family dynamic he now sees as flawed and dishonest.

Because ethnicity is closely tied to our concept of social identity, so too is our surname. A crisis of identity can also accompany a Y-DNA result indicating that the surname you carry and that defines you is not the one your genetic profile dictates. Misattributed parentage from the eighteenth century can still have emotional ramifications today.

Epigenetics

Epigenetics is defined as "the study of the process by which genetic information is translated into the substance and behavior in an organism: specifically, the study of the way in which expression of heritable traits is modified by environmental influences or other mechanisms without a change to the DNA sequence."[6] Epigenetics tells us that we are given genetic information from each of our parents, but the expression of those genes can be influenced by stresses and trauma inflicted on the ancestors that came before us.[7]

While DNA surprises inflict a psychological toll on those who are subject to these surprises, it is entirely possible that they may have also inherited other effects related to the psychological well-being of their ancestors. Rachel Yehuda, director of the Traumatic Stress Studies Division at the Mount Sinai School of Medicine, has spent decades researching the neurobiology of post-traumatic stress syndrome (PTSD) as well as the intergenerational effects of traumatic stress. Her published works includes measurable correlations between PTSD experienced by pregnant mothers exposed to the World Trade Center attacks and their babies, as well as that between Holocaust survivors and their children.[8]

5. Petter Weinreich, "The operationalization of identity theory in racial and ethnic relations," in John Rex and David Mason, *Theories of Race and Ethnic Relations* (Cambridge, Mass.: Cambridge University Press, 1986).

6. Epigenetics, *Dictionary.com* (https://www.dictionary.com/browse/epigenetics).

7. "Epigenome," *International Society of Genetic Genealogists Wiki* (https://isogg.org/wiki/Epigenome).

8. "Rachel Yehuda," *Wikipedia* (https://en.wikipedia.org/wiki/Rachel_Yehuda).

In short, exposure to unexpected DNA results does not merely create a logistic nightmare in reconfiguring one's family tree. It can create real psychological problems in those presented with these results. Test takers not only have to deal with the real shift in identity, they have to process effects this discovery may have on them. A genealogical line filled with octogenarians and nonagenarians may be suddenly substituted with a family plagued with the genetics for breast cancer, heart disease, or other health risks. Not only do we inherit those genes from a plethora of new ancestors, we may very well inherit the stressors they experienced. It is not uncommon that those faced with DNA surprises set out on their own personal journey to redefine their self. They seek the biological connections that have been taken from them by a single test.

Decision Making

> *A good decision is based on knowledge and not on numbers.* —Plato
> (https://www.brainyquote.com/quotes/plato_400440)

Rarely discussed in human-interest stories regarding reunions of biological siblings and parents, or in *Facebook* groups dedicated to finding those missing biological links, are the ramifications of such a search. The previous paragraphs certainly portray a convincing argument for seeking such a connection, and the overwhelming majority of people interviewed for this chapter immediately undertook such a search when presented with their unexpected DNA results.

This is not an entirely new phenomenon. Adoptees have long sought the stories of their origins. Many feel a loss of self- and social-identity by not knowing these details. The American Adoption Congress has been publishing guidelines for making contact with newly discovered family members since 1993, long before the advent of DNA testing.[9]

Nonetheless, adoption is a definable act that legally separates a bloodline. Presumably both birthparents know a child exists, and that adopted child knows another set of parents exists. DNA surprises often yield a misattributed paternity that nobody but the mother may have known about. Often those involved with the conception and rearing of a child are dead. No answers can be sought for questions, and no paper file exists to be examined for details. Many biological fathers identified via DNA are just as surprised by the results as the children seeking them.

DNA test takers with unexpected results are presented with a dilemma. How can they resolve their identity crisis without creating one for others? They cannot. The biological relatives they seek may not want to be found. More often, they do not know the seeker exists. The very existence of this subjective chapter in an otherwise objective book is borne of this problem.

9. "Making Contact," 1993, revised 2008, American Adoption Congress (https://www.americanadoptioncongress.org/pdf/making_contact.pdf).

Discussions regarding the search for familial answers and origins via DNA testing bring up the ubiquitous phrases, such as the "right to know" and the "right to act" and the "right to privacy." Each of these rights is valid, and each of them generates lists of pros and cons and dos and don'ts that will not be discussed here in detail. The unfortunate reality is that these rights must coexist and be exercised within the constraints of each other. This balancing act is impossible.

If the right to know a birth father's identity yields five brothers as candidates via DNA research, what is our right to act in contacting all five of them to determine which is the father we seek? What is their right to privacy in resisting your enquiries? When is it morally and ethically right to resolve your crisis while creating one for others?

Rachel Yehuda summarizes this dilemma nicely:

> In theory, people should have the right to be protected from disruptive information, especially information that they did not seek. In practice, this is very difficult to implement. If someone learns that he has a different biological father, he becomes the owner of that information. The owner technically has the right to share that information and create any havoc that arises from that knowledge.
>
> In our society, it is difficult to make the nuanced argument that one should not always exercise one's right in order to consider the impact on others or the greater good. Moreover, in our society, privacy is no longer an expectation that people have, and therefore, respecting someone's privacy no longer seems to be part of the cultural value system. We are constantly and unwittingly providing a lot of private information about ourselves to others, so it is difficult to draw the line and tell someone that they should think twice about disclosing their personal information because it violates someone's privacy and may have untoward circumstances.[10]

The decision to seek answers to the questions raised by unexpected DNA results is not one made only by the test taker. Decisions are made by all those involved and by all those contacted as a result of such a search, and all the decisions made by these parties are valid. Those who have undergone DNA testing do so for a number of reasons. They are not obligated to respond to inquiries associated with their results. Many adoptees have embraced DNA testing to give them a sense of cultural identity or to provide familial medical history of which they have no knowledge. One test taker conceived by sperm donation found a deeper sense of historical connection when his test revealed he was 50% Ashkenazi Jew, born of his Irish mother. Another adoptee confided in me: "I did [DNA testing] mostly for medical reasons. I was a blank slate. I wanted to know whatever the test would tell me as I age. I personally have no desire to know my bio parents. And I would never disrupt their

10. Rachel Yehuda, to Michael D. Lacopo, email, 13 November 2017; privately held by Lacopo.

lives by seeking them out. If they contacted me, I'd probably ignore [them]. They're strangers to me. They owe me nothing. I owe them nothing. This should not be interpreted as bad feelings. The genetic relationship matters not to me. It's who raised you that matters to me."

We must remember that recipients of unexpected DNA results are well within their rights to seek the answers to the questions raised by these results. We must also remember that these searches enmesh others within this crisis of identity that can in turn be faced with the same shock and surprise. Their decisions to respond, act, and assist must be respected even when it can create a roadblock in the searcher's quest for a "new" biological family.

Privacy

> *When it comes to privacy and accountability, people always demand the former for themselves and the latter for everyone else.* —David Brin
> (https://www.brainyquote.com/quotes/david_brin_389049)

As Rachel Yehuda pointed out, we live in a society that has experienced a rapid change in the perception of privacy brought on by the computer age. The popularity of DNA testing alone shows us that millions of people are willing to share a very personal part of themselves within the databases of the testing companies. Over a million such test takers are willing to place their results on third-party sites such as *GEDmatch* for the general public to peruse.[11]

Privacy is a tricky paradox amongst genealogists, and, sadly, one that many perceive as a one-way street. When seeking personal stories from test takers who had received surprises as a result of their DNA tests, I was asked in public forums no less than *twenty* times if my future commentary would guarantee their privacy. Yet these same genealogists, in a variety of online forums, strongly advocate for immediate and direct contact with paternal candidates and effortlessly ask those candidates about their sexual habits from decades before. They express anger and outrage when other test takers decline to reply. One genealogist who fiercely protects her locked family tree online, publicly stating she makes the decisions about who does or does not see her years of research, also routinely posts angry tirades regarding DNA test takers who do not immediately respond to her questions regarding adoptee and birth parent searches she pursues as a volunteer. In November 2017, I was removed and barred from a *Facebook* DNA group by an administrator who deemed my search for those willing to share their DNA stories "too personal and inappropriate." This same administrator allowed another post regarding the graphic sexual proclivities of a test taker's paternal grandmother go uncontested when DNA showed that the woman's son was not conceived of her husband. My request for those willing to voluntarily share their stories was met with a significant amount of venom and vitriol; most of them stating "it was none of my business." Many of these same people can be found advocating for the right to know the answers to unexpected

11. Larkin, "Database Sizes—September 2018 Update." Official size of GEDmatch database as of September 2018 was 1,080,000 test takers.

DNA results at all costs and with complete disregard for the feelings of those they contact. The irony is not lost on me, nor should it be lost on readers.

Again, from Rachel Yehuda: "Confronting people with information that will shake up their world needs to be done with an abundance of care and thought, but generally this kind of information is disclosed impulsively, with a high level of emotion, and very little planning of the next steps."

Outcomes

> *Knowing the truth changes nothing, yet it changes everything.* —anonymous test taker who discovered the man who raised her was not her father

The crisis of identity has already been discussed. All those interviewed who had this sense of loss and confusion immediately set out on a journey of discovery to seek answers to questions they previously had no reason to ask. Unfortunately for many, the answers they seek will never be found. Often all the parties involved are gone—their stories taken to their graves. For those presented with misattributed paternities, a common denominator revealed in dozens of interviews was an immediate sense of betrayal by the affected persons' mothers. Their secrets resulted in a child denied the knowledge of half their biological identity. Yet others commiserate.

> On a hard day, I feel heartbroken about my mom's secret. I also have empathy for her, especially after reflecting on the fact that she faced the decision whether to terminate the pregnancy and then carried the burden of the secret of my paternity for the rest of her life.[12]

> My feelings towards my mother (who is still alive) range from hurt, betrayal, anger, and sometimes sympathy.[13]

Conversely, many mothers confronted with these results reacted with the same shock as their children although they were obviously well aware of their past infidelities. They convinced themselves that the child was that of the husband. Call it willful denial or wishful thinking, but the shock waves reverberate from the test taker in unexpected ways.

As millions test their DNA, the media has become saturated with stories of tearful reunions and happy endings. Biological fathers who had no idea they had fathered children embrace their newfound offspring. Large extended families happily accept the new half-sibling, nephew, niece, grandchild, or cousin they never knew existed. TLC network's *Long Lost Family* premiered in 2016 to capitalize on the tear-jerking reunions of long-lost biological family

12. Phil Noble, "Letters: "My Dad Was Not My Biological Father," *The Atlantic,* 18 August 2018 (https://www.theatlantic.com/letters/archive/2018/08/letters-my-dad-was-not-my-biological-father/566903/).

13. Anonymous, to Michael D. Lacopo, email, 10 November 2017; privately held by Lacopo.

members. All the major DNA-testing companies either actively air stories of reunions facilitated by their DNA tests on their internet platforms or sponsor videos of this nature on other forms of media.

The outcomes of a DNA surprise can be many and varied. Support groups for those affected by unexpected results have sprung up on social media as a platform for test takers to share their stories, experiences, and outcomes.[14] A narrow range of psychologists in clinical practice were interviewed for this chapter. None had been personally presented with a patient with this problem, yet all expressed a concern about the psychiatric welfare of those involved in such a scenario, including not just those tested but also those contacted as a result of testing. In the words of a medical doctor who also holds a doctorate in psychiatry: "Oh my, this is a situation that definitely needs to be addressed!"

For many test takers presented with DNA surprises, the outcomes can be worse than the shock of the test results. Rejection, indifference, and denial are commonplace. "Sometimes, even when the evidence is present, people cannot deal with the truth and call others liars."[15] Children conceived of incest or violence or betrayal are often refused access to their biological families so as to minimize the trauma of discovery for the families involved. Sometimes contact is rebuked merely because the family is disinterested or feels that the past should be left in the past. All of these are appropriate responses, yet are often met with aggressive reactions from those seeking a "happily ever after" reunion that will never materialize.

The outcomes of receiving unexpected test results are varied and unpredictable. My own search for my maternal grandfather described in the blog *Hoosier Daddy?* began as an academic exercise in identifying one-quarter of my missing ancestry. Its loss was revealed by my mother's surprising DNA results.[16] My maternal grandmother had been long deceased. Had she been alive, she would have been nearly a century old when my search started. It was presumed that the man with whom she conceived my mother would also be deceased. He was not. The subsequent meeting and irrational and selfish acts of the daughter he never knew existed resulted in his death and the complete alienation of many members of my family.

My story is not an isolated story. The sudden appearance of previously unknown half-siblings and children has broken marriages and alienated families. For all the joyous reunions and happy endings portrayed in the media, there are an equal number of individuals and families negatively affected by the searches borne out of unexpected DNA results. This hearkens back to the right to know versus the right to act. We all are entitled to the basic right to know our beginnings, to know our interconnectivity to mankind, and to have knowledge of those who share our DNA. We do not have an inherent right to insert ourselves into the lives of others. We must think before we act.

---

14. Two such groups are *NPE Friends Fellowship* (https://www.npefellowship.org) and the closed *Facebook* group *DNA NPE Gateway* (https://www.facebook.com/groups/NPEGateway/).

15. Anonymous, to Michael D. Lacopo, email, 10 November 2017.

16. Michael Lacopo, *Hoosier Daddy?* (http://roots4u.blogspot.com/).

Of the nearly twenty million people who have had their autosomal DNA tested, only a very small percentage of them are genealogists skilled in genetics who can immediately and fully grasp the test taker's results. Many test takers have DNA surprises that they are blissfully unaware of. They do not understand the matching algorithms, nor do they know the centimorgan requirements to define a relationship. During a DNA lecture, I was asked by an audience member who had attended with her sister why the two of them did not seem to be classified as siblings by the company with which they tested. Deferring the question to a private consultation after the lecture, it was revealed that they were half-siblings. Ignorance of the mechanics of genetic genealogy resulted in a very public and very unexpected emotional reaction from these two women.

This also raises the question of when it is appropriate for test takers to reveal unexpected DNA results to their matches. Many of these matches may be completely unaware that their family tree is incorrect. If your DNA analysis indicates that one of your matches is almost certainly the child of a relative and not the parents listed on their public tree, is it your responsibility to reveal this information to a faceless individual of whom you know nothing? You are not personally affected by this reveal; are you presenting this stranger with a gift of knowledge or a crisis of identity?

Similarly, many test takers confused by results or unable to understand why they are close matches to people whose names are not familiar to them are turning to social media for answers. While crowdsourcing has become a vital part of our research and discovery in all aspects of genealogy, many test takers have naively released the names of previously unknown biological parents and siblings into a public forum. Many learn life-altering truths via the flippant response of a stranger. Some are bombarded with unsolicited advice, remonstrations, and derision before they can even process the information laid before them.

Furthermore, accusations are made and people contacted based on erroneous assumptions made from incorrect DNA analysis. One man interviewed for this chapter was repeatedly harassed by a test taker who was convinced that the interviewee was his biological father. The man had lived in the same town as the test taker's mother at the time of his conception. Another DNA cousin pointed the finger at the man in question because he had "a history of being quite a womanizer in his younger days." The man seeking his biological father and the man he accused of impregnating his mother did share DNA in common—62 cM. This obvious genetic incompatibility with the definition of a father-son relationship did not stop the test taker's persistence. His ignorance of mathematics and probability of DNA matching created a nightmare for an innocent man. The affordability of DNA testing has put the results of such into the hands of the hobbyist while at the same time creating dilemmas for those ill-equipped to handle them.

Conclusion

> *Humans aren't as good as we should be in our capacity to empathize with feelings and thoughts of others, be they humans or other animals on Earth. So maybe part of our formal education should be training in empathy. Imagine how different the world would be if, in fact, that were 'reading, writing, arithmetic, empathy.'* —Neil deGrasse Tyson (interview with PETA President Ingrid E. Newkirk, https://www.peta.org/features/dr-neil-degrasse-tyson-interview/

The title of this section is a bit of a misnomer, as we can never bring this topic to a conclusion. This chapter reads more like an editorial and opinion piece on the hazards of DNA testing, which stands in stark contrast to most of the content in this book. Working with DNA is rewarding because it is factual and often irrefutable. The ramifications of making these concrete conclusions are less so.

This chapter may also sound overwhelmingly negative. That is not the intention. The popular news media overflows with the feel good stories regarding DNA reunions. It rarely delves into the darker side, nor does it adequately caution test takers against the obstacles they may face on their DNA journey.

If we return to the so-called rights mentioned earlier in the chapter, the right to know is the most accepted. Proponents for open adoption records have been using the same terminology long before the advent of DNA testing. When presented with unexpected DNA results, especially those that alter the sense of identity for the test taker, it is without question that person is entitled to seek clarification and resolution to the crisis unexpectedly thrust upon them.

Unfortunately, to gain that knowledge, one must act. To identify and contact the people who hold the answers to the questions of identity, one must invade the perceived privacy of others. There is no guarantee of one right without infringing upon the others.

One right that test takers do not have, and which is lamented upon vehemently in DNA forums, is the right to a relationship. Nary a day passes that I do not see a heated discussion within a DNA forum where a test taker has confirmed his or her paternity, reached out to the reputed father, and been rebuked even when the numbers are conclusive. Without fail, people respond with violently scornful derision toward the unseen and unknown father. Test takers have a right to know the truth behind their paternity. They never have to right to force a relationship with the person they identify. One may argue the morality of shunning an unknown child, often unwittingly brought into the world through their willful actions. While they are certainly entitled to empathy, they are not entitled to an interpersonal relationship. The very definition of relationship implies a connection between two or more persons. When one party is unwilling to partake in such a connection, regardless of how desperately the other party wants it, such a relationship can only exist in the definition of the biological relationship and no more.

As genealogists we seek to solve problems. We repeatedly talk in terms of brick walls and proof arguments. We seek to define relationships. The advent of DNA testing has changed the way we assess and evaluate our evidence. Mathematically, it is virtually impossible to have a family tree complete to eight generations and *not* have an ancestor with misattributed parentage within your pedigree.[17] DNA testing, when used as an active tool within the genealogist's toolbox, will help identify these anomalies. As databases grow, identifying them will become easier and more obvious. Many of these unexpected relationships will personally affect people living today. Do not fool yourself that this will not happen to you. It will. The anecdotes that opened this chapter should be evidence enough that it can happen to all of us. Be prepared.

Genealogists are analytical thinkers. DNA findings that do not match the presumed paper trail only propel us deeper into problem-solving mode. It is wise to remember that solving problems brought forth by unexpected DNA results is not equivalent to solving a Rubik's Cube. Both are complex problems that require skill, time, and analysis to solve. One results in a uniformly colored cube and a sense of accomplishment. The other has the potential to change the lives of others in a ripple effect that quickly spreads beyond our immediate control.

The conclusion therefore is that there is no conclusion. There are no right or wrong answers, and there are no recommendations for what one should or should not do. Each situation is unique. This chapter is merely presented to readers and to those who embark upon DNA testing to inform them that this is a dialog that needs to be had, and it is one that is largely unaddressed.

When DNA uncovers family secrets there must be a balance between being a fact-finding genealogist and an empathetic human being. That balance may be difficult, if not impossible, but it is a balance that must be considered, attempted, and respected.

17. Leah Larkin, "MPEs, Probabilities, and Why You Need DNA, Even if You Think You Don't," *The DNA Geek: Mixing Science and Genealogy,* 5 October 2017 (http://thednageek.com/mpes-probabilities-and-why-you-need-dna-even-if-you-think-you-dont/).

The Promise and Limitations of Genetic Genealogy

Debbie Kennett, MCG

DNA testing is a powerful tool for family history research. Despite the advances of the last two decades, the tests still have many limitations. It is important to consider these uncertainties when incorporating DNA evidence into genealogical research. As the cost of sequencing comes down, microarray tests will be replaced by whole genome sequencing. With hundreds of millions of people in the databases, scientists will need to develop sophisticated algorithms and machine learning tools to analyze the data. This chapter quantifies the limitations of the existing tests and takes a look at exciting developments expected in the coming years—the era of genomics and big data.

The Promise and Limitations of Y-DNA Testing

A Y-DNA test can be taken only by males and provides matches with genetic cousins on the patriline. The first commercial Y-DNA tests became available in 2000. In those early days, only low-resolution Y-STR tests were available at 10 or 12 markers. In the intervening years, the cost of the tests has come down, while the range and resolution of the tests have improved. Standalone Y-STR tests can be purchased for 37, 67, and 111 markers. We also have learned about the importance of SNP testing and the limitations of drawing conclusions based on STRs alone. Y-SNPs are included on some microarrays, and it is also possible to purchase single SNPs or SNP pack tests. The first commercial Y chromosome sequencing tests using next generation sequencing (NGS)[1] became available in 2013 with the launch of

---

All URLs were accessed 18 September 2018 unless otherwise indicated.

Master Craftsman of the Guild of One-Name Studies (MCG).

1. For differences in sequencing and older testing see "Y-DNA next generation sequencing," *International Society of Genetic Genealogists (ISOGG) Wiki* (https://isogg.org/wiki/Y-DNA_next_generation_sequencing) and "Y-DNA STR testing comparison chart," *ISOGG Wiki* (https://isogg.org/wiki/Y-DNA_STR_testing_comparison_chart).

the Big Y test from FamilyTreeDNA and the Y Elite from Full Genomes Corporation. Now it is also possible to purchase commercial whole genome sequencing (WGS) tests providing better coverage than the standalone Y-sequencing tests.

Genetic genealogy is all about the comparison process. For Y-DNA results to be useful they need to be compared with other test takers, and this is usually done in a matching database. The matching databases were historically based on STRs. With the increasing popularity of autosomal testing, many of the companies that previously sold STR tests are no longer in existence. FamilyTreeDNA has always been the market leader for STR testing and they are now the only company that provides an STR matching database suitable for use by genealogists.

Y chromosome sequencing provides the most accurate estimates of relationships. Depending on the test taken, a new SNP occurs, statistically, every 80 to 130 years.[2] SNPs are much more stable than STRs and can be used to construct a robust tree right down to the present day. Alex Williamson's Big Tree incorporates NGS data for the P312 and U106 subclades of haplogroup R1b; it shows the subclades with all the associated surnames.[3] It provides a fine example of how NGS testing is bringing SNPs into the surname era and is defining recent branches of the Y-tree. Similarly-detailed trees are produced by some other haplogroup projects.

However, NGS tests are still relatively expensive. Upgrade prices at FamilyTreeDNA start at several hundred dollars, depending on the prior STR testing level. A new test starts at a few hundred more than an upgrade though big savings are sometimes available with sale prices. Hence, the number of NGS results available for comparison is still quite small. There are probably only around 30,000 people with NGS results across all the companies. At FamilyTreeDNA, the Big Y results are included in a matching database. The other vendors do not have matching databases, but raw data can be uploaded for a fee to YFull, where results can be compared. Because of the higher prices, NGS tests are currently the preserve of advanced genealogists who are using the tests for SNP discovery, though there are some well-funded surname projects that have made good use of NGS tests.[4]

This means that, at present, genealogists will be mostly working with Y-STR results and not Y chromosome sequences. Unfortunately, Y-STR tests are poor predictors of relatedness. If enough markers are tested, they can tell us that two men are related, but they cannot tell us precisely how or when they are related. For example, a father and son might only match on 65 of 67 markers whereas two men who share a common ancestor in the 1700s might match on 67 of 67 markers. Matches can be refined by upgrading to 111 markers. According to FamilyTreeDNA's guidelines, there is a 95% chance that an exact match at 111 markers will

2. James Kane, "Next Generation Sequencing Statistics," *Haplogroup R* (http://www.haplogroup-r.org/stats.html).

3. Alex Williamson, "The Big Tree," *The Big Tree* (https://www.ytree.net/).

4. See the "Y-DNA Analysis for a Family Study" chapter of this book for examples.

share a common ancestor within the last five generations. This also means that there is still a 5% chance that a match at this level will be beyond five generations.[5] To make an upgrade worthwhile, the people in your match list must also have upgraded to the same testing level, but only a small percentage have done so. Only around 26% of the 681,000 men in the FamilyTreeDNA database have taken the 67-marker test.[6] The percentage who have upgraded to 111 markers is likely to be much smaller.

Calculations about relatedness using Y-STRs rely upon knowledge of the mutation rates of the different markers, but there is considerable uncertainty about the mutation rates and different methods produce different dates. The most reliable estimates of the time to the most recent common ancestor (TMRCA, MRCA) are obtained by combining Y-STR and Y-SNP data.[7] FamilyTreeDNA's guidelines are based only on STR data and they have not published their methodology.

STR results are always best interpreted within the context of a surname project and in combination with all the available documentary evidence. The aim should be to test two or more representatives of each of the known lineages for the surname of interest in order to verify that the results are consistent with the documented genealogies. If a framework of results for known lineages can be established, when someone tests who has limited or no information about his family tree he can be matched to a genetic cluster or family. It can be inferred that he must fit somewhere into the documented tree for that cluster. It is then back to the documentary research to try and identify the connection. The methodology is described in two articles about the Pomeroy DNA Project.[8]

With the growing availability of SNP data from SNP packs and NGS testing, it has become apparent that a number of STR matches are false positives. SNP testing can sometimes indicate that two men who appear to have a close match at 37 markers belong to different subclades or even different haplogroups.[9] This phenomenon is known as convergence.[10]

5. "Paternal Lineages Tests," *Family Tree DNA Learning Center* (https://www.familytreedna .com/learn/dna-basics/ydna/).

6. "About the Family Tree DNA Database.," *Family Tree DNA* (https://www.familytreedna.com/why-ftdna.aspx).

7. O. Balanovsky, "Toward a Consensus on SNP and STR Mutation Rates on the Human Y-Chromosome," *Human Genetics* 136 (2017): 575–90 (https://doi.org/10.1007/s00439-017-1805-8).

8. Chris Pomery, "The Advantages of a Dual DNA-Documentary Approach," *Journal of Genetic Genealogy* 5 (2009): 86–95 (http://jogg.info/pages/52/files/Pomery.htm). See also Chris Pomery, "Defining a Methodology to Reconstruct the Family Trees of a Surname Within a DNA/Documentary Dual Approach Project," *Journal of Genetic Genealogy* 6 (2010) (http://jogg.info/pages/62/index.html).

9. Maarten H. D. Larmuseau et al., "Recent Radiation within Y-Chromosomal Haplogroup R-M269 Resulted in High Y-STR Haplotype Resemblance," *Annals of Human Genetics* 78 (1 March 2014): 92–103 (https://doi.org/10.1111/ahg.12050). See also Neus Solé-Morata et al., "Recent Radiation of R-M269 and High Y-STR Haplotype Resemblance Confirmed," *Annals of Human Genetics* 78 (July 2014): 253–54 (https://doi.org/10.1111/ahg.12066).

10. Maurice Gleeson, "Convergence - What Is It?," *The Gleason / Gleeson DNA Project*, 20 May 2017 (http://gleesondna.blogspot.com/2017/05/convergence-in-practice.html).

The full scale of the problem is not yet known, though it appears to be less of an issue with higher-resolution tests. A good indicator of convergence is when a person has a large number of matches with a variety of different surnames. For example, I have one project member who has over 1,500 matches at 67 markers, only eleven of which share the same surname or variant as him. At 111 markers the number of matches is reduced to seven, though most of his matches have not upgraded to this level. SNP testing has shown that two of these 111-marker matches are in different subclades of the R1b-M222 tree. Convergence is more of a problem with matches with different surnames but, even with matches of the same surname, STRs can give a false sense of relatedness. This is particularly the case when trying to interpret results in a surname project for trees that date back to the 1600s and earlier, when the documentary evidence is also very sparse.

In addition to ruling out false positive matches, SNPs are also very useful for defining branches within genetic clusters, where STRs have not been able to provide sufficient resolution. In these scenarios, Y chromosome sequencing can pay dividends.[11]

However, unless next generation Y chromosome sequencing is included in a structured testing plan, generally as part of a surname project, it is not likely to provide genealogically useful information. It is, as always, the comparison process that is important. You might learn from your Y chromosome sequence that you have fifteen novel SNPs that have never been seen before, but to make use of those SNPs you need to find other people who share those SNPs with you.

With NGS tests, information is reported for around 500 STRs, now 700 with the Big Y-700. This discussion focuses on the Big Y-500.[12] The STR information was reported from the outset by Full Genomes. NGS results can also be uploaded to the third-party website YFull for an analysis that includes a list of STRs. FamilyTreeDNA began reporting STRs from NGS data in April 2018. The FamilyTreeDNA NGS product is now known as the Big Y-500.[13] It includes the full 111-marker panel, and FamilyTreeDNA guarantees that the test will include at least 389 additional STRs.[14] In practice the number is much higher. One comparison found that an average of 435 STRs were reported for 2,118 tests.[15] However, not all of the STRs in the 111-marker panel are amenable to NGS testing. Some, if not all, of these STRs are also tested using conventional methods, though the details have not been revealed.

11. Linda Jonas, "The Amazing Power of Y-DNA," *The Ultimate Family Historians*, 17 April 2018 (http://ultimatefamilyhistorians.blogspot.com/2018/04/the-amazing-power-of-y-dna.html).

12. See 'Big Y-700' in the "Y-DNA Analysis for a Family Study" chapter for more information on Big Y-700.

13. Linda Jonas, "Announcing Family Tree DNA's New Big Y-500 Test," *The Ultimate Family Historians*, 20 April 2018 (http://ultimatefamilyhistorians.blogspot.com/2018/04/announcing-family-tree-dnas-new-big-y.html).

14. "Big Y-500," *Family Tree DNA Learning Center* (https://www.familytreedna.com/learn/y-dna-testing/big-y/big-y/).

15. Iain McDonald, "How Many Y-STRs Can Currently Be Tested?," R1b1c U106-S21 Haplogroup Forum, *Yahoo Groups*, 30 August 2018 (https://groups.yahoo.com/neo/groups/R1b1c_U106-S21/conversations/messages/53984).

The additional STR results beyond the first 111 markers were not originally included in FamilyTreeDNA's matching database but are now.

While the companies are not yet combining the SNP and STR data into their predictions about relatedness, citizen scientists have taken the initiative to fill the gap. Four genetic genealogists have set up the Y-DNA Data Warehouse, which is collecting raw Y-DNA data from individual test takers in haplogroup R1b. The goal is to produce a detailed haplotree with age analysis based on both SNPs and STRs and with ancient DNA data from published papers incorporated.[16]

Despite all the technological advances of the last two decades, there is still much to be learned. The Y chromosome is nearly 60 million bases long, but the current Y-sequencing tests can only report results for up to fifteen million positions.[17] New methods are being developed to provide better coverage. Chromium sequencing from 10x Genomics provides long-range information from short reads. This process tags the DNA fragments with molecular barcodes that can then be used to link the reads together.[18] Full Genomes now offers a whole genome test using Chromium technology. This test provides the best available Y chromosome coverage, providing results for up to 20 megabases (20 million positions) from a good-quality sample, but at a premium price of several thousand dollars.[19]

The Y chromosome has proved difficult to decipher because it includes many long repetitive regions. Next generation sequencing breaks the genome up into short 100 to 600 base pair fragments. These are then re-assembled and mapped onto a reference genome. This methodology cannot be used for the repetitive areas.[20] Companies such as Oxford Nanopore and PacBio have developed long-read technologies that can read thousands of bases in a single run, making it possible to map these repetitive regions. Jain and coauthors used nanopore long-read sequencing to resolve the structure of the centromere of the Y chromosome.[21]

The current NGS tests provide results for around 500 (or 700) Y-STRs, but over 4,500 Y-STRs have been identified.[22] The commercially available NGS tests have read lengths of between 100 and 250 bases, but this is not sufficient to capture the longer STRs. There is a significant correlation between marker size and mutation rates. Y-STRs with more repeats

16. "Y-DNA Data Warehouse Submission," *Haplogroup R* (http://www.haplogroup-r.org/submit_data.php).

17. James Kane, "Next Generation Sequencing Statistics." *Haplogroup R* (https://haplogroup-r.org/stats.html).

18. Dianna, 10x, "A Basic Introduction to Linked-Reads," *10x Genomics Blog*, 26 September 2016 (https://community.10xgenomics.com/t5/10x-Blog/A-basic-introduction-to-linked-reads/ba-p/95).

19. Justin Loe, to Debbie Kennett, personal communication, 5 September 2018.

20. Mark A. Jobling and Chris Tyler-Smith, "Human Y-Chromosome Variation in the Genome-Sequencing Era," *Nature Reviews Genetics* 18 (August 2017): 485–97 (https://doi.org/10.1038/nrg.2017.36).

21. Miten Jain et al., "Linear Assembly of a Human Centromere on the Y Chromosome," *Nature Biotechnology* 36 (April 2018): 321–23 (https://doi.org/10.1038/nbt.4109).

22. Thomas Willems et al., "Population-Scale Sequencing Data Enable Precise Estimates of Y-STR Mutation Rates," *American Journal of Human Genetics* 98 (May 2016): 919–33 (https://doi.org/10.1016/j.ajhg.2016.04.001).

tend to be much more variable.[23] Beyond the basic 111-marker panel, the additional STRs provided with the NGS tests are generally shorter STRs that have little variation. Hence, they are not as useful for defining male lineages. Long-read sequencing would be able to capture the longer STRs. It is possible that a better base panel of STRs could be developed that would improve TMRCA calculations. However, there will always be a need to maintain backward compatibility with the existing database. Many of the people in the database are no longer active or are now deceased so their results cannot be upgraded.

While technological advances are always welcome, the one improvement that is likely to have the biggest impact is the growth of the database. The number of men who have taken a Y-DNA test is still very small as a proportion of the global population. Also, there is a strong Euro-American bias in the database. FamilyTreeDNA estimates that about 70% of their customers are in the U.S., though they do not provide any regional breakdowns. Y-DNA testing is also popular in the U.K., Ireland, New Zealand, Scandinavia, Finland, and the Middle East, with smaller numbers testing in other European countries. There appears to be a low rate of Y-DNA testing in Africa and Asia. As of 1 September 2018, there were 681,745 Y-DNA records in the FamilyTreeDNA database, 10,115 group projects, and 593,690 unique surnames.[24] However, there are millions of different surnames. The World Names Public Profiler has surname data from just twenty-six countries but has identified over eight million surnames.[25] While many variant surname spellings are related to each other and will fall into the same genetic clusters, there are still many surnames that are not yet represented in the database. The sampling in many surname projects is incomplete. There are disproportionate numbers of test takers from the U.S. and few from the country or region where the surnames originated. Surnames are not used in all countries and in others they were adopted relatively recently; in these countries the results are generally coordinated in geographical projects.

The growth of the database will be facilitated as the cost of testing comes down and the current complex range of testing choices is replaced by a single low-cost sequencing test. That test will provide reports on SNPs and STRs in the same package that can be added to a genetic genealogy database. In the genomic era, the Y chromosome information will be included as a matter of routine with a WGS test. The dream would be to have a Y chromosome signature associated with all the patrilines in our family trees. Then one day we can look forward to a fully delineated Y-tree cross-referenced with all the available genealogical records.

---

23. Sofie Claerhout et al., "Determining Y-STR Mutation Rates in Deep-Routing Genealogies: Identification of Haplogroup Differences," *Forensic Science International. Genetics* 34 (May 2018): 1–10 (https://doi.org/10.1016/j.fsigen.2018.01.005).

24. "About the Family Tree DNA Database," *Family Tree DNA* (https://www.familytreedna.com/why-ftdna.aspx).

25. "Frequently Asked Questions about Names," *Worldnames Public Profiler,* University College London (http://worldnames.publicprofiler.org/FAQ.aspx).

Ancient DNA is beginning to transform our understanding of the past. We now know that migrations have been a constant feature throughout human history and that the DNA of living people cannot be reliably used to make inferences about past populations.[26] However, with the growing body of ancient DNA samples, we have direct DNA evidence of our distant ancestors' presence in specific locations at different times in the past. There is currently no easy way to search this ancient DNA data, but we can expect to see the establishment of online ancient DNA databases where we can look for Y-DNA matches with our distant ancestors and cousins from pre-history.

The Promise and Limitations of mtDNA Testing

A mitochondrial (mtDNA) test can be taken by both males and females and provides matches with genetic cousins only on the matriline. The first commercial mtDNA tests went on the market in the year 2000. The early tests sequenced all or part of the hypervariable region (HVR), also known as the control region, of the mitochondrial genome. This is the part of mtDNA that is most prone to variation, but it represents just 7% of the mtDNA genome. The first commercial full mtDNA sequence tests became available in 2005. The price has come down in recent years to a more affordable level, though it is still relatively expensive compared to autosomal DNA testing and entry-level Y-DNA tests. A full mtDNA sequence is at present the only complete DNA test you can take. Once you have had your mtDNA genome sequenced you will not need to get it done again because there is nothing left to be sequenced.

The mtDNA genome is easy to sequence in its entirety because it is so small. It has just 16,569 base pairs compared to nearly sixty million in the Y chromosome. Because of its small size, and the fact that the mutation rate of mtDNA SNPs is very slow, mtDNA lacks discriminatory power. Even if two people have an exact full sequence match, the common ancestor could still have lived hundreds of years ago.[27] Nevertheless, mtDNA testing has been successfully used as a tool, in combination with other evidence, to answer genealogical and historical questions.[28]

As with any other DNA test, to use an mtDNA test for genealogy you need to be in a matching database. FamilyTreeDNA is currently the only company providing such a service. As of

26. Joseph K. Pickrell and David Reich, "Toward a New History and Geography of Human Genes Informed by Ancient DNA," *Trends in Genetics: TIG* 30 (September 2014): 377–89 (https://doi.org/10.1016/j.tig.2014.07.007).

27. Debbie Parker Wayne, "Using Mitochondrial DNA for Genealogy," *National Genealogical Society Magazine* 39 (October-December 2013): 26–30 (http://debbiewayne.com/pubs/pub_NGSMag_201308_mtDNA_ALL.pdf).

28. Turi E. King et al., "Identification of the Remains of King Richard III," *Nature Communications* 5 (2 December 2014) (https://doi.org/10.1038/ncomms6631). See also Elizabeth Shown Mills, "Testing the FAN Principle Against DNA: Zilphy (Watts) Price Cooksey Cooksey of Georgia and Mississippi," *National Genealogical Society Quarterly* 102 (June 2014): 129–52 (https://www.historicpathways.com/download/ZilphyArticle072915.pdf).

10 September 2018, FamilyTreeDNA had over 311,000 mtDNA results in its database. That is less than half the number of results in the company's Y-DNA database. The mtDNA database is still very small compared with the global population. It is heavily biased toward test takers from Europe and especially North America. However, the database has grown rapidly in recent years as test prices have decreased.

A full mtDNA sequence is already included when ordering a whole genome sequence and, in future, a standalone mtDNA test will no longer be necessary. The dream would be to have a mtDNA signature associated with all the matrilines in our family trees. For this we need millions of people testing and not thousands. The mtDNA genome will never provide the precision we are beginning to get from Y chromosome sequencing, but it can be used to rule specific lineages in or out. It will also be very useful as complementary evidence when mtDNA results are fully integrated with autosomal DNA results. We can also expect to see ancient mtDNA samples included in ancient DNA databases, which will allow us to search for mtDNA matches with our distant ancestors and cousins from pre-history.

The Promise and Limitations of Autosomal DNA Testing

The first autosomal DNA (atDNA) test for genealogy purposes became available in 2009 when 23andMe introduced its Relative Finder service, now known as DNA Relatives. Since then, relative-matching services have been introduced by three other companies: FamilyTreeDNA, AncestryDNA, and MyHeritage. At the time of writing, a fifth company, Living DNA, is beta testing a similar service known as Family Networks. As the cost of testing has decreased and marketing budgets have increased, the databases have grown exponentially. In 2017, more people took a genetic ancestry test than in all the previous years combined and it was atDNA testing that dominated the sales.[29]

An atDNA test can be taken by both males and females and provides matches with genetic cousins on all ancestral lines. The amount of DNA we inherit from our ancestors is halved with each generation. We inherit 50% of our DNA from each of our parents but only 25%, on average, from each of our grandparents. There are, however, wide variations in the range of DNA inherited from our ancestors. Because of the random nature of inheritance, once we go back about six to seven generations, we may have ancestors from whom we have inherited no DNA at all. This also means that we will inherit proportionally more DNA from the remaining genetic ancestors represented in our tree. In order to have a match with a more distant relative, both cousins need to have inherited some DNA from the ancestor of interest. They must also share DNA on the same segment of a specific chromosome in order to show up as a match. The chances of this happening rapidly diminish beyond the fourth and fifth cousin level.[30] We can, however, increase the chances of finding matches

29. Antonio Regalado, "2017 Was the Year Consumer DNA Testing Blew Up," *MIT Technology Review*, 12 February 2018 (https://www.technologyreview.com/s/610233/2017-was-the-year-consumer-dna-testing-blew-up/).

30. "Cousin Statistics," *ISOGG Wiki* (https://isogg.org/wiki/Cousin_statistics).

with cousins who descend from ancestors of interest by testing other family members. Siblings or cousins will have inherited different segments of DNA than we have and will have matches that we do not have.

Our genome contains a mosaic of pieces of DNA that have been inherited from many different ancestors, both recent and ancient. Because of the progressive loss of our genetic ancestors with each succeeding generation, most of our DNA is inherited from only a subset of our more distant ancestors.[31] Theoretically, the number of our ancestors doubles with every generation. If we go back thirty generations to the year 1000 AD, using thirty-three years per generation, we each have over one billion ancestors. Yet the world population in the year 1000 AD was estimated to be just 253 to 283 million.[32] This is known as the ancestor paradox. The reason for the discrepancy is, of course, that our ancestors do not all occupy separate positions on our family trees. Pedigree collapse, which occurs when marriages take place between people who are already related to each other, causes the same ancestor to appear in our tree more than once. Computer models suggest that all humans alive today share the same genealogical ancestors from just a few thousand years ago.[33] A genetic analysis of 2,257 Europeans showed that all Europeans are genealogically related to each other within a very recent time frame and that most people alive today in Europe share the same set of ancestors from just 1,000 years ago.[34] We are, therefore, all connected to each other through a complicated network of relationships via multiple ancestral pathways. The degree of pedigree collapse varies between and within populations. It will be more pronounced in endogamous populations, where people have historically married within the same community. This can happen, for example, in island populations or in cultures where people are encouraged to marry someone with the same religion or heritage. Few of us can trace all of our ancestors on all our lines back for more than a few generations. The vast majority of this tangled network will be invisible to the genealogist, though there are many genealogists who have been able to find recent evidence of pedigree collapse in their family trees as a result of ancestors marrying known cousins.

The limitations of our family tree research, the complexities of our shared population history, and the random nature of the inheritance process mean that the application of atDNA for genealogical research is fundamentally restricted. There will always come a point where a genetic match cannot be confidently ascribed to a specific ancestor or ancestral couple. We cannot rule out the possibility that the DNA is shared, not by the relationship we have

31. Graham Coop, "How Many Genetic Ancestors Do I Have?," *gcbias: The Coop Lab Blog*, 11 November 2013 (https://gcbias.org/2013/11/11/how-does-your-number-of-genetic-ancestors-grow-back-over-time/).

32. Scott Manning, "Year-by-Year World Population Estimates: 10,000 B.C. to 2007 A.D.," *Historian on the Warpath*, 12 January 2008 (https://scottmanning.com/content/year-by-year-world-population-estimates/).

33. Douglas L. T. Rohde et al., "Modelling the Recent Common Ancestry of All Living Humans," *Nature* 431 (30 September 2004): 562–66 (https://doi.org/10.1038/nature02842).

34. Peter Ralph and Graham Coop, "The Geography of Recent Genetic Ancestry across Europe," *PLOS Biology* 11 (7 May 2013) (https://doi.org/10.1371/journal.pbio.1001555). See also Peter Ralph and Graham Coop, "European Genealogy FAQ," *gcbias: The Coop Lab Blog*, 25 April 2013 (https://gcbias.org/european-genealogy-faq/).

identified in our family tree, but through another more distant relationship that is beyond the reach of genealogical records. That point will vary depending on the availability of genealogical records and the history of the ancestral population. In practice, atDNA tests can generally only be reliably used in combination with documentary records to identify relationships within the last five or six generations.

There are also limitations with the atDNA tests that are currently on the market. The microarray chips used by the testing companies cover between 600,000 and 930,000 autosomal SNPs scattered across the genome, depending on the chip used by the company.[35] This represents just a tiny fraction of the 10 million or more SNPs that have been identified in the human genome.[36] Each company uses its own proprietary methodology to predict relationships. They look for consecutive runs of matching SNPs to infer shared segments of DNA. These predictions become increasingly less reliable for more distant relationships, and, especially, when the amount of sharing falls below 15 cMs.[37] In my own research, I have found that a sizeable percentage of my matches do not match either of my parents. At FamilyTreeDNA, 26% of my matches do not appear on my parents' match lists, compared with 36% at AncestryDNA, and 29% at MyHeritage.[38] These results are not directly comparable because each company uses different thresholds for matching.[39] While some of these missing matches might be false negatives, it seems likely that many are false positives as a result of the limitations of the tests. The processes of statistical phasing and imputation, along with differences in SNP coverage on the chips, can introduce errors. If parent-child phasing is not used, there is a high error rate, especially for smaller segments under 15 cMs.[40] Some inferred segments might fall in SNP-poor areas. This could potentially result in small segments being artificially joined together to make a larger segment because the SNPs that would break up the run are not included on the chip. Over-matching is a good indicator of problem areas, which are known colloquially as pileups.[41] Haplotype frequency, along with segment size, is used for detecting identical by descent (IBD) segments—those likely to have

35. "Autosomal SNP Comparison Chart," *ISOGG Wiki* (https://isogg.org/wiki/Autosomal_SNP_comparison_chart).

36. "What Are Single Nucleotide Polymorphisms (SNPs)?," *Genetics Home Reference,* National Institutes of Health, U.S. Library of Medicine (https://ghr.nlm.nih.gov/primer/genomicresearch/snp).

37. Blaine T. Bettinger, "A Small Segment Round-Up," *The Genetic Genealogist*, 29 December 2017 (https://thegeneticgenealogist.com/2017/12/29/a-small-segment-round-up/).

38. Debbie Kennett, "Comparing Match Tallies for Family Members with Family Tree DNA's Family Finder Test," *Cruwys News*, 29 July 2017 (https://cruwys.blogspot.com/2017/07/comparing-match-tallies-for-family.html). See also Debbie Kennett, "Comparing Parent and Child Matches at AncestryDNA," *Cruwys News*, 6 August 2017 (https://cruwys.blogspot.com/2017/08/comparing-parent-and-child-matches-at.html). See also Debbie Kennett, "MyHeritage DNA Updates Announced at Rootstech," *Cruwys News*, 4 March 2018 (https://cruwys.blogspot.com/2018/03/myheritage-dna-updates-announced-at.html).

39. "Autosomal DNA Match Thresholds," *ISOGG Wiki* (https://isogg.org/wiki/Autosomal_DNA_match_thresholds).

40. For more on phasing see "Phasing," *ISOGG Wiki* (https://isogg.org/wiki/Phasing).

41. Debbie Kennett, "Small Segments and Pile-Ups - a Visualisation," *Cruwys News*, 22 January 2018 (https://cruwys.blogspot.com/2018/01/small-segments-and-pile-ups.html).

been inherited from a common ancestor.[42] AncestryDNA has an algorithm known as Timber that is used to downweight the cM count for high-frequency matches.[43] 23andMe has a proprietary list of problem areas for which it makes adjustments. DNA.Land uses allele frequencies to distinguish between recent and "ancient" shared segments. Otherwise, very limited use is made of haplotype frequency. This is an area that is ripe for further study and which would lend itself to citizen science investigations.

Within the genetic genealogy community, we now have a large amount of empirical evidence from parent-child comparisons and extended family studies about the effectiveness of atDNA testing. The Shared cM Project, a citizen-science study initiated by Blaine T. Bettinger, is collecting data on the amount of DNA shared for known genealogical relationships.[44] The most recent update of the project provided information for over 25,000 known relationships.[45] Because of the amount of data collected, outliers can easily be detected and removed from the dataset. The data can be used to check if a relationship falls within the expected range. We can also look at the amount of DNA shared and generate a list of the likely relationships. This process is now facilitated by the introduction of the "Shared cM" tool and the "What are the Odds?" tool on the DNA Painter website.[46]

Experience has shown that the currently available atDNA tests can be effectively used in combination with documentary records to confirm close relationships up to the second cousin level. With high matches for close relationships, there are only a limited number of possibilities and genealogical research can be used to identify the precise relationship. All relationships up to the second cousin level can be reliably detected through DNA. Theoretically, there is a tiny possibility that two second cousins would not share sufficient DNA for a match to be detectable,[47] but there are no proven examples where this has happened. The further back in time to a common ancestor, the more difficult it is to predict the relationships. Problems with false matches and the wider range of possible relationships contribute to this problem. Our match lists at all the testing companies are dominated by matches with more distant cousins, with whom we generally share just one segment of DNA, but there is currently no way of determining how far back we are related. They could potentially be our fifth cousins or they could be very distant cousins.[48] Because we have so many more distant

42. Sharon R. Browning and Brian L. Browning, "Identity by Descent between Distant Relatives: Detection and Applications," *Annual Review of Genetics* 46 (2012): 617–33 (https://doi.org/10.1146/annurev-genet-110711-155534).

43. Julie M. Granka, "Filtering DNA Matches at AncestryDNA with Timber" *Ancestry Blog*, 8 June 2015 (https://blogs.ancestry.com/ancestry/2015/6/8/filtering-dna-matches-at-ancestrydna-with-timber/).

44. Blaine T. Bettinger, "The Shared cM Project: A Demonstration of the Power of Citizen Science," *Journal of Genetic Genealogy* 8 (2017): 38–42 (http://jogg.info/pages/vol8/editorial/bettinger/bettinger-sharedcMProject.html).

45. Blaine T. Bettinger, "August 2017 Update to the Shared CM Project," *The Genetic Genealogist*, 26 August 2017 (http://thegeneticgenealogist.com/2017/08/26/august-2017-update-to-the-shared-cm-project/).

46. *DNA Painter* (https://dnapainter.com/tools).

47. "Cousin Statistics," *ISOGG Wiki* (https://isogg.org/wiki/Cousin_statistics).

48. Peter Ralph and Graham Coop, "Identification of Genomic Regions Shared between Distant Relatives," *gcbias; The Coop Lab Blog*, 10 May 2013 (https://gcbias.org/2013/05/10/identification-of-genomic-regions-shared-between-distant-relatives/).

cousins, it is more likely that the relationship will be very distant, and, hence, beyond the reach of documentary records.[49]

Relationship predictions are more difficult if there are recent consanguineous marriages in your family tree (marriages between close relatives such as first or second cousins) or if a person is descended from a highly endogamous population. In these situations, two people will share an inflated amount of DNA because of the double or multiple relationships. They may also have some fully identical regions where they match on both the maternal and paternal chromosomes.[50] Not all of the testing companies currently take these regions into account when making relationship predictions. There is great scope for improvement in this area.

The growth of the autosomal databases

By September 2018, over seventeen million people had taken an atDNA test at the four major testing companies. There are no official breakdowns by country, but I would estimate that around 80% of those test takers reside in the U.S. This means that around one in twenty-five Americans are now represented in the databases. The databases are effectively at critical mass for most Americans, and particularly those with deep roots in the U.S. It is taking longer for other countries to catch up, partly because the tests are often much more expensive than they are in the U.S. Also, AncestryDNA, the company with the largest database, launched its autosomal test in the U.S. three years before making it available elsewhere. While FamilyTreeDNA, MyHeritage, and Living DNA sell their tests in most countries of the world, the AncestryDNA test is only available in thirty-five countries,[51] and the 23andMe test is only available in fifty-four countries.[52] Most of the people in the atDNA databases are from North America and Europe. The highest representation outside the U.S. appears to be in Australia, Canada, Ireland, New Zealand, and the U.K., but DNA testing is also very popular in Sweden, Norway, and Finland. It is the growth of the databases, particularly outside the U.S., that is likely to have the biggest impact on our genealogical research.

The promise of autosomal STRs

Autosomal STRs are routinely used in forensic science for identification purposes. Forensic tests typically use small panels of up to 27 STRs that have been specially chosen for their high discriminatory power.[53] However, over one million STRs have been identified in the

49. Steve Mount, "Genetic Genealogy and the Single Segment," *On Genetics*, 19 February 2011 (http://ongenetics.blogspot.com/2011/02/genetic-genealogy-and-single-segment.html).

50. Paul Woodbury, "Endogamy Part 1: Exploring Shared DNA" *Legacy Tree*, 13 October 2016 (https://www.legacytree.com/blog/dealing-endogamy-part-exploring-amounts-shared-dna).

51. "AncestryDNA® Availability by Country," *Ancestry* Support (https://support.ancestry.com/s/article/AncestryDNA-Availability).

52. "What Countries Do You Ship To?," *23andMe*, Customer Care (http://eu.customercare.23andme.com/hc/en-us/articles/204712980-What-countries-do-you-ship-to).

53. Nicole M. M. Novroski, August E. Woerner, and Bruce Budowle, "Potential Highly Polymorphic Short Tandem Repeat Markers for Enhanced Forensic Identity Testing," *Forensic Science International: Genetics* 37 (1 November 2018): 162–71 (https://doi.org/10.1016/j.fsigen.2018.08.011).

human genome.[54] They are not currently used for genealogy or forensics. The Sorenson Molecular Genealogy Foundation (SMGF) experimented with the use of autosomal STR data combined with pedigree data for genealogical purposes in the early years of DNA testing. SMGF scientists presented an interesting poster at the 2003 meeting of the American Society of Human Genetics. They showed how autosomal haplotypes of three STRs on chromosome 2 were transmitted in a six-generation pedigree.[55] By August 2009, SMGF had tested 78,568 samples for an average of 68.5 autosomal STRs. They had also tested 73,394 samples for thirteen X chromosome STRs.[56] The DNA assets of SMGF were purchased by Ancestry.com in 2012.[57] Ancestry launched its autosomal SNP product in the U.S. that same year. The autosomal STR data appears to have been put on the backburner and has never been made publicly available or published. Autosomal STRs are currently a neglected resource, and there is a wealth of potential information that is yet to be exploited. STRs are routinely used in combination with SNPs to increase the power of Y-DNA testing and are likely to have a similar application for autosomal DNA testing.

The Promise of Whole Genome Sequencing

Direct-to-consumer whole genome sequencing (WGS) tests have been available for a number of years, and a WGS test can now be purchased for under $500. It can only be a matter of time before the dream of the $100 genome is realized. Once this tipping point is reached, we can expect to see the genetic genealogy companies move from microarrays to whole genome sequencing. In consumer genomics, it is predicted that sixty million Americans will have had their DNA sequenced by 2025.[58]

WGS data is being generated on a large scale for research purposes. The fruits of this research will inform the development of the commercial tests. The Broad Institute of MIT and Harvard announced on DNA Day in April 2018 that they had sequenced 100,000 genomes; they estimated that almost one million genomes had been sequenced worldwide.[59] Other large-scale sequencing projects are under way or in the pipeline. In April 2018, U.K.

54. Melissa Gymrek et al., "Interpreting Short Tandem Repeat Variations in Humans Using Mutational Constraint," *Nature Genetics* 49 (11 September 2017): 1495–1501 (https://doi.org/10.1038/ng.3952).

55. S. R. Woodward et al., "Large Scale DNA Variation as an Aid to Reconstruction of Extended Human Pedigrees," American Society of Human Genetics, 2003 (https://web.archive.org/web/20080626062033/http://www.smgf.org/resources/papers/PosterASHG2003.pdf).

56. CeCe Moore, "An Important Update on SMGF from Dr. Tim Janzen," *Your Genetic Genealogist*, 30 June 2012 (http://www.yourgeneticgenealogist.com/2012/06/important-update-on-smgf-from-dr-tim.html).

57. CeCe Moore, "Ancestry.Com Buys GeneTree (Another Competitor) and Launches Their New Autosomal DNA Product to Subscribers," *Your Genetic Genealogist*, 3 May 2012 (http://www.yourgeneticgenealogist.com/2012/05/ancestrycom-buys-genetree-and-launches.html).

58. Razib Khan and David Mittelman, "Consumer Genomics Will Change Your Life, Whether You Get Tested or Not," *Genome Biology* 19 (20 August 2018): 120 (https://doi.org/10.1186/s13059-018-1506-1).

59. Broad Communications, "Broad Institute Sequences Its 100,000th Whole Human Genome on National DNA Day," Broad Institute News and Media, 25 April 2018 (https://www.broadinstitute.org/news/broad-institute-sequences-its-100000th-whole-human-genome-national-dna-day).

Biobank announced plans to sequence all 500,000 participants.[60] The U.S. National Institutes of Health announced in May 2018 the launch of the All of Us research program. It aims to sequence the DNA of one million Americans.[61]

The advent of the WGS era will bring exciting possibilities for genealogists but is also likely to present significant challenges. The microarrays used by the consumer testing companies cover up to one million SNPs. A whole genome sequence will, in theory, provide results for all three billion base pairs. This will require a massive increase in computer processing power. Relative matching is computationally intensive. Each person in the database has to be compared one to one to everyone else in the database. The number of pairwise comparisons increases exponentially.[62] The promise of WGS can only be fulfilled if computing power and storage can be increased to cope with the large datasets. The data file from a WGS, even in compressed format, is over 100 gigabytes in size.

WGS should improve the accuracy of relationship predictions and reduce the number of false positive matches, especially on smaller segments. However, the current methods of IBD detection are based on the number and size of shared segments of DNA, and the returns are not likely to be that great for genealogical purposes. A 2014 study looked at relationship predictions from WGS in a dataset of 258 individuals in 30 families with 490 known relationships. The authors found that by moving from SNP microarrays to WGS they were able to detect "all 1st through 6th degree relationships and 55% of 9th through 11th degree relationships in the 30 pedigrees." There was "a 5% to 15% increase in power for distant relationships between 7th and 11th degree."[63]

The promise of using WGS for relationship predictions possibly lies in the use of alternative or complementary methods of estimating relatedness. WGS will allow researchers to detect all variants in the genome and not just the known SNPs included on the microarrays. Information about rare variants could be combined with information about the size and frequency of the shared segments. This would allow us to make a better estimate of their age and ancestral history. It is likely that we will be able to identify rare variants that are unique to certain families or lineages. We will also be able to ascertain the order in which the variants were inherited and construct a phylogeny, in much the same way that we currently do for Y-SNPs and mtDNA SNPs. A team of researchers from the University of Toledo in Ohio were able to detect distant relationships up to the eighth and ninth degree by counting the number of rare shared variants identified through WGS. The number of rare variants

60. "Whole Genome Sequencing Will 'Transform the Research Landscape for a Wide Range of Diseases,'" U.K. *Biobank*, 5 April 2018 (http://www.ukbiobank.ac.uk/2018/04/whole-genome-sequencing-will-transform-the-research-landscape-for-a-wide-range-of-diseases/).

61. Dina Fine Maron, "Can the U.S. Get 1 Million People to Volunteer Their Genomes?," *Scientific American* (https://www.scientificamerican.com/article/can-the-u-s-get-1-million-people-to-volunteer-their-genomes/).

62. Leah LaPerle Larkin, "Fun With Factorials," *The DNA Geek*, 7 March 2018 (http://thednageek.com/fun-with-factorials/).

63. Hong Li et al., "Relationship Estimation from Whole-Genome Sequence Data," *PLOS Genetics* 10 (30 January 2014) (https://doi.org/10.1371/journal.pgen.1004144).

per individual varies from one geographical region to another, so it is also important to take into account the underlying population structure.[64] In an ancient DNA study, Schiffels and coauthors developed a tool known as "rarecoal" that looked at the sharing of rare variants to infer relationships within and between populations.[65]

In order to make use of rare variants, it will be important to determine the phase of the sequence so that we can work out whether the variants are on the maternal or paternal chromosomes. Phasing is difficult with short-read technologies. Because of the cost and complexity, researchers have historically worked with consensus sequences where the parental source of the variant is not known. There are a number of different strategies for phasing whole genomes. A summary of the different approaches is provided by Choi and coauthors.[66] These methods will improve over time as larger reference panels become available and as scientists refine their algorithms.

The first draft of the human genome reference sequence was published in 2003. Great progress has been made in the last fifteen years, especially since the introduction of next generation sequencing. The Genome Reference Consortium publishes updates to the reference sequence. The most recent update, GRCh38 (Build 38), was published in 2017.[67] However, there are still many unresolved areas in the human genome, particularly in the repetitive regions such as the centromeres and telomeres. These areas cannot be reliably mapped with short-read technologies. New long-read technologies from PacBio and Oxford Nanopore are helping to decipher these intractable regions, but the long reads come at the expense of reduced accuracy.[68] As the technology evolves, long-read sequencing will improve mapping and phasing. The new technology will also make it easier to detect structural variants such as insertions, deletions, and copy number variants.[69] It is possible that some of these structural variants will help to define family lineages. The accuracy of the relationship predictions is likely to improve as more of the human genome is deciphered and mapped. It is possible that some of the problems we are seeing with pileups might be resolved as these occur at a much higher rate in or near the poorly mapped regions.[70]

---

64. Ahmed Al-Khudhair et al., "Inference of Distant Genetic Relations in Humans Using '1000 Genomes,'" *Genome Biology and Evolution* 7 (7 January 2015): 481–92 (https://doi.org/10.1093/gbe/evv003).

65. Stephan Schiffels et al., "Iron Age and Anglo-Saxon Genomes from East England Reveal British Migration History," *Nature Communications* 7 (19 January 2016) (https://doi.org/10.1038/ncomms10408).

66. Yongwook Choi et al., "Comparison of Phasing Strategies for Whole Human Genomes," *PLOS Genetics* 14 (5 April 2018) (https://doi.org/10.1371/journal.pgen.1007308).

67. Valerie A. Schneider et al., "Evaluation of GRCh38 and de Novo Haploid Genome Assemblies Demonstrates the Enduring Quality of the Reference Assembly," *Genome Research* 27 (1 May 2017): 849–64 (https://doi.org/10.1101/gr.213611.116).

68. Adam M. Phillippy, "New Advances in Sequence Assembly," *Genome Research* 27 (1 May 2017): xi–xiii (https://doi.org/10.1101/gr.223057.117).

69. Katherine Miller, "Beyond Short Reads: New Genomics Technologies Reveal Previously Unseen Structural Variants," *Clinical OMICs*, 11 April 2018 (https://clinicalomics.com/articles/beyond-short-reads-new-genomics-technologies-reveal-previously-unseen-structural-variants/1581).

70. Hong Li et al., "Relationship Estimation from Whole-Genome Sequence Data." *PLOS Genetics* 10, no. 1 (30 January 2014): e1004144 (https://doi.org/10.1371/journal.pgen.1004144).

In the short term, the major genetic genealogy companies will likely offer WGS as a standalone service for users interested in health and trait reports. This would also be useful for those who wish to sequence their genome for citizen-science research. A WGS test is likely to be a better value in the long term than a dedicated Y chromosome sequencing. It seems likely that WGS testing will completely replace the existing Y chromosome sequencing products. Eventually, there will be no need to order separate Y-DNA, mtDNA and atDNA tests. You will get all the information needed from a single WGS test, perhaps obtaining different reports on a pay-as-you-go basis. In future, it is likely that entire populations will be routinely sequenced as part of an integrated healthcare service. Sequences could then be uploaded to consumer WGS databases for further analysis. The competition will then be between the companies that can provide the most meaningful reports from the raw data and are able to attract the largest number of users to their matching databases.

Compatibility between tests run at different companies will not be an issue with WGS. However, imputation will still be needed for backwards compatibility with existing results where the customers are not willing or able to upgrade. We may see mergers or acquisitions and new companies entering the arena. Once the WGS databases have reached a reasonable size, the companies will be able to start pilot studies for relative matching. 23andMe has already experimented with WGS through its African American Sequencing Project, though this project is aimed at advancing health research and not genealogy research.[71] At the beginning of September 2018, 23andMe was gauging interest in a premium whole-genome service for less than one thousand dollars, though they do not appear to have any immediate plans to launch a WGS test.[72]

Advanced genetic genealogists have been among the early adopters of WGS. Some genealogists have participated in the Personal Genome Project, which provides for open sharing of WGS data to advance scientific research.[73] Some of the genetic genealogists who have taken a WGS test are mostly interested in the Y chromosome information but have also wanted to learn about the additional information from the rest of the genome. Although there are no matching databases available, there are tools that allow genealogists to extract a SNP file for upload to *GEDmatch*. Other genealogists have ordered direct-to-consumer WGS tests, either out of curiosity or to investigate health issues. Debbie Parker Wayne and Louis Kessler are writing about the WGS journey on their blogs.[74] Otherwise, the results are only

71. Turna Ray, "23andMe to Sequence Genomes of 925 African American Customers, Share Results With Researchers," *GenomeWeb*, 10 November 2016 (https://www.genomeweb.com/sequencing/23andme-sequence-genomes-925-african-american-customers-share-results-researchers).

72. Christina Farr, "23andMe is Gauging Interest in a $749 'Premium' Service that Would Offer Deeper Health Data," *CNBC*, 5 September 2018 (https://www.cnbc.com/2018/09/05/23andme-considering-749-dollar-premium-service.html).

73. *Personal Genome Project* (https://www.personalgenomes.org/).

74. Debbie Parker Wayne, "Whole Genome Sequence (Part 1)," *Deb's Delvings in Genealogy*, 25 April 2018 (http://debsdelvings.blogspot.com/2018/04/whole-genome-sequence-part-1.html). See also Debbie Parker Wayne, "Whole Genome Sequence (Part 2) - Analysis Tools," *Deb's Delvings in Genealogy*, 15 May 2018 (http://debsdelvings.blogspot.com/2018/05/whole-genome-sequence-part-2-analysis.html). See also Louis Kessler,

just starting to be explored, and little has been written. There is great scope for genealogists to research their own genomes and identify their own rare variants. We can expect to see advanced genetic genealogists doing comparative analyses of WGS versus microarray data. This will allow us to get a better understanding of the accuracy of the existing relationship predictions and a taste of things to come. Genealogists will need to be prepared for the possibility that their previous conclusions will be overturned once the rich genome data is available. Some apparently matching segments, especially the smaller ones, may well turn out to be false matches.

The Promise of Ancestral Reconstruction

Chromosome mapping is a well-established technique used by many advanced genealogists. It involves comparing DNA results from known relatives in order to assign particular segments of DNA to a specific ancestor or ancestral couple. If enough of your relatives have tested, a significant percentage of your chromosomes can be assigned in this way. The knowledge that a specific segment has been handed down from a particular ancestor means that, in theory, we will be able to identify the inheritance path of particular genes. This raises the tantalizing possibility that we might be able to make inferences about the physical features of our ancestors, based purely on the DNA of their living descendants.

The scientists at AncestryDNA announced in 2014 the results of a pilot study in which they were able to partially reconstruct the genome of a man by the name of David Speegle. Speegle lived in Alabama in the early 1800s. He was the focus of the research because he had two wives, Winifred Crawford and Nancy Garren, by whom he had many children. Those children, in turn, had many living descendants. There was, therefore, a good chance that fragments of his DNA would still be found in many of his descendants. The scientists were able to reconstruct about 50% of the genomes of Speegle and his two wives and identify segments that could be traced back to David Speegle. They were also able to identify portions of the genome containing SNPs associated with male pattern baldness and blue eyes. The study provided proof of concept for Ancestry's DNA Circles feature, but the results have not been published in a peer-reviewed journal.[75]

The first ancestral-reconstruction study published in the scientific literature appeared in 2018 in *Nature Genetics*. An international team of researchers reconstructed part of the genome of Hans Jonatan, who is thought to be the first person of African ancestry to settle in Denmark. Jonatan was born in the Caribbean in 1784. Historical sources indicated that his father was European and his mother was an enslaved African. Jonatan moved to Iceland in 1802, where he married and had two children. The researchers were able to trace 788 of

"My Whole Genome Sequencing. The VCF File," *BEHOLD Genealogy*, n.d. (http://www.beholdgenealogy.com/blog/?p=2879).

75. Thomas MacEntee, "AncestryDNA Reconstructs Partial Genome of Person Living 200 Years Ago," *GeneaPress*, 16 December 2014 (http://www.geneapress.com/2014/12/ancestrydna-reconstructs-partial-genome.html). See also Catherine Ball and Julie Granka, *AncestryDNA Reconstructs Partial Genome of Person Living 200 Years Ago, Ancestry* (https://www.youtube.com/watch?v=xiApcdRuWsM).

his descendants, 182 of whom were genotyped on a SNP chip, and twenty of whom had their whole genomes sequenced. They were able to reconstruct 38% of Jonathan's maternal genome, including 64% of his maternal chromosome 3. They were also able to retrieve 9% of his X-chromosome. The task was made easier because of the distinctive nature of the African chromosomal segments. The researchers did not make any inferences about Jonatan's phenotype, but it is possible that this could be the subject of further research.[76]

It would be the dream of all genealogists to generate a picture of their ancestors' faces from the DNA of their descendants, but is this ever likely to be a reality, or is it just a science fiction fantasy? Predicting appearance from DNA is also potentially useful in criminal investigations and there has been considerable research in this area. While it is self-evident that physical characteristics like hair, eye color, and skin color are highly heritable, the inheritance of such traits is not as simple as once thought. Nevertheless, forensic scientists have identified a number of SNPs associated with visual traits. It is now possible to make probabilistic predictions about some features. Researchers at the Erasmus Medical Center in Rotterdam have developed "HIrisPlex-S," a webtool that can be used to predict eye, hair, and skin color with varying degrees of accuracy.[77] The Visible Attributes through Genomics (VISAGE) Consortium in Europe is in the process of developing a next generation sequencing toolkit to predict traits for forensic purposes.[78] The third-party website *GEDmatch* has an eye color predictor tool using forty-one SNPs. It works best on European populations, though with variable results.

A company called Parabon NanoLabs, based in Virginia in the U.S., has developed a test for forensic purposes known as the Snapshot Forensic DNA Phenotyping System. This test provides a composite profile of an individual's facial features based on their DNA. Genotyping is done on an Illumina Cyto-SNP chip covering around 850,000 SNPs.[79] Parabon claims that its test "accurately predicts genetic ancestry, eye color, hair color, skin color, freckling, and face shape in individuals from any ethnic background, even individuals with mixed ancestry." However, the methods have not been published and subjected to peer review, and they have not been forensically validated. Critics claim that much of the information could just as easily be gleaned by looking at a person's ancestry and gender.[80]

76. Anuradha Jagadeesan et al., "Reconstructing an African Haploid Genome from the 18th Century," *Nature Genetics* 50 (February 2018): 199–205 (https://doi.org/10.1038/s41588-017-0031-6). See also Gisli Palsson, *The Man Who Stole Himself: The Slave Odyssey of Hans Jonathan* (Chicago, Ill.: University of Chicago Press, 2016).

77. "HIrisPlex-S Eye, Hair and Skin Colour DNA Phenotyping Webtool," Department of Genetic Identification of Erasmus MC, Rotterdam, The Netherlands (https://hirisplex.erasmusmc.nl/).

78. Turna Ray, "European Consortium to Validate NGS Toolkit for Analyzing Phenotypes in Forensics," *GenomeWeb*, 4 May 2018 (https://www.genomeweb.com/research-funding/european-consortium-validate-ngs-toolkit-analyzing-phenotypes-forensics).

79. Rachel Wiley et al., "Blind Testing and Evaluation of a Comprehensive DNA Phenotyping System," in Poster Presented at the *International Symposium on Human Identification* (http://docs.parabon.com/pub/Parabon_Snapshot_Scientific_Poster-ISHI_2016.pdf).

80. Laura DeFrancesco, "DNA Makes an Appearance," News, *Nature Biotechnology,* 10 January 2018 (https://doi.org/10.1038/nbt.4057). See also Sense About Science, "Making Sense of Forensic Genetics" (London: Sense About Science, 2017).

Another company called Human Longevity has also been researching phenotype prediction. It published some preliminary results in 2017 in the *Proceedings of the National Academy of Sciences*. The company used whole genome sequencing on 1,061 individuals of diverse ancestry and developed algorithms to re-identify these individuals using various biometric traits.[81] However, the research was widely criticized by other scientists, including one of the coauthors of the paper who no longer works for the company. Critics again claimed that the same results could be achieved simply by knowing a person's sex, age, and ethnicity.[82]

More promising is a 2018 study from Crouch and coauthors on the genetics of the human face. The researchers used samples from the People of the British Isles study and a U.K. twins cohort. They were able to identify three genetic variants that affected the eyes and facial profiles.[83]

Other traits such as height have turned out to be even more complex. It is clear that there are hundreds or thousands of places in the genome associated with any given trait, but each of these variants only has a very tiny effect. It is possible to come up with a scoring system, known as the polygenic score, to predict the likelihood that an individual will have a particular trait or a disease risk; at present these are only weak predictors.[84] Trait reports are provided by 23andMe. AncestryDNA has started offering trait reports to customers in the U.S. A limited selection of free trait reports is available from *DNA.Land*, if you agree to share your data for scientific research.

Current studies already use hundreds of thousands of samples, mostly in Europeans. As more people have their DNA sequenced, it will be possible to look for effects in sample sizes running into the millions and in many different populations. This will allow scientists to discover many more variants, and to predict traits and disease risk with greater accuracy. However, environmental factors such as diet and socioeconomic status also have an important role to play. There are many other complicating factors. Our understanding will improve in the coming decades, but it is likely that it will never be possible to separate out all the environmental and cultural differences from the genetic effects.[85] Genealogists could potentially have an important role to play in this type of research. Multi-generation family studies could be undertaken where genealogists have genealogical and genetic records combined with

81. Christoph Lippert et al., "Identification of Individuals by Trait Prediction Using Whole-Genome Sequencing Data," *Proceedings of the National Academy of Sciences* 114 (19 September 2017): 10166–71 (https://doi.org/10.1073/pnas.1711125114).

82. Sara Reardon, "Geneticists Pan Paper That Claims to Predict a Person's Face from Their DNA," *Nature News* 549 (8 September 2017): 139 (https://doi.org/10.1038/nature.2017.22580).

83. Daniel J. M. Crouch et al., "Genetics of the Human Face: Identification of Large-Effect Single Gene Variants," *Proceedings of the National Academy of Sciences* 115 (23 January 2018): E676–85 (https://doi.org/10.1073/pnas.1708207114).

84. Brian Resnick, "How scientists are learning to predict your future with your genes," *Vox*, 23 August 2018 (https://www.vox.com/science-and-health/2018/8/23/17527708/genetics-genome-sequencing-gwas-polygenic-risk-score).

85. Graham Coop, "Polygenic Scores and Tea Drinking," *gcbias; The Coop Lab Blog*, 14 March 2018 (https://gcbias.org/2018/03/14/polygenic-scores-and-tea-drinking/).

medical records. Family photographs could also be used to track the inheritance of visual traits across the generations. Ultimately, it should then be possible to make inferences about the phenotype of our ancestors from the sections of their genome that we have been able to reconstruct, though the process is never likely to be completely reliable.

DNA from Our Ancestors

The process of reconstructing the genomes of our ancestors would be greatly helped if we could get direct DNA evidence from our forebears. The mass exhumation of the graves of our ancestors is not likely to be a realistic prospect for obvious reasons, but it is possible to extract DNA from hair, teeth, postage stamps, and other items. This type of testing is mostly done in forensic labs, but some companies are starting to offer tests aimed at the genealogy market. An Australian company called Totheletter DNA[86] provides DNA testing from artifacts. They produce results that they claim can be uploaded to *GEDmatch*'s Genesis database, though at the time of writing no one had reported receiving any results. Living DNA was able to successfully extract DNA from 30-year-old envelopes to confirm the identity of a foundling's father. They now offer such testing on a one-to-one basis.[87] However, such tests are still very expensive and, even with advances in technology, the success rate is likely to be very low. There is also the problem of contamination. For example, there is more DNA on the outside of an envelope than under the stamp or the envelope flap. There is also no guarantee that the stamp or envelope has been licked by the sender. It is generally only mtDNA that can be obtained from hair samples, though it is possible to extract nuclear DNA if the hair follicle or root is available.[88]

The Power of Big Data

It has long been the genealogist's dream to have "one world family tree" that integrates both genealogical and genetic records. In an ideal world, all the information in the trees would be supported by source citations. While that dream is not likely to be realized for many years, there are already a number of initiatives that are attempting to aggregate family tree data into giant databases. Some of these also provide opportunities for collaborative work. About 95% of the population of Iceland from the last three centuries is documented in an online version of the Íslendingabók–the *Book of Icelanders*; this has proved to be a valuable resource for geneticists.[89] The *FamilySearch* Family Tree aims to provide "One complete, accurate re-

86. *Totheletter DNA* (https://www.totheletterdna.com/).

87. "Living DNA Provide Closure on Lifetime Search for Biological Father," *Living DNA*, 19 March 2018 (https://livingdna.com/news/living-dna-provide-closure-on-lifetime-search-for-biological-father).

88. Caroline Hughes, "Challenges in DNA Testing and Forensic Analysis of Hair Samples," *Forensic Magazine*, 2 April 2013 (https://www.forensicmag.com/article/2013/04/challenges-dna-testing-and-forensic-analysis-hair-samples).

89. Olga Khazan, "How Iceland's Genealogy Obsession Leads to Scientific Breakthroughs," *The Atlantic*, 7 October 2014 (https://www.theatlantic.com/health/archive/2014/10/how-icelands-genealogy-obsession-leads-to-scientific-breakthroughs/381097/).

cord for each person who has lived on the earth."[90] By the end of 2017, the tree contained records for over 1.2 billion people.[91] *Wikitree* is a free website that uses both DNA and genealogical records to produce a single collaborative tree. As of 8 September 2018, the tree included 18,322,357 profiles and, of these, 4,414,853 had DNA test connections.[92]

Ancestry.com is working on an aggregated Big Tree that will include everyone in its database.[93] This tree powers some of the company's features such as its We're Related app.[94] Ancestry has a collection of over 100 million trees,[95] but no statistics are available on the composition of the Big Tree. It is used internally and is not accessible to users, but as the amount of genealogical data increases and the machine learning improves, the Big Tree is likely to drive the development of interesting new automated features.

Purists might scoff at the idea of genealogy by algorithm, but there are many new insights to be derived from analyzing giant aggregated data sets. The scientists from MyHeritage published a fascinating study in which they analyzed family trees containing millions of people at their sister site *Geni.com*. They used sophisticated algorithms to clean up the trees and fix errors, leaving them with 5.3 million independent family trees. The largest of these included 13 million individuals covering an average of eleven generations. From this refined dataset, they looked at longevity, migration, and marriage patterns.[96]

Population geneticists are also looking at new ways of interpreting big data. The concept of something known as the ancestral recombination graph (ARG) was developed in the 1990s. The ARG has enormous potential for the interpretation of genetic sequence data and has been described as the Holy Grail of evolutionary genetics. The idea is that you can use recombination points (the places where segments get broken up) and coalescent events (the points where lineages join together) to draw up an evolutionary tree for a set of samples. The method can be used to explore the history of an entire chromosome, a section of a chromosome, or a single gene, and will produce the "best fit" tree from the data.[97] We could perhaps think of it as a type of chromosome mapping on steroids. The work is mathemat-

90. "Contributing to the Family Tree - FamilySearch Developers," *FamilySearch* (https://www.familysearch.org/developers/docs/guides/scoe-model).

91. "FamilySearch 2017 Genealogy Highlights," *FamilySearch*, 11 January 2018 (https://media.familysearch.org/familysearch-2017-genealogy-highlights).

92. *WikiTree* (https://www.wikitree.com/).

93. Doug Black, "Where Do We Come From? Ancestry.Com's 19B-Record 'Big Tree' Database May Have the Answer," *EnterpriseTech*, 20 December 2016 (https://www.enterprisetech.com/2016/12/20/24561/).

94. Randy Seaver, "Is There Really an Ancestry.Com 'Big Tree?,'" *Genea-Musings* 9 March 2017 (https://www.geneamusings.com/2017/03/is-there-really-ancestrycom-big-tree.html).

95. "The many ways our family has grown," *Ancestry.com* (https://www.ancestry.com/corporate/about-ancestry/company-facts).

96. Joanna Kaplanis et al., "Quantitative Analysis of Population-Scale Family Trees with Millions of Relatives," *Science* (1 Mar 2018): 171–75 (https://doi.org/10.1126/science.aam9309).

97. Adam Siepel, "Our Paper: Genome-Wide Inference of Ancestral Recombination Graphs," *Haldane's Sieve*, 2 July 2013 (https://haldanessieve.org/2013/07/02/our-paper-genome-wide-inference-of-ancestral-recombination-graphs/).

ically challenging and computationally intensive, but potentially will have many applications in genetic genealogy.

Conclusion

DNA testing has been a transformative tool for family history research, but we have so far only scratched the surface of its potential. We can expect to see the databases continue to grow at a steady pace. In the future, with hundreds of millions of people from around the world in the DNA databases, everyone will have the opportunity to make exciting discoveries. We will also have ancient DNA databases where we can search for matches with our distant ancestors from pre-history. As we enter the genomic era, we will be able to purchase a single test that will give us fully integrated results and matches for Y-DNA (for males), mtDNA, X-DNA, and atDNA.

The quantity and quality of genealogical data will continue to improve. Advances in machine learning and sophisticated algorithms will allow the data to be mined in innovative ways. We will learn much more about the effects of population structure on the matching process. Companies will be able to develop bespoke matching algorithms that take into account the amount of endogamy and pedigree collapse in a specific population. At the click of a button, we will be able to see which ancestors contributed particular traits to our genome. The faces of our ancestors will be reconstructed through the DNA of their descendants. The time will come when it will be possible to take a DNA test, put your name into a database, and produce an instant, fully sourced family tree, complete with family photographs and composite facial reconstructions. It is likely to be many years before that happens and something we will never see in our own lifetime, but in the meantime, we can all join in on the journey toward the dream.

Glossary

Ahnentafel number: A genealogical numbering system for listing a person's direct ancestors. The subject is number 1; the subject's father is number 2, mother is number 3, the grandparents are numbers 4 to 7, and so on, back through the generations. Male ancestors are even numbers; females are odd numbers.

allele: An inherited variant (A, C, G, or T) occupying a specific spot on a chromosome that often results from a mutation.

ancestor paradox Because the number of ancestors doubles with each generation back in time (two parents, four grandparents, eight great-grandparents, and so on), at some generations back a person theoretically has more ancestors than the number of people who were living at that time. See also pedigree collapse.

autosomal DNA (atDNA): The chromosomes numbered 1 through 22 are the autosomes or the autosomal DNA.

base pair: See megabase pair.

CA: See Common Ancestor.

centimorgan (cM): A measure of genetic distance; the distance between two points on a chromosome for which there is an average of one crossover in a single generation. The cM is the most common method for describing (or "measuring") a DNA segment.

centromere: A connection/linkage point for sister chromatids during meiosis and displayed in chromosome diagrams as constricted or pinched areas. This area may contain fewer known variants for testing.

chain of custody: In forensics and in legal contexts, an unbroken trail that records the collection, possession, and sequence in the handling of evidence.

chip: See microarray.

chromatid: One half of a chromosome that has undergone replication initially into two identical copies. Those that are joined by the centromere are called sister chromatids. Non-sister chromatids are not identical and represent the product of the homologous chromosome in the pair.

chromosome mapping: A technique used in atDNA analysis to assign DNA segments to the ancestor from which the segment was inherited.

clade: A Y-DNA or mtDNA haplogroup. See also subclade.

cM: See centimorgan.

Common Ancestor (CA): An ancestor or ancestral couple shared by researchers.

coding region: An area of mtDNA that codes for genes.

control region: An area of mtDNA that contains non-coding DNA (not genes).

convergence: The process that results in two unrelated people sharing DNA that has mutated separately to the same or similar values.

copy number variant: Variations among individuals in the numbers of copies or repeating sequences of DNA.

cover (or coverage): In atDNA analysis, triangulated groups "cover" an area, chromosome, or an entire genome, when the groups are adjacent to each other, and the end point of one is the start point of the next triangulated group. In relation to Next Generation Sequencing, coverage refers to how many times a specific location is read with higher numbers of reads providing higher quality results with fewer miscalls or no calls.

crossover points: The points on a chromosome at which the parent's two chromosomes exchange DNA segments resulting in a single chromosome usually with portions of both to be passed on to the next generation.

CSV file: A Comma Separated Values file that stores letters and numbers in plain text. Raw atDNA data, mtDNA reference differences, and Y-DNA SNP and STR values are often supplied as CSV files.

deletion: A DNA mutation where genetic material has been lost.

Direct To Consumer (DTC): A product or service offered directly to a consumer without going through an intermediary. A test taker can order directly from a company without a doctor or prescription.

documentary research (paper-based or event-based research): Traditional genealogical research that does not include DNA test results.

downstream: A DNA mutation (or SNP variant) that occurred after (later in time) an upstream SNP.

DTC: See Direct to Consumer.

endogamy: The practice of individuals marrying within the same community or group over a long period encompassing numerous generations. This genetic isolation eventually results in many, if not all, members of the community being distantly related to each other, often in multiple ways.

excess IBD: A matching segment of DNA shared by many people originating from a distant, shared ancestral population. See Identical By Descent.

Fibonacci sequence: A sequence of numbers introduced to Europeans by Leonardo Fibonacci in 1202, but is actually much older than that and has its origins in India. Every number in the sequence after the first two is the sum of the two preceding numbers. The number of individuals in each generation in the X chromosome pattern of inheritance adds up to a Fibonacci number.

FIR: See Fully Identical Region.

Fully Identical Region (FIR): A DNA segment on both the maternal and paternal chromosomes shared by two people. This is most often seen in full siblings or double cousins, but may also be seen in endogamous populations.

GEDCOM: An acronym standing for GEnealogical Data COMmunication which is a specification for transferring data between genealogy programs.

gene: A region of DNA that codes for a function such as a trait, characteristic, or biological function.

genealogy triangulation: (1) Shared match analysis. (2) At least three matches in a triangulated group who all agree on the same MRCA or the same ancestral line. Some researchers recommend that there should be as many matches in agreement on the MRCA line as there are "greats" back to the MRCA—for example, six matches if the MRCA is a sixth-great-grandparent.

genetic genealogists: Genealogists who use DNA tests results to answer genealogical questions.

genome: A complete set of DNA. See Whole Genome Sequence.

genotype: The allele values inherited from both parents at a specific DNA location. For example, if an A was inherited from one parent and a G from the other parent at the same location, the genotype is AG.

Half Identical Region (HIR): A DNA segment on either the maternal or paternal chromosome, but not both, shared by two people.

haplogroup: A group sharing a common ancestor on the Y-DNA line or mtDNA line. Separate phylogenetic or haplogroup trees exists for Y-DNA and mtDNA. See also clade or subclade.

haplotype: A set of DNA variations, or polymorphisms, that tend to be inherited together.

haplotype block: A group of markers that occur sequentially that are inherited together from a parent.

HIR: See Half Identical Regions.

HLA Region: A region of DNA called human leukocyte antigen (HLA) where excess IBD matches are seen.

heterozygous: Different alleles (A, C, G, or T) were inherited from each parent at a particular location. See homozygous.

homologous/homologue: A homologue is a member of a pair of chromosomes. For example, the pair of chromosomes labeled number one are considered an homologous pair.

homozygous: The same alleles (A, C, G, or T) were inherited from both parents at a particular location. See heterozygous.

IBD: See Identical By Descent.

IBS: See Identical By State and excess IBD.

ICW: See In Common With.

Identical By Descent (IBD): A non-recombined matching segment of DNA, shared by two or more people that was inherited from a common ancestor.

Identical By State (IBS): A matching segment of shared DNA that was *not* inherited from a recent common ancestor, but is likely because of a shared distant ancestral population. See excess IBD.

Illumina: Ilumina, Inc. is an American company that develops, manufactures and markets systems for the analysis of genetic material. The most common autosomal DNA test for genealogy use Illumina chips or microarrays.

imputation: A computational technique used to infer (impute) gaps in tested DNA sequences.

In Common With (ICW): A list of shared matches found in the list of multiple test takers. An ICW list is a subset of two people's match lists containing only the people found on both lists. ICW does not imply triangulation: some segments will triangulate and some will not.

insertion: A DNA mutation where genetic material has been added.

interference: During recombination there is a decreased likelihood of two crossovers in nearby locations on the same chromosome due to interference.

locus (pl. loci): A specific physical position or point on the chromosome.

matriline: The line of inheritance through which mtDNA passes; the line from a person to his or her mother, to her mother (the maternal grandmother), to her mother, and so on.

Mbp/Mb: Megabases / megabase pairs. See megabase pairs.

megabase pair (Mbp/Mb): Megabase pair or 1,000,000 base pairs, where each one is a paired set of the building blocks of our DNA. This gives us a physical "address" or location for the start and end points of shared segments. The differences in the start and end points give a measure of the physical length of a segment. One Mbp is roughly equal to 1 cM, but not exactly the same. The term megabase (Mb) also refers to a specific location on a chromosome, often when discussing one base and not the pair.

meiosis: The cell-division process through which egg and sperm are created in reproduction. It reduces the parental chromosomes by half. The resulting chromosome passed to a child from each pair is usually a distinct combination of the two but may be passed unmixed.

microarray: The "chips" used for the autosomal DNA tests used by genealogists since about 2009 that determine the allele values of several hundred-thousand SNPs scattered across the chromosomes. A collection of DNA sequences are attached to a solid surface used for genotyping single nucleotide variants. Only the variants of interest are tested.

misattributed parentage (MPE): A term describing a scenario where the biological parent is not the expected or documented parent; also referred to as a non-paternal event. Also see NPE.

miscall: Incorrect identification of an allele because of an error while sequencing the DNA.

missing teeth pattern: The pattern, resembling missing teeth, of missing DNA segments seen in a chromosome browser when certain close family members are compared. This was described by Bennett Greenspan when Family Finder was introduced by FamilyTreeDNA.

mitochondrial DNA (mtDNA): DNA passed from a mother to all of her children; males inherit mtDNA from their mothers but do not pass it to their children; females inherit mtDNA from their mothers and pass it to their male and female children.

mitochondria: Organelles in every cell containing the only DNA found outside of the nucleus.

mitochondrion: The singular form of mitochondria.

Most Recent Common Ancestor (MRCA): Most Recent Common Ancestor or Ancestral couple.

MPE: See misattributed parentage.

MRCA: See Most Recent Common Ancestor.

Next Generation Sequencing (NGS): A massively parallel sequencing process. A newer generation of test type than the microarray typically used for autosomal DNA tests.

NGS: See Next Generation Sequencing.

no-call: An allele that cannot be read while sequencing the DNA; typically identified as "- -" (hyphen hyphen) in atDNA data.

NPE: Not the Parent Expected or Non-Parental Event, which is a misnomer. See misattributed parentage.

nucleotide: A building block of DNA that includes a base A, C, G, or T making up one half of a base-pair. The structural unit of DNA.

oocyte: An egg cell in a female prior to maturation.

orientation: Each chromosome has two orientations depending on which strand (plus or minus) in the double helix is being read. If a marker on a new chip or a new build does not conform to the expected orientation, an A will be read as a T and vice versa and a C will be read as a G and vice versa. Care must be taken when comparing DNA between individuals to always use the same orientation, either plus or minus (also known as the forward or reverse strand).

PAR: See pseudoautosomal region.

parental type of DNA: Refers to a chromosome (or in some cases a section between two markers on a chromosome) that has not undergone noticeable recombination.

patriline: The line of inheritance through which Y-DNA passes; the line from a man to his father, to his father (the paternal grandfather), to his father, and so on.

pedigree collapse: When cousins marry cousins, a pedigree chart can have multiple ancestor slots, each represented by a different Ahnentafel number, occupied by the same individual. This results in fewer total ancestors in a given generation than there will be with no cousin marriage. See ancestor paradox.

phased: DNA data that has been processed in a way to assign alleles to either the paternal or maternal chromosome. This reduces the likelihood of false positive matches.

phasing: The process of assigning alleles to the paternal or maternal chromosome.

phenotype: An observable genetic trait, such as hair or eye color, that is partially determined by a genotype.

phylogeny: A graphical representation displaying relationships of the human family tree. Also called a cladogram.

proof argument: A documented narrative explaining why the answer to a complex genealogical problem should be considered acceptable and which may be a stand-alone product, like a case study, journal article, or report, or may appear within a chapter, family history, or other genealogical work in print, online, or elsewhere.[1]

PseudoAutosomal Region (PAR): The regions at each end of the X and Y chromosome that interact in the same manner as autosomal DNA. The pseudoautosomal regions (PAR1 at the beginning and PAR2 at the end) of the human X and Y chromosomes pair and recombine during meiosis.

raw data: For atDNA, a file containing a list of the allele values or DNA bases (A, C, G, or T) on both chromosomes at each tested location; for mtDNA, the values contained in a FASTA file that represent the DNA chemicals detected at each location tested; for Y-STR tests, the marker name and value of the marker (the number of repeats at that location); for Y-SNP tests, the marker name and whether the SNP is ancestral (not mutated, sometimes represented by a hyphen or minus sign) or derived (mutated, sometimes represented as a plus sign); some data file formats also indicate the quality of the sequenced data, the DNA chemical detected at each location, or both.

recombination: The process in which segments of DNA inherited from a person's parents are exchanged, creating unique chromosomes to be passed to the next generation in the sperm or egg.

segment triangulation: The process of comparing atDNA segments from three or more people to verify they all share DNA along the same or an overlapping segment of a particular chromosome, thereby forming a triangulated group; shared DNA segments of a significant size indicate all may have inherited that DNA segment from a common ancestral line. Match A matches Match B on a segment; Match B matches Match C on the same or overlapping segment; and Match C matches Match A on the same or overlapping segment. The matches should be at least as distant as first cousins. A parent and a child or grandchild are only considered as one leg of the three needed for triangulation. Segment overlap of at least 7 cM is preferred by many researchers.

sex chromosomes: The X and the Y chromosome pair. Most biological females carry XX and most biological males carry XY. A gene on the Y chromosome determines maleness.

Short Tandem Repeat (STR): A short, repeating pattern of DNA at consecutive rungs of the DNA ladder.

side: Matches sharing DNA that a person inherited from the father are matches on the "paternal side." Matches sharing DNA that a person inherited from the mother are matches on the "maternal side." A shared segment could be on one person's maternal side and be on the second person's paternal side.

Single Nucleotide Polymorphism (SNP): A common genetic variation among people. A variation, or mutation, at one nucleotide (one base).

1. From Thomas W. Jones, *Mastering Genealogical Proof* (Arlington, Va.: National Genealogical Society, 2013), 137. Used here with permission.

Single Nucleotide Variant (SNV): A variation similar to a SNP occurring through mutation, but not limited by thresholds of frequency.

SNP: See Single Nucleotide Polymorphism.

SNV: Single Nucleotide Variant. See Single Nucleotide Polymorphism and variant.

STR: See Short Tandem Repeat.

subclade: A subgroup of a Y-DNA or mtDNA haplogroup; a branch that is downstream (later in time). See also clade.

telomere: DNA found at the ends of the chromosomes which protects the chromosome and helps it maintain its function.

TG: See Triangulated Group.

Triangulated Group (TG): A group of at least three triangulated segments. The "group" can be thought of as the segment or as the matches who share the segments. Matches in one triangulated group may share other segments and sometimes be related through more than one MRCA.

upstream: A DNA mutation (or SNP) that occurred before (earlier in time than) a downstream SNP.

visual phasing: A methodology for assigning the test taker's segments to specific parents and grandparents based on all known crossover points of at least three siblings as determined through the comparison of matches in a chromosome browser. This technique can be employed when parents and grandparents are not available for testing. It uses cousin matches to determine pathways of inheritance. The mapping to each of four grandparents involves colored (hence the term "visual") haplotype blocks that cover the distance along each chromosome through the process of elimination. In some circumstances, two siblings can be used through the identification of all crossover points as obtained by comparison with multiple relatives.

Whole Genome Sequencing (WGS): A process in which the entire genome is tested, as opposed to a microarray test which tests several hundred-thousand locations.

X Chromosome (often referenced as X-DNA): One of the two sex chromosomes. The Y is the other sex chromosome. The X chromosome pairs with a second X in females and with the Y in males to form the 23rd pair of human chromosomes.

Y chromosome (often referenced as Y-DNA): One of the two sex chromosomes. The X is the other sex chromosome. The Y chromosome is found only in males and pairs with an X to form the 23rd pair of male human chromosomes.

Recommended Reading

All URLs were accessed 2 February 2019 unless otherwise indicated.

Genetic and Genealogical Standards

Board for Certification of Genealogists. "Ethics and Standards." http://bcgcertification .org/ethics-super-parent/ethics-standards/.

______. *Genealogy Standards*. Second edition. Nashville, Tenn.: Ancestry.com, 2019.

______. *Genealogy Standards*. Nashville, Tenn.: Ancestry.com, 2014.

Genetic Genealogy Standards. http://www.geneticgenealogystandards.com/.

General DNA and Genealogy Sources

23andMe. "Privacy Highlights." https://www.23andme.com/about/privacy/.

American Adoption Congress. https://www.americanadoptioncongress.org/state.php.

AncestryDNA. "Privacy for Your AncestryDNA Test." http://www.ancestry.com/cs/legal/PrivacyForAncestryDNATesting/.

Bartlett, Jim. *Segment-ology*. http://segmentology.org/.

Bettinger, Blaine T. and Debbie Parker Wayne. *Genetic Genealogy in Practice*. Falls Church, Va.: National Genealogical Society, 2016.

Bettinger, Blaine T. "Evaluating Genealogical Conclusions using DNA." *Paths to Your Past: 2018 Program Syllabus*. Grand Rapids, Michigan, 2018. Also, audio recording of the same, available at *Playback NGS*. https://www.playbackngs.com/.

———. "The Danger of Distant Matches." National Genealogical Society, *Paths to Your Past: 2018 Program Syllabus*. Grand Rapids, Michigan, 2018. Also, audio recording of the same, available at *Playback NGS*. https://www.playbackngs.com/.

______. *The Family Tree Guide to DNA Testing and Genetic Genealogy*. Blue Ash, Ohio: Family Tree Books, 2016.

———. "The Recombination Project: Analyzing Recombination Frequencies Using Crowdsourced Data." *The Genetic Genealogist*. 20 February 2017. https://thegeneticgenealogist.com/wp-content/uploads/2017/02/Recombination_Preprint.pdf.

______. "The Shared cM Project," version 3.0. *The Genetic Genealogist*. https://www.thegeneticgenealogist.com/.

______. "Unlocking the Genealogical Secrets of the X Chromosome." *The Genetic Genealogist*. 21 December 2008. https://thegeneticgenealogist.com/2008/12/21/unlocking-the-genealogical-secrets-of-the-x-chromosome.

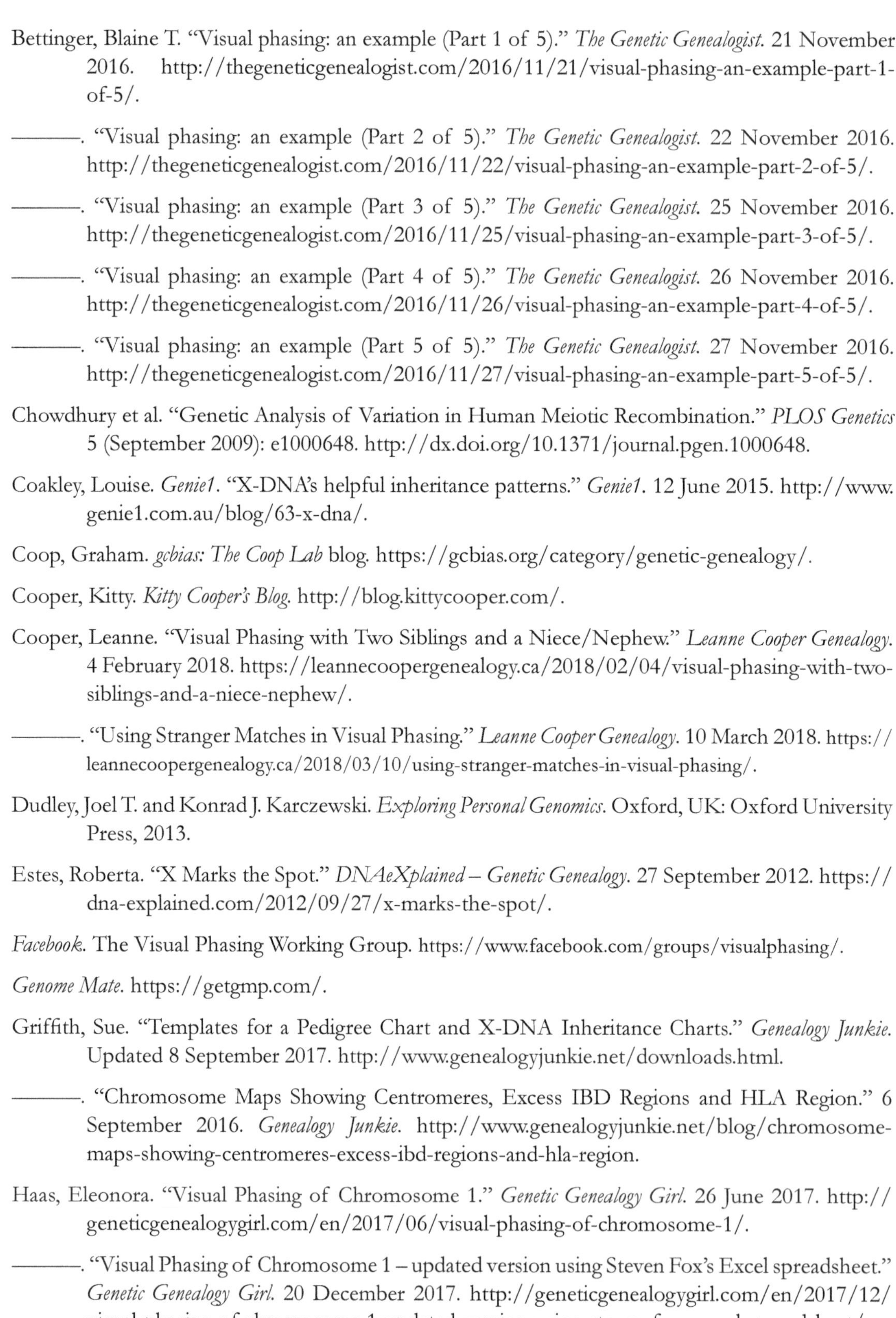

Bettinger, Blaine T. "Visual phasing: an example (Part 1 of 5)." *The Genetic Genealogist.* 21 November 2016. http://thegeneticgenealogist.com/2016/11/21/visual-phasing-an-example-part-1-of-5/.

———. "Visual phasing: an example (Part 2 of 5)." *The Genetic Genealogist.* 22 November 2016. http://thegeneticgenealogist.com/2016/11/22/visual-phasing-an-example-part-2-of-5/.

———. "Visual phasing: an example (Part 3 of 5)." *The Genetic Genealogist.* 25 November 2016. http://thegeneticgenealogist.com/2016/11/25/visual-phasing-an-example-part-3-of-5/.

———. "Visual phasing: an example (Part 4 of 5)." *The Genetic Genealogist.* 26 November 2016. http://thegeneticgenealogist.com/2016/11/26/visual-phasing-an-example-part-4-of-5/.

———. "Visual phasing: an example (Part 5 of 5)." *The Genetic Genealogist.* 27 November 2016. http://thegeneticgenealogist.com/2016/11/27/visual-phasing-an-example-part-5-of-5/.

Chowdhury et al. "Genetic Analysis of Variation in Human Meiotic Recombination." *PLOS Genetics* 5 (September 2009): e1000648. http://dx.doi.org/10.1371/journal.pgen.1000648.

Coakley, Louise. *Genie1*. "X-DNA's helpful inheritance patterns." *Genie1*. 12 June 2015. http://www.genie1.com.au/blog/63-x-dna/.

Coop, Graham. *gcbias: The Coop Lab* blog. https://gcbias.org/category/genetic-genealogy/.

Cooper, Kitty. *Kitty Cooper's Blog.* http://blog.kittycooper.com/.

Cooper, Leanne. "Visual Phasing with Two Siblings and a Niece/Nephew." *Leanne Cooper Genealogy.* 4 February 2018. https://leannecoopergenealogy.ca/2018/02/04/visual-phasing-with-two-siblings-and-a-niece-nephew/.

———. "Using Stranger Matches in Visual Phasing." *Leanne Cooper Genealogy.* 10 March 2018. https://leannecoopergenealogy.ca/2018/03/10/using-stranger-matches-in-visual-phasing/.

Dudley, Joel T. and Konrad J. Karczewski. *Exploring Personal Genomics.* Oxford, UK: Oxford University Press, 2013.

Estes, Roberta. "X Marks the Spot." *DNAeXplained – Genetic Genealogy.* 27 September 2012. https://dna-explained.com/2012/09/27/x-marks-the-spot/.

Facebook. The Visual Phasing Working Group. https://www.facebook.com/groups/visualphasing/.

Genome Mate. https://getgmp.com/.

Griffith, Sue. "Templates for a Pedigree Chart and X-DNA Inheritance Charts." *Genealogy Junkie.* Updated 8 September 2017. http://www.genealogyjunkie.net/downloads.html.

———. "Chromosome Maps Showing Centromeres, Excess IBD Regions and HLA Region." 6 September 2016. *Genealogy Junkie.* http://www.genealogyjunkie.net/blog/chromosome-maps-showing-centromeres-excess-ibd-regions-and-hla-region.

Haas, Eleonora. "Visual Phasing of Chromosome 1." *Genetic Genealogy Girl.* 26 June 2017. http://geneticgenealogygirl.com/en/2017/06/visual-phasing-of-chromosome-1/.

———. "Visual Phasing of Chromosome 1 – updated version using Steven Fox's Excel spreadsheet." *Genetic Genealogy Girl.* 20 December 2017. http://geneticgenealogygirl.com/en/2017/12/visual-phasing-of-chromosome-1-updated-version-using-steven-foxs-excel-spreadsheet/.

Hartley, Joel. "My big fat chromosome 20." *Hartley DNA & Genealogy*. 12 January 2016. http://www.jmhartley.com/HBlog/?p=462.

Hill, Richard. *Finding Family: My Search for Roots and the Secrets in My DNA*. Sanger, Calif.: Familius, paperback edition 2017.

Hutchison, Luke A. D., Natalie M. Myres, and Scott R. Woodward. "Growing the Family Tree: The Power of DNA in Reconstructing Family Relationships." *Proceedings of the First Symposium on Bioinformatics and Biotechnology (BIOT-04, Colorado Springs)*, (September 2004): 42-49.

International Society of Genetic Genealogists. *Wiki*. "Identical by Descent—Excess IBD Sharing." https://isogg.org/wiki/Identical_by_descent.

______. "Recombination." *ISOGG Wiki*. https://isogg.org/wiki/Recombination.

______. "Visual Phasing." *ISOGG Wiki*. https://isogg.org/wiki/Visual_phasing.

JoGG: Journal of Genetic Genealogy. http://jogg.info/.

Johnston, Kathy. "Segment Matching with Grandparents Using Crossover Lines." *FamilyTreeDNA* Forums. http://forums.familytreedna.com/showthread.php?t=36812 (registration required).

Kilpatrick, Holly. "Chromosome Maps." *Tilley Pearsal Genealogy Database*. http://www.stonecropicelandics.com/Tilley_Pearsall-o/up/chr.htm.

Martin, Lars. "Visual Phasing of chromosome 20 using Excel." *YouTube*. 24 February 2017. https://www.youtube.com/watch?v=neB4oFlwJHA.

Owston, Jim. "Phasing the X-Chromosome." *The Lineal Arboretum*. 21 November 2012. http://linealarboretum.blogspot.com/2012/11/phasing-x-chromosome.html.

———. "Pruning the Family Tree with DNA Evidence." *The Lineal Arboretum*. 13 October 2010. http://linealarboretum.blogspot.com/2010/10/pruning-family-tree-with-dna-evidence.html.

Pike, David. Autosomal Utilities. http://www.math.mun.ca/~dapike/FF23utils/.

Raymont, Ann. "Chromosome mapping with siblings – part 1." *DNA Sleuth*. 13 May 2016. https://dnasleuth.wordpress.com/2016/05/13/chromosome-mapping-with-siblings-part-1/.

———. "Chromosome mapping with siblings – part 2." *DNA Sleuth*. 1 June 2016. https://dnasleuth.wordpress.com/2016/06/01/chromosome-mapping-with-siblings-part-2/.

Speed, D. and DJ Balding. "Relatedness in the post-genomic era: is it still useful?" *Nature Reviews Genetics* (January 2015): 33-44. https://doi.org/10.1038/nrg3821 is behind a paywall. Author's copy online at http://dougspeed.com/wp-content/uploads/nrg_relatedness.pdf.

Stanbary, Karen. "Can a DNA Relationship be Proved by DNA Alone?" *BCG Springboard*. 17 January 2018. https://bcgcertification.org/can-a-genetic-relationship-be-proved-by-dna-alone/.

Sweeney, Deborah. "Down the DNA Rabbit Hole – Visual Phasing with Two Siblings." *Genealogy Lady*. 2 May 2017. https://genealogylady.net/2017/05/02/down-the-dna-rabbit-hole-visual-phasing-with-two-siblings/.

Turner, Ann. "Ahnentafel numbers of ancestors who could contribute a segment on the X chromosome." *Internet Archive*. https://web.archive.org/web/20170607171833/http://dnacousins.com/AHN_X.TXT.

Wayne, Debbie Parker. "X-DNA Inheritance Charts." *Deb's Delvings in Genealogy.* 25 October 2013. https://debsdelvings.blogspot.com/2013/10/x-dna-inheritance-charts.html.

Whited, Randy. "Reconstructing Grandparent DNA Using Sibling Results." SCGS Jamboree 2016 Session #TH 023. http://www.myconferenceresource.com/products/45-01-scgs-genetic-dna-jamboree-conference-2016.aspx.

WikiTree. https://www.wikitree.com/.

Guides to Crafting Case Studies and Proof Arguments

Bettinger, Blaine T. "Evaluating a Genetic Genealogy Proof Argument." *Association of Professional Genealogists Quarterly* 30 (September 2015): 162–64.

Jones, Thomas W. *Mastering Genealogical Proof.* Arlington, Va.: National Genealogical Society, 2014. Also available in a Kindle edition.

———. "Reasoning from Evidence" and "Proof Arguments and Case Studies." In Elizabeth Shown Mills, ed. *Professional Genealogy: Preparation, Practice, and Standards.* Baltimore, Md.: Genealogical Publishing, 2018.

Little, Barbara Vines. "Skillbuilding: It's Not That Hard to Write Proof Arguments." *OnBoard: Newsletter of the Board for Certification of Genealogists* 15 (September 2009): 20–23. Online at *Board for Certification of Genealogists.* https://bcgcertification.org/learning/skills/onboard/.

Mills, Elizabeth Shown. "Proof Arguments and Case Studies." In *Professional Genealogy: A Manual for Researchers, Writers, Editors, Lecturers, and Librarians.* Baltimore, Md.: Genealogical Publishing, 2001.

Guides to Writing and Documentation

Bates, Jefferson D. *Writing with Precision: How to Write So that You Cannot Possibly Be Misunderstood.* New York: Penguin, 2000.

Blake, Gary, and Robert W. Bly. *The Elements of Technical Writing: The Essential Guide to Writing Clear, Concise Proposals, Reports, Manuals, Letters, Memos, and Other Documents in Every Technical Field.* New York: Longman, 1993.

Cheney, Theodore A. Rees. *Getting the Words Right: How to Rewrite, Edit and Revise.* Cincinnati, Ohio: Writer's Digest Books, 1983.

The Chicago Manual of Style, 17th edition. Chicago: University of Chicago Press, 2017.

The Chicago Manual of Style Online, 17th edition. http://www.chicagomanualofstyle.org/home.html.

Clark, Roy Peter. *Writing Tools: 50 Essential Strategies for Every Writer.* New York: Little, Brown, 2006.

Fiske, Robert Hartwell. *The Dictionary of Concise Writing: 10,000 Alternatives to Wordy Phrases.* Oak Park, Ill.: Marion Street Press, 2002.

Garner, Bryan A. *The Chicago Guide to Grammar, Usage, and Punctuation.* Chicago: University of Chicago Press, 2016.

Jones, Thomas W. *Mastering Genealogical Documentation.* Arlington, Va.: National Genealogical Society, 2017. Also available in a Kindle edition.

Mills, Elizabeth Shown. *Evidence Explained: Citing History Sources from Artifacts to Cyberspace,* 3rd edition revised. Baltimore, Md.: Genealogical Publishing, 2017. Also available in a Kindle edition. The essential guide to citing genealogical sources.

———. *Evidence Explained: Historical Analysis, Citation, and Source Usage.* http://www.evidenceexplained.com/. Online forum addressing citation issues.

———. *QuickSheet: Citing Genetic Sources for History Research; Evidence Style,* laminated folder. 2nd ed. Baltimore, Md.: Genealogical Publishing, 2017. Also available in a Kindle edition.

———. Editor. "Writing and Compiling." In *Professional Genealogy: A Manual for Researchers, Writers, Editors, Lecturers, and Librarians.* Baltimore, Md.: Genealogical Publishing, 2001.

Ross-Larson, Bruce. *Edit Yourself: A Manual for Everyone Who Works with Words.* New York: W. W. Norton, 1982.

Strunk, William, Jr., and E. B. White. *The Elements of Style.* New York: Macmillan, 1959.

Zinsser, William. *On Writing Well: The Classic Guide to Writing Nonfiction.* New York: Collins, 2006.

Selected Blind-Reviewed Genetic-Genealogy Case Studies in Genealogical Journals

Devine, Donn. "Sorting Relationships among Families with the Same Surname: An Irish-American DNA Study." *National Genealogical Society Quarterly* 93 (December 2005): 283–93. [Uses results of Y-chromosome DNA testing]

Dunn, Victor S. "Determining Origin with Negative and Indirect Evidence: Cylus H. Feagans of Virginia and West Virginia. *National Genealogical Society Quarterly* 105 (March 2017): 5–18. [Uses results of autosomal DNA testing]

Eagleson, Pamela Stone. "Parents for Robert Walker of Rockingham County, North Carolina, and Orange County, Indiana." *National Genealogical Society Quarterly* 102 (September 2013): 189–99. [Uses results of Y-chromosome DNA testing]

Fein, Mara. "A Family for Melville Adolphus Fawcett." *National Genealogical Society Quarterly* 104 (June 2016): 107–24. [Uses results of autosomal DNA testing]

Fox, Judy Kellar. "Documents and DNA Identify a Little-Known Lee Family in Virginia." *National Genealogical Society Quarterly* 102 (June 2011): 85–96. [Uses results of Y-chromosome DNA testing]

Green, Shannon. "Connecting William W. Hawkins of Newark, New Jersey, and William Wallace Hawkins of New York City." *New York Genealogical and Biographical Record* 148 (October 2017): 265–77. [Uses results of Y-chromosome DNA testing]

Hobbs, Patricia L. "DNA Identifies a Father for Rachel, Wife of James Lee of Huntingdon County, Pennsylvania." *National Genealogical Society Quarterly* 105 (March 2017): 43–56. [Uses results of autosomal DNA testing]

Hollister, Morna Lahnice. "Goggins and Goggans of South Carolina: DNA Helps Document the Basis of an Emancipated Family's Surname." *National Genealogical Society Quarterly* 102 (September 2014): 165–76. [Uses results of autosomal and Y-chromosome DNA testing and a DNA test of ethnic origin]

Jackson, B. Darrell. "George Craig of Howard County, Missouri: Genetic and Documentary Evidence of His Ancestry." *National Genealogical Society Quarterly* 99 (March 2011): 59–72. [Uses results of Y-chromosome DNA testing]

Jones, Thomas W. "Too Few Sources to Solve a Family Mystery? Some Greenfields of Central and Western New York." *National Genealogical Society Quarterly* 103 (June 2015): 85–103. [Uses results of autosomal DNA testing]

Lustenberger, Anita A. "David Meriweather: Descendant of Nicholas[1] Meriweather? A DNA Study." *National Genealogical Society Quarterly* 93 (December 2005): 269–82. [Uses results of Y-chromosome DNA testing]

Mills, Elizabeth Shown. "Testing the FAN Principle against DNA: Zilphy (Watts) Price Cooksey Cooksey of Georgia and Mississippi." *National Genealogical Society Quarterly* 102 (June 2014): 129–52. [Uses results of autosomal and mitochondrial DNA testing and a DNA test for Native American ancestry]

———. "Frontier Research Strategies—Weaving a Web to Snare a Birth Family: John Watts (ca. 1749–ca. 1822)." *National Genealogical Society Quarterly* 104 (September 2015): 165–90. [Uses results of Y-chromosome DNA testing]

Moneta, Daniela. "Identifying the Children of David Pugh and Nancy Minton of Virginia, Kentucky, and Tennessee." *National Genealogical Society Quarterly* 96 (March 2008): 13–22. [Uses results of Y-chromosome DNA testing]

———. "Virginia Pughs and North Carolina Wests: A Genetic Link from Slavery in Kentucky." *National Genealogical Society Quarterly* 97 (September 2009): 179–94. [Uses results of Y-chromosome DNA testing]

Morelli, Jill. "DNA Helps Identify "Molly" (Frisch/Lancour) Morelli's Father." *National Genealogical Society Quarterly* 106 (December 2018): 293–306. [Uses results of autosomal and X-chromosome DNA testing]

Ouimette, David S. "Proving the Parentage of John Bettis: Immigrant Ancestor of Bettis Families in Vermont." *National Genealogical Society Quarterly* 98 (September 2010): 189–210. [Uses results of Y-chromosome DNA testing]

Pratt, Warren C. "Finding the Father of Henry Pratt of Southeastern Kentucky." *National Genealogical Society Quarterly* 100 (June 2012): 85–103. [Uses results of Y-chromosome DNA testing]

Ralls, Stephen Alden. "The Lost Second Family of Colonel Hugh[2] McGary Jr. (Hugh[1]); Part II: Autosomal DNA Evidence." *The Genealogist* 31 (Fall 2017): 182–90. [Uses results of autosomal DNA testing]

Stanbary, Karen. "Rafael Arriaga, A Mexican Father in Michigan: Autosomal DNA Helps Identify Paternity." *National Genealogical Society Quarterly* 104 (June 2016): 85–98. [Uses results of autosomal and X-chromosome DNA testing]

Guides to Ethical Considerations and Family Dynamics

Boyle, Alan. "Who's keeping your genetic keys? Questions about DNA testing, confidentiality and ethics." *NBCNews.com.* 16 January 2002. http://www.nbcnews.com/id/3077152/ns/technology_and_science-science/.

Clyde, Linda. "DNA Genealogy: Making Contact with Biological Relatives." *RootsTech blog.* 12 April 2017. https://www.rootstech.org/blog/dna-genealogy-making-contact-with-biological-relatives.

Council for Responsible Genetics. "Genetic Testing, Privacy and Discrimination." http://www.councilforresponsiblegenetics.org/projects/PastProject.aspx?projectId=1.

Estes, Roberta. "No (DNA) Bullying," *DNAeXplained.com — Genetic Genealogy.* 15 May 2013. http://dna-explained.com/2013/05/15/no-dna-bullying/.

FamilyTreeDNA. "Common Misconceptions About DNA Privacy." https://www.familytreedna.com/common-misconceptions.aspx.

———. "Group Project Administrator Terms & Policies." https://www.familytreedna.com/legal/terms/group-project-administrator.

———. "Privacy Statement." https://www.familytreedna.com/legal/privacy-statement.

———. "Terms of Service." https://www.familytreedna.com/legal/terms-of-service.

Fox, Judy Kellar. "Respecting the Privacy of DNA Test Takers." *BCG SpringBoard* blog. 8 October 2015. https://bcgcertification.org/respecting-the-privacy-of-dna-test-takers/.

Fronczak, Paul Joseph, and Alex Tresiowski. *The Foundling: The True Story of a Kidnapping, a Family Secret, and My Search for the Real Me.* New York, NY: Simon & Schuster, 2018.

"Genetic Information Nondiscrimination Act of 2008." Pub. L. 110–233, 122 Stat. 881 (May 21, 2008). https://www.gpo.gov/fdsys/pkg/PLAW-110publ233/pdf/PLAW-110publ233.pdf.

Griffeth, Bill. *The Stranger in My Genes: A Memoir.* Boston, Mass.: New England Historic Genealogical Society, 2016.

Hughes, Virginia. "23 and You: Does that commercial DNA test you just bought violate somebody else's privacy?" *Matter.* 8 Dec 2013. https://medium.com/matter/66e87553d22c.

———. "It's Time To Stop Obsessing About the Dangers of Genetic Information." *Slate.com.* 7 January 2013. http://www.slate.com/articles/health_and_science/medical_examiner/2013/01/ethics_of_genetic_information_whole_genome_sequencing_is_here_and_we_need.html.

Humbert, Mathias, lead author. De-anonymizing Genomic Databases Using Phenotypic Traits. *Proceedings on Privacy Enhancing Technologies* 2015 (2): 99–114. https://petsymposium.org/2015/papers/07_Humbert.pdf.

International Society of Genetic Genealogy. "Ethics, guidelines and standards." Resource list, rev. 15 December 2017. http://www.isogg.org/wiki/Ethics,_guidelines_and_standards.

———. "ISOGG Project Administrator Guidelines." Rev. 18 March 2016. https://isogg.org/wiki/ISOGG_Project_Administrator_Guidelines.

Kaminer, Ariel. "Sizing Up the Family Gene Pool." *The New York Times Magazine*, 24 February 2012. http://www.nytimes.com/2012/02/26/magazine/ethicist-dna.html.

Krimsky, Sheldon, and David Cay Johnston. *Ancestry DNA Testing and Privacy: A Consumer Guide.* Washington, D.C.: Council for Responsible Genetics, 2017. http://www.councilfor responsiblegenetics.org/img/Ancestry-DNA-Testing-and-Privacy-Guide.pdf.

National Genealogical Society. "Guidelines for Sharing Information with Others." https://www.ngsgenealogy.org/wp-content/uploads/NGS-Guidelines/Guidelines_SharingInfo2016-FINAL-30Sep2018.pdf.

National Human Genome Research Institute. "Genetic Discrimination" (including discussion of federal genetic nondiscrimination statute). http://www.genome.gov/10002077.

Presidential Commission for the Study of Bioethical Issues. Blog, now archived. https://bioethicsarchive.georgetown.edu/. Various dates including:

Donnelly, John. "Ethical questions around genetic testing." 2 February 2012. https://bioethicsarchive.georgetown.edu/pcsbi/blog/2012/02/02/ethical-questions-around-genetic-testing/index.html.

Klotz, Dan. "Do privacy concerns follow the coffee cup?" 1 August 2012. https://bioethicsarchive.georgetown.edu/pcsbi/blog/2012/08/01/do-privacy-concerns-follow-the-coffee-cup/index.html.

———. *Privacy and Progress in Whole Genome Sequencing.* October 2012. https://bioethics archive.georgetown.edu/pcsbi/sites/default/files/PrivacyProgress508_1.pdf.

Russell, Judy G. *The Legal Genealogist* blog, various dates including:

"Games grandparents play." 29 Sep 2013. http://www.legalgenealogist.com/2013/09/29/games-grandparents-play/.

"Protection for genetic privacy." 21 Apr 2013. http://www.legalgenealogist.com/2013/04/21/protection-for-genetic-privacy/.

"The ethics of DNA testing." 18 Nov 2012. http://www.legalgenealogist.com/2012/11/18/the-ethics-of-dna-testing/.

———. "Skillbuilding: The Ethics of DNA Testing," *OnBoard* 21 (January 2015): 1–2, 7. https://bcgcertification.org/skillbuilding-the-ethics-of-dna-testing/.

Sandberg, Anders. "Caught in the genetic social network." *Practical Ethics.* 3 July 2013. http://blog.practicalethics.ox.ac.uk/2013/07/caught-in-the-genetic-social-network/.

Shringarpure, Sutash S., and Carlos D. Bustamante. "Privacy Risks from Genomic Data-Sharing Beacons." *American Journal of Human Genetics* 97 (November 2015): 631-646. http://www.cell.com/ajhg/fulltext/S0002-9297(15)00374-2.

Tanner, Adam. "The Promise & Perils of Sharing DNA." *UnDark.* 13 September 2016. https://undark.org/article/dna-ancestry-sharing-privacy-23andme/.

Weinreich, Petter. "The operationalization of identity theory in racial and ethnic relations." In John Rex and David Mason, *Theories of Race and Ethnic Relations.* Cambridge, Mass.: Cambridge University Press, 1986.

Recommended Reading

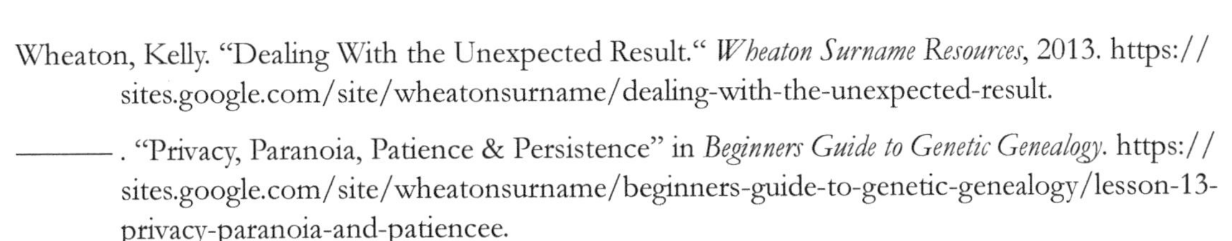

Wheaton, Kelly. “Dealing With the Unexpected Result.“ *Wheaton Surname Resources*, 2013. https://sites.google.com/site/wheatonsurname/dealing-with-the-unexpected-result.

———. “Privacy, Paranoia, Patience & Persistence” in *Beginners Guide to Genetic Genealogy*. https://sites.google.com/site/wheatonsurname/beginners-guide-to-genetic-genealogy/lesson-13-privacy-paranoia-and-patiencee.

Index

This indexes pages where an important concept is discussed; every page where a phrase is used in a minor context is not indexed. Names of real persons and places are indexed. Genetic analysis concepts may use slightly different phrases in different chapters. When scanning a page for indexed concepts look for similar keywords, not exact phrases.

Index

Chapter images provided from screen shots and tools used by authors. Additional credit information may be in the text. Images from screen shots and tool displays of companies and websites include

23andMe: 3.17, 8.4, 8.8
AncestryDNA: 3.9, 3.13,
ConnectedDNA: 6.14, 6.15, 7.6
DNA Painter: 2.29, 2.30, 2.31
Exploring Family Trees: 6.2
FamilyTreeDNA: 3.14, 4.3, 4.4, 4.5, 4.6, 4.7, 4.8,
GEDmatch: 2.1, 2.2, 2.5, 2.7, 2.12, 2.13, 3.2, 3.4, 3.5, 3.7, 3.10, 3.11, 3.12, 6.4, 8.2, 8.3, 8.9, 8.10
Genome Mate Pro: 6.9
MyHeritage: 6.10, 6.11
Shared cM Project: 5.3, 5.4

CPSIA information can be obtained
at www.ICGtesting.com
Printed in the USA
LVHW072259250319
611832LV00023B/129/P

9 781733 694902